Advanced Lectures in Mathematics
Volume XV

An Introduction to Groups and Lattices:
Finite Groups and Positive Definite Rational Lattices

by Robert L. Griess, Jr.

International Press
www.intlpress.com

高等教育出版社
HIGHER EDUCATION PRESS

Advanced Lectures in Mathematics, Volume XV
An Introduction to Groups and Lattices:
Finite Groups and Positive Definite Rational Lattices

by Robert L. Griess, Jr.

2010 Mathematics Subject Classification. 11H56, 20C10, 20C34, 20D08.

Copyright © 2011 by International Press, Somerville, Massachusetts, U.S.A., and by Higher Education Press, Beijing, China.

This work is published and sold in China exclusively by Higher Education Press of China.

All rights reserved. Individual readers of this publication, and non-profit libraries acting for them, are permitted to make fair use of the material, such as to copy a chapter for use in teaching or research. Permission is granted to quote brief passages from this publication in reviews, provided the customary acknowledgement of the source is given. Republication, systematic copying, or mass reproduction of any material in this publication is permitted only under license from International Press. Excluded from these provisions is material in articles to which the author holds the copyright. (If the author holds copyright, notice of this will be given with article.) In such cases, requests for permission to use or reprint should be addressed directly to the author.

ISBN: 978-1-57146-206-0

Printed in the United States of America.

15 14 13 12 11   1 2 3 4 5 6 7 8 9

Advanced Lectures in Mathematics

Executive Editors

Shing-Tung Yau
Harvard University

Lizhen Ji
University of Michigan, Ann Arbor

Kefeng Liu
University of California at Los Angeles
Zhejiang University
Hangzhou, China

Editorial Board

Chongqing Cheng
Nanjing University
Nanjing, China

Zhong-Ci Shi
Institute of Computational Mathematics
Chinese Academy of Sciences (CAS)
Beijing, China

Zhouping Xin
The Chinese University of Hong Kong
Hong Kong, China

Weiping Zhang
Nankai University
Tianjin, China

Xiping Zhu
Zhongshan University
Guangzhou, China

Tatsien Li
Fudan University
Shanghai, China

Zhiying Wen
Tsinghua University
Beijing, China

Lo Yang
Institute of Mathematics
Chinese Academy of Sciences (CAS)
Beijing, China

Xiangyu Zhou
Institute of Mathematics
Chinese Academy of Sciences (CAS)
Beijing, China

Contents

1 **Introduction** 1
   1.1 Outline of the book 2
   1.2 Suggestions for further reading 3
   1.3 Notations, background, conventions 5

2 **Bilinear Forms, Quadratic Forms and Their Isometry Groups** .. 7
   2.1 Standard results on quadratic forms and reflections, I 9
      2.1.1 Principal ideal domains (PIDs) 10
   2.2 Linear algebra 11
      2.2.1 Interpretation of nonsingularity 11
      2.2.2 Extension of scalars 13
      2.2.3 Cyclicity of the values of a rational bilinear form 13
      2.2.4 Gram matrix 14
   2.3 Discriminant group 16
   2.4 Relations between a lattice and sublattices 18
   2.5 Involutions on quadratic spaces 19
   2.6 Standard results on quadratic forms and reflections, II 20
      2.6.1 Involutions on lattices 20
   2.7 Scaled isometries: norm doublers and triplers 23

3 **General Results on Finite Groups and Invariant Lattices** 25
   3.1 Discreteness of rational lattices 25
   3.2 Finiteness of the isometry group 25
   3.3 Construction of a G-invariant bilinear form 26
   3.4 Semidirect products and wreath products 27
   3.5 Orthogonal decomposition of lattices 28

4 **Root Lattices of Types A, D, E** 31
   4.1 Background from Lie theory 31
   4.2 Root lattices, their duals and their isometry groups 32
      4.2.1 Definition of the A_n lattices 33

		4.2.2	Definition of the D_n lattices	34
		4.2.3	Definition of the E_n lattices	34
		4.2.4	Analysis of the A_n root lattices	34
		4.2.5	Analysis of the D_n root lattices	37
		4.2.6	More on the isometry groups of type D_n	39
		4.2.7	Analysis of the E_n root lattices	41

5 Hermite and Minkowski Functions 49
    5.1 Small ranks and small determinants......................... 51
        5.1.1 Table for the Minkowski and Hermite functions 52
        5.1.2 Classifications of small rank, small determinant lattices 53
    5.2 Uniqueness of the lattices E_6, E_7 and E_8 54
    5.3 More small ranks and small determinants...................... 57

6 Constructions of Lattices by Use of Codes 61
    6.1 Definitions and basic results 61
        6.1.1 A construction of the E_8-lattice with the binary $[8, 4, 4]$ code 62
        6.1.2 A construction of the E_8-lattice with the ternary $[4, 2, 3]$ code 64
    6.2 The proofs... 64
        6.2.1 About power sets, boolean sums and quadratic forms 64
        6.2.2 Uniqueness of the binary $[8, 4, 4]$ code 65
        6.2.3 Reed-Muller codes.. 66
        6.2.4 Uniqueness of the tetracode.............................. 67
        6.2.5 The automorphism group of the tetracode 67
        6.2.6 Another characterization of $[8, 4, 4]_2$ 69
        6.2.7 Uniqueness of the E_8-lattice implies uniqueness of the
              binary $[8, 4, 4]$ code 69
    6.3 Codes over $\mathbb{F}_7$ and a (mod 7)-construction of E_8 70
        6.3.1 The A_6-lattice 71

7 Group Theory and Representations 73
    7.1 Finite groups 73
    7.2 Extraspecial p-groups 75
        7.2.1 Extraspecial groups and central products 75
        7.2.2 A normal form in an extraspecial group 77
        7.2.3 A classification of extraspecial groups 77
        7.2.4 An application to automorphism groups of extraspecial
              groups 79
    7.3 Group representations....................................... 79
        7.3.1 Representations of extraspecial p-groups.................. 80
        7.3.2 Construction of the BRW groups 82
        7.3.3 Tensor products... 85
    7.4 Representation of the BRW group G 86
        7.4.1 BRW groups as group extensions 88

Contents

8 Overview of the Barnes-Wall Lattices 91
   8.1 Some properties of the series 91
   8.2 Commutator density 93
      8.2.1 Equivalence of 2/4-, 3/4-generation and commutator density for Dih_8 93
      8.2.2 Extraspecial groups and commutator density 96

9 Construction and Properties of the Barnes-Wall Lattices 99
   9.1 The Barnes-Wall series and their minimal vectors 99
   9.2 Uniqueness for the BW lattices 101
   9.3 Properties of the BRW groups 102
   9.4 Applications to coding theory 103
   9.5 More about minimum vectors 104

10 Even unimodular lattices in small dimensions 107
   10.1 Classifications of even unimodular lattices 107
   10.2 Constructions of some Niemeier lattices 108
      10.2.1 Construction of a Leech lattice 109
   10.3 Basic theory of the Golay code 111
      10.3.1 Characterization of certain Reed-Muller codes 111
      10.3.2 About the Golay code 112
      10.3.3 The octad Triangle and dodecads 113
      10.3.4 A uniqueness theorem for the Golay code 116
   10.4 Minimal vectors in the Leech lattice 116
   10.5 First proof of uniqueness of the Leech lattice 117
   10.6 Initial results about the Leech lattice 118
      10.6.1 An automorphism which moves the standard frame 118
   10.7 Turyn-style construction of a Leech lattice 119
   10.8 Equivariant unimodularizations of even lattices 121

11 Pieces of Eight 125
   11.1 Leech trios and overlattices 125
   11.2 The order of the group $O(\Lambda)$ 128
   11.3 The simplicity of M_{24} 130
   11.4 Sublattices of Leech and subgroups of the isometry group 132
   11.5 Involutions on the Leech lattice 134

References 137

Index 143

Appendix A   The Finite Simple Groups 149

Appendix B   Reprints of Selected Articles 153
   B.1   Pieces of Eight: Semiselfdual Lattices and a New Foundation for
         the Theory of Conway and Mathieu Groups..................... 155
   B.2   Pieces of 2^d: Existence and Uniqueness for Barnes-Wall and
         Ypsilanti Lattices... 181
   B.3   Involutions on the Barnes-Wall Lattices and Their Fixed Point
         Sublattices, I. 223

1

Introduction

Rational lattices occur naturally in many areas of mathematics, such as number theory, geometry, combinatorics, representation theory, discrete mathematics, finite groups and Lie theory.

The main goal of this book is to explain methods for construction and analysis of positive definite rational lattices and their finite groups of isometries.

It seems that many lattices of great interest are related to finite groups and vice versa. One thinks of root lattices, the Barnes-Wall lattices, the Leech lattice and others which occur as sublattices or overlattices of these. The Leech lattice is closely related to twenty of the twenty-six sporadic simple groups. Many lattices with relatively high minimum norms have interesting finite isometry groups.

Materials in this book are similar to that in graduate courses we gave during the 2000s decade at the University of Michigan in Ann Arbor, USA and Zhejiang University in Hangzhou, China. We present group theory and lattice theory as closely interrelated subjects.

Many topics in the theory of lattices and the theory of groups shall be treated from first principles and proofs will be self-contained. Our presentation is more classroom style or conversational than encyclopedic. We try to provide clear introductions, give examples and indicate directions. If a full treatment would be long and is otherwise available in publications, we may refer to outside sources.

We shall assume the basic knowledge in graduate algebra and introduce more specialized results as we go along. Elementary linear algebra (Jordan and rational canonical forms, multilinear algebra and tensors, modules over a principal ideal domain) is necessary. Elementary representation theory for groups and algebras over fields is assumed, e.g., [23, 52]. Integral representation theory is less well-known, so we shall cover some basics on this topic. Group cohomology theory will be quoted as needed. The knowledge of root systems for the finite dimensional Lie algebras would be helpful but not absolutely necessary.

We thank the Center for Mathematical Research at Zhejiang University in Hangzhou, China, for an invitation to teach a course on Groups and Lattices in winter, 2008. Also, we thank the University of Michigan and the National Science Foundation of the United States for financial support during this period.

1.1 Outline of the book

The goal is an introduction to groups, positive definite rational lattices and their interactions.

Chapter 2 covers the basic algebra associated to rational lattices, such as integrality, the dual, Gram matrices and relations between a lattice and a sublattice. Definitions for quadratic spaces and their isometries are treated with some generality. Particular attention is paid to involutions.

Chapter 3 deals with rational lattices invariant under a given finite group and with finiteness of the isometry group of a given rational lattice. An orthogonal decomposition of a lattice into orthogonally indecomposable summands indicates certain decompositions of its isometry group.

Chapter 4 deals with root lattices of types ADE. These lattices and closely related ones occur widely and are an important part of basic vocabulary in this subject. We give detailed analysis of these lattices, their duals and isometry groups.

Chapter 5 discusses the two inequalities of Hermite and Minkowski which say that, given integers n and d, there is a number $f(n,d)$ so that a lattice of rank n and determinant d has a nonzero vector such that the absolute value of its norm is at most $f(n,d)$. This technique is important for starting structure analyses of lattices for which n and d are not too large. An application is given to uniqueness of the exceptional root lattices E_6, E_7 and E_8 and other cases.

Chapter 6 introduces elementary theory of error correcting codes and their role in building lattices. Applications are given to root lattices (e.g. several different constructions of E_8).

Chapter 7 begins with a review of representation theory of finite groups then specializes to extraspecial p-groups and groups obtained from them by extending upwards by subgroups of the outer automorphism group. In particular, we construct the Bolt-Room-Wall groups. Such groups play important roles in the theory of lattices, as explained in the next chapter.

Chapters 8 and 9 are about an inductive construction of the family of Barnes-Wall lattices, in ranks 2^d. We sketch how to get the rank 2^d case by starting with the rank 2^{d-1} case and using integral representation theory of a dihedral group of order 8. The concepts of 2/4-generation, 3/4-generation and commutator density are developed in generality then specialized to the Barnes-Wall constructions. Applications are given, including a description of minimal vectors and indication of how the Reed-Muller binary codes occur within the Barnes-Wall lattices.

Chapter 10 is about the even unimodular integral lattices in dimensions 8, 16 and 24. The number of isometry types are, respectively, 1, 2 and 24. We describe many of these and devote a lot of attention to the Golay code and the Leech lattice, the unique even unimodular integral lattices in dimension 24 which has no norm 2 vectors. Its isometry group is a remarkable finite groups whose quotient by the center is simple. We sample the rich combinatorics and group theory.

Chapter 11 gives a new treatment of existence and uniqueness for the Leech lattice. It has many logical advantages over the past treatments. For example, it implies existence and properties of the Golay code and Mathieu groups, rather than using these respective theories.

An appendix gives a table of orders for the finite simple groups.

Three articles of this author are reprinted (one in revised form) to supplement treatments in the text.

1.2 Suggestions for further reading

Chapter 2: Bilinear forms, quadratic forms and their isometry groups
For basics about integer quadratic forms, see [17, 55, 58].

Chapter 3: General results on finite groups and invariant lattices

There are many good texts on basic representation theory of finite groups over fields, e.g. [1, 2, 23, 24, 25, 26, 29, 30, 52]. The term "modular representation theory" refers to representations of a finite group over fields of positive characteristic which divides the group order. In this case, the group algebra has a nonzero radical and finite dimensional modules are not completely reducible. Integral representation theory is not as widely treated as representation theory over fields. Aspects are treated in the abovementioned texts.

The text [59] studies many interesting integral lattices which have strong connections to Lie algebras and finite groups.

Chapter 4: Root lattices of types A, D, E

In this text, consider the classification of root systems as given. For an axiomatic treatment, see [11, 50]; the appendices in [11] are quite useful and have become a standard reference.

Chapter 5: Hermite and Minkowski functions

We emphasize the Hermite function because in low ranks, it gives better results than the Minkowski function. For larger ranks, the Minkowski function is much better than the Hermite function, but does not seem to be strong enough for practical use in classification results, such as the ones at the end of this chapter. For a proof of the Minkowski result, see [80].

We have wondered if there is a generalization of such functions to the following situation. We are given two rational lattices L and M, where $\text{rank}(M) \leq \text{rank}(L)$. For each integer $r \geq 1$, one would like some estimate of the number of embeddings of $\sqrt{r}\, M$ into L. Perhaps a sharper question would get a better answer, such as one about preferred bases of M: given a finite set of vectors with Gram matrix G can one estimate the number of embeddings in L of a set of vectors with Gram matrix rG, for $r \geq 1$. Some kinds of estimate in the case of L a rootless rank 24 even unimodular lattice and M the E_8-lattice might be useful for a new uniqueness proof of the Leech lattice along the lines of [38].

The problem of determining the minimum norm of a given lattice is generally hard. Some techniques are given in [21, 57].

Chapter 6: Constructions of lattices by use of codes

For basic coding theory, see [64, 68, 77, 79] and for an extensive report, see [69]. This text gives only the simplest constructions of lattices from sublattices and glue codes. For a greater range of such constructions, see the systematic expositions in [21].

Chapter 7: Group theory and representations

Extraspecial p-groups got a lot of attention from the work of [47] and they frequently played roles throughout the development of finite group theory and the classification of simple groups. For example, see [31, 37, 51]. In [74], Theorem B of [47] is used a lot.

The result on character values of an element in a BRW group (7.4.6) was reported in [35] but it may be older; we do not know the earliest references.

Chapter 8: Overview of the Barnes-Wall Lattices

This short exposition shows that the Barnes-Wall series involves some familiar lattices, covered earlier in the text. The commutator density theory from [40] seems to be new. There are homological issues for representations of a group over $\mathbb{Z}$ which are trivial for the corresponding rational representations. For background on homological algebra, see [6, 46, 49, 70].

Chapter 9: Construction and properties of the Barnes-Wall lattices

These lattices were first described in [5]. In fact, [5] describes more lattices than we consider in this book. Their lattices depend on a set of parameters. For each of those lattices, the isometry group has a subgroup G for which $G/O_2(G)$ is some $GL(m, 2)$. Certain values of the parameters give "the" Barnes-Wall lattices, the ones we treat in this book, and for these lattices, the BRW group is properly larger than the preceding group G. Shortly after [5] appeared, there came several articles describing lattices like Barnes-Wall for odd primes and their groups [7, 8, 9].

The Barnes-Wall lattices were discovered independently in [15], which defines each as an ascending chain of lattices, depending on a sequence of Reed-Muller binary error-correcting codes. The authors give a lot of group theoretic information. See also the earlier articles [12, 13, 14]. Their viewpoint is more group-theoretic than that of [5].

Chapter 10: Even unimodular lattices in small dimensions

The classifications of rank 2 lattices and even unimodular lattices of ranks 8, 16 and 24 are well known. In other low dimensions, there are a few results for cases of interest. See the books [21] and [58] and the article [44]. Some such characterizations are found in Subsection 5.3.

Chapter 11: Pieces of eight

The early constructions of the Leech lattice were done by first creating a Golay code, then using it to make glue vectors over a square lattice with minimum norm 4 [20, 62]. Uniqueness of the Leech lattice is deduced from uniqueness of the Golay code. In [37], this program is described in detail. The approach of Borcherds [10] is based on hyperbolic lattices and so is quite different. He proves existence of a rootless Niemeier lattice but his proof indicates no properties of such a lattice. Uniqueness is proven using analysis of roots in the hyperbolic overlattice of rank 26. The Pieces of Eight [38] approach gets structure theory of the Leech lattice and its isometry group by the uniqueness viewpoint. The ideas in [38] led this author to [40].

1.3 Notations, background, conventions

The conventions in this book will be similar to that of [31]. However, we shall use mostly left actions of groups and rings, though in a few situations we use right actions.

Since we use n-tuples a lot, it is often more convenient to write row vectors for arguments with linear combinations, whereas with matrix work, we may apply a matrix on the left to a column vector. Conjugation and commutation follow the style of [31], e.g. x^y means $y^{-1}xy$ and $[x,y] = x^{-1}y^{-1}xy$ so that $x^y = x[x,y]$.

We tend to write $A \leq B$ when A is a subobject of B in an algebraic category (groups, rings, etc.).

Set theoretic notations include $A \setminus B$ for set-theoretic difference, i.e., $\{x \in A \mid x \notin B\}$, $A+B$ for Boolean sum ($A+B := A \cup B \setminus A \cap B = A \setminus B \cup B \setminus A$) and $A \sqcup B$ for disjoint union.

See the book index for a list of notations which occur in the text.

2
Bilinear Forms, Quadratic Forms and Their Isometry Groups

In this section we cover a few basic definitions which shall apply throughout this book. We include the definition of rational lattice.

Definition 2.0.1. *Let R be a commutative ring and let M and N be R-modules. An R-bilinear map $f : M \times M \to N$ is a function which satisfies*

$$f(x + x', y) = f(x, y) + f(x', y), \quad f(rx, y) = rf(x, y);$$
$$f(x, y + y') = f(x, y) + f(x, y'), \quad f(x, ry) = rf(x, y)$$

for all $x, x', y, y' \in L$ and all $r \in R$.

Such a map is called symmetric if $f(x, y) = f(y, x)$ for all $x, y \in M$. It is called alternating if $f(x, x) = 0$ for all $x, y \in M$.

The alternating property implies that $f(x, y) = -f(y, x)$, but the converse statement is not generally true. Consider the map $f(x, x) = x$, for $x \in \mathbb{F}_2$.

Definition 2.0.2. *Let R be a commutative ring and let M, N be R-modules with respective bilinear forms $f : M \times M \to R$ and $g : N \times N \to R$. An isometry from M to N is an R-linear isomorphism of modules $T : M \to N$ which preserves the bilinear forms, i.e., for all $x, y \in M$, $g(Tx, Ty) = f(x, y)$. When such a T exists, we say that M and N are isometric. In case $M = N$, the set of such maps forms a group under composition, called the isometry group of f. We may write $O(M, f)$ or $O(f)$ for this group.*

Definition 2.0.3. *Let R be a commutative ring and A, B modules over R. A quadratic map is a function $Q : A \to B$ such that,*
  (1) $f(x, y) := Q(x + y) - Q(x) - Q(y)$ *is bilinear;*
  (2) $Q(cx) = c^2 Q(x)$ *for all $c \in R$ and $x \in A$.*
The function f is called the bilinear function associated to Q. A quadratic form is a quadratic map $Q : A \to R$. We say that Q is nondegenerate iff f is nondegenerate.

Remark 2.0.4. (1) *The definition implies that $Q(0) = 0$ and the bilinear form f associated to Q is symmetric.*
  (2) *If $g : A \times A \to R$ is a bilinear form, then $Q(x) := g(x, x)$ is a quadratic form. The bilinear form f associated to Q is in this case equal to $g(x, y) + g(y, x)$.*

If g is symmetric, this is $2g(x,y)$. In case 2 is a unit in R, we derive that $g = \frac{1}{2}f$. However, it is possible that $2R = 0$ and that Q is nonzero. This means that Q is not obtained from a symmetric bilinear form in such a way.

(3) The map {quadratic forms}⟶{symmetric bilinear forms} *given by the definition of quadratic form is many-to-one in general. Consider the next example.*

Example 2.0.5. Let $R = \mathbb{F}_2$, $A = \mathbb{F}_2^2$, and $a \in A$. We define $Q_0 : A \to R$ by $Q_0(0) = 0$ and $Q_0(x) = 1$ if $x \neq 0$. For $a \neq 0$, we define $Q_a : A \to R$ by
$$Q_a(x) := \begin{cases} 0 \text{ if } x = a \\ 1 \text{ otherwise.} \end{cases}$$
It is straightforward to check that all four of these functions are quadratic forms and all of them have the same associated bilinear form f, which satisfies $f(x,y) = 1$ if $x, y \in A$ are linearly independent, and $f(x,y) = 0$ otherwise.

Lemma 2.0.6. *Suppose that $c_i \in R$ for and $v_i \in A$, for $i = 1, \ldots, n$. Then $Q(\sum_i c_i v_i) = \sum_i c_i^2 Q(v_i) = \sum_{i<j} c_i c_j f(v_i, v_j)$.*

Proof. Induction. □

However, we can do something with nonsymmetric forms.

Proposition 2.0.7. *Suppose that A is a free R-module with basis $\{v_i \mid i \in I\}$. Well-order the index set, I. Let $Q : A \to R$ be a quadratic form. Define a bilinear map $g : A \times A \to R$ by declaring its values on pairs of basis elements to be $g(v_i, v_i) := Q(v_i)$, $g(v_i, v_j)$ arbitrary if $i < j$ and if $i > j$, define $g(v_i, v_j) := Q(v_i + v_j) - Q(v_i) - Q(v_j) - g(v_j, v_i)$. Then for all $u, v \in A$, $g(u,v) + g(v,u) = f(u,v)$ and $Q(v) = g(v,v)$.*

Proof. The first statement is clear upon evaluating on pairs of basis elements. For the second statement, write v as a unique linear combination $v = \sum_i c_i v_i$, then expand according to the axioms in the definition of quadratic form. □

Definition 2.0.8. *A pair (A, Q) consisting of an R-module A and a quadratic map Q is called a quadratic space. In a quadratic space we say that $0 \neq a \in A$ is singular if $Q(a) = 0$, and nonsingular if $Q(a) \neq 0$.*

We say $A_1 \leq A$ is anisotropic if $Q(a) \neq 0$ for all $0 \neq a \in A_1$. We call A_1 totally singular if $Q(a) = 0$ for all $a \in A$, and totally isotropic if $f\mid_{A_1 \times A_1} \equiv 0$.

When R is a field and A is finite dimensional over R, the Witt index is the largest dimension of a totally singular subspace. This number is at most $\frac{1}{2}dim(A)$ when the form is nondegenerate. When furthermore the field is finite, this number is $\frac{1}{2}(dim(A) - 1)$ when $dim(A)$ is odd and it is $\frac{1}{2}dim(A)$ or $\frac{1}{2}dim(A) - 1$ when $dim(A)$ is even.

Definition 2.0.9. *A lattice is a finitely generated free abelian group L, with a symmetric bilinear form $f : L \times L \to \mathbb{Q}$, the rational numbers.*

For the moment, the signature of the form, in the sense of Sylvester, is arbitrary. Later in this book we shall assume that L positive definite: $f(x,x) > 0$ if $0 \neq x \in L$.

Notation 2.0.10. *Let L be a lattice with a bilinear form f. We often write (x,y) for the value of f on the pair $x, y \in \mathbb{Q}$. This may cause some confusion with the ordered pair (x,y), but context should indicate the difference.*

We call (x, x) the *norm* of the vector x.

In case L is positive definite (so that we can think of it contained in Euclidean space), the norm is the square of the *length*: $(x, x) = |x|^2$. We shall often use the *triangle inequality* ($|x + y| \leq |x| + |y|$) and the *Cauchy-Schwarz inequality* ($|(x, y)| \leq |x||y|$).

Definition 2.0.11. *A vector x in a lattice L is* primitive *if whenever n is a nonzero integer and $y \in L$ such that $ny = x$, then $n = \pm 1$.*

In $\mathbb{Z}^2$, $(4, 9)$ is primitive while $(2, 4) = 2(1, 2)$ is not.

2.1 Standard results on quadratic forms and reflections, I

We shall state a few general results on quadratic forms without proofs. See [3, 28, 61].

Theorem 2.1.1. (Witt) *Let F be a field, V a finite dimensional vector space over F and Q a nondegenerate quadratic form. Suppose U_1, U_2 are subspaces of V and there is an isometry $\alpha : (U_1, Q|_{U_1}) \to (U_2, Q|_{U_2})$. Then there is an isometry $\sigma \in O(V, Q)$ such that $\sigma|_{U_1} = \alpha$.*

As far as we know, all proofs for Witt's theorem involve special calculations in small dimensions.

There are versions of Witt's theorem for alternating and Hermitian forms.

Theorem 2.1.2. *Let F be a field, V a finite dimensional vector space over F and Q a nondegenerate quadratic form. Suppose the characteristic of the field F is not 2. Then every isometry is a product of reflections.*

Recall that we say a field F is perfect if all roots of the polynomial $x^p - a$ are in F for any $a \in F$, where p is the characteristic of F.

Theorem 2.1.3. *Let F be a field, V a finite dimensional vector space over F and Q a nondegenerate quadratic form. Suppose F is perfect with characteristic 2. Then every isometry is a product of transvections (i.e., $x \mapsto x - \frac{(x,v)}{Q(v)}v$), except when F is the two-element field $\mathbb{F}_2$, $\dim(V) = 4$ and Q has "maximal Witt index": there exists a subspace $U \leq V$ with $\dim(U) = 2$ and $Q|_U \equiv 0$ (i.e., U is totally singular of dimension 2).*

Definition 2.1.4. *Suppose that F is a field and that we have a quadratic form Q on the finite dimensional vector space V over F. Let f be the associated symmetric bilinear form. The* radical *is $R(f) := \{x \in V \mid f(x, V) = 0\}$, a subspace of V. Now suppose that $char(F) = 2$. It is possible that $Q|_{R(f)}$ is not identically 0. This restriction gives a map to F whose image is a subspace with respect to the field F^2, the subfield of squares in F. If F is perfect, this subfield is F and the* rank *of Q is defined to be $\dim(V) - \dim(Ker(Q|_{R(f)}))$. The* defect deficiency *of Q is defined to be $\dim(R(f))$. See [28], Section 16.*

Example 2.1.5. *Assume the notations of* (2.1.5) *and that F is perfect of characteristic* 2. *Suppose that V has basis $v_0, \ldots, v_{2m}$ and that $Q(\sum_i c_i v_i) := c_0^2 + \sum_{j=1}^m c_{2j-1} c_{2j}$. Then Q has rank $2m+1$ and is degenerate of defect 1.*

Theorem 2.1.6. *Let F be a perfect field of characteristic 2, V an n dimensional vector space over F and Q a nondegenerate quadratic form on V. Given $n = 2m$, there are, up to isometry, just two quadratic spaces of dimension n. They are distinguished by having Witt index $m-1$ or m (the dimension of the maximal totally singular subspace). (See* (2.0.8) *for the definition of Witt index.)*

Proposition 2.1.7. *Let q be a power of 2.*

(1) *The order of the orthogonal group $O^\varepsilon(2n, q)$ is twice the order of $\Omega^\varepsilon(2n, 2)$, which equals $q^{n(n-1)}(q^n - \varepsilon 1) \prod_{i=1}^{n-1}(q^{2i} - 1)$.*

(2) *The order of $Sp(2n, q) \cong O(2n+1, q)$ is $q^{n^2} \prod_{i=1}^n (q^{2i} - 1)$.*

2.1.1 Principal ideal domains (PIDs)

The term PID means principal ideal domain, an integral domain in which every ideal is principal. For work on lattices, we need to deal with abelian groups, which are just free modules over the ring $\mathbb{Z}$, a PID. Occasionally, we shall work with other rings which are PIDs, namely certain rings of integers in number fields.

In this section, we review a few basic points of PID theory. The reader is referred to [48, 61].

Let R be a PID. If F is a finitely generated free module and E is a submodule, then E is free and $\mathrm{rank}(E) \leq \mathrm{rank}(F)$. Also, $\mathrm{rank}(E) = \mathrm{rank}(F)$ if and only if E/F is torsion.

Given a matrix $A_{m \times n}$ over R, there exist two invertible matrices $P_{m \times m}$ and $Q_{n \times n}$ over R such that

$$PAQ = D = \begin{pmatrix} d_1 & & & \\ & d_2 & & \\ & & \ddots & \end{pmatrix},$$

and $d_1 \mid d_2, d_2 \mid d_3, \ldots$. If P', Q' are invertible matrices over R such that $P'AQ' = D'$ and

$$D' = \begin{pmatrix} d_1' & & & \\ & d_2' & & \\ & & \ddots & \end{pmatrix},$$

with $d_1' \mid d_2', d_2' \mid d_3', \ldots$, then for all i, $d_i \sim d_i'$, i.e., they are associates in R, meaning there is a unit $u_i \in R^\times$ so that $d_i' = u_i d_i$.

Such a diagonal matrix is called a *Smith canonical form* for A. The sequence $d_1, d_2, \ldots$ is called the *Smith invariant sequence* or the *fundamental invariant sequence*.

We add that the matrices P, Q are a particular type of invertible matrix. They are products of diagonal matrices, permutation matrices and so-called elementary matrices (matrices of the form $I + aE_{ij}$ for $i \neq j$ and scalar a, where I is the identity and E_{ij} is all zero except for a 1 in position i,j). Actually permutation matrices may be omitted here since they are products of matrices of the other two types.

This theory has a lot of applications, including:

(1) Expressions of a finitely generated module as a direct sum of cyclic modules.

(2) Compatibility of bases: given finitely generated free modules $E \leq F$, there exists a basis $v_1, \ldots, v_n$ of F such that there exist scalars $d_1, d_2, \ldots, d_m$ that $d_1 | d_2, d_2 | d_3, \ldots$ and the nonzero members of the sequence $d_1 v_1, \ldots, d_m v_m$ form a basis for E.

Exercise 2.1.8. (1) *Let $p > 2$ be a prime. The integer matrix* $\begin{pmatrix} 2 & 1 \\ 1 & \frac{p+1}{2} \end{pmatrix}$ *has determinant p. Since the product if its Smith invariants $d_1 d_2$ equals p (up to a unit), the only possibility for Smith invariants is $d_1 = 1, d_2 = p$ (up to units). What happens if we assume only that p is an odd number?*

(2) *Determine the Smith normal form for* $\begin{pmatrix} 1 & 1 & 1 & 1 \\ 1 & 1 & -1 & -1 \\ 1 & -1 & 1 & -1 \\ 1 & -1 & -1 & 1 \end{pmatrix}$.

2.2 Linear algebra

In this section, bilinear forms are arbitrary (so are not assumed to be symmetric or alternating). The scalars are from a commutative ring.

Definition 2.2.1. *Let K be a commutative ring and V a module over K. Let f be a K-bilinear form $V \times V \to K$. The left-annihilator of $X \subseteq V$ is $\{v \in V \mid f(v,x) = 0, \text{ for all } x \in X\}$. The left-radical of f is the left annihilator of V. Similarly we define right annihilator and right radical.*

Notation 2.2.2. *We shall use the notation $\text{ann}_V(X)$ for the left or right annihilator of X in V in case the left and right annihilators coincide (they do coincide if the form is symmetric or alternating).*

We say that an ordered pair of sequences $v_1, v_2, \ldots, v_n$ and $w_1, w_2, \ldots, w_n$ is *in duality with respect to the bilinear form g* if $g(v_i, w_j) = \delta_{ij}$ for all $i, j = 1, \ldots, n$. In case g is not symmetric, we may speak of a basis and its left dual basis and right dual basis.

2.2.1 Interpretation of nonsingularity

The following is a standard result about bilinear forms over fields.

Theorem 2.2.3. *Let K be a field and V a finite dimensional vector space over K. Let f be a K-bilinear form $V \times V \to K$.*

The following are equivalent.

(1) *The left radical of f is 0;*

(2) *The right radical of f is 0;*

(3) *For any subspace W of V, the natural map $V \to Hom(W, K)$ defined by $v \mapsto [x \mapsto f(v, x)]$ is onto;*

(4) *For any subspace W of V, the natural map $V \to Hom(W, K)$ defined by $v \mapsto [x \mapsto f(x, v)]$ is onto;*

(5) *For any basis $v_1, \ldots, v_n$ of V, the matrix $(f(v_i, v_j))$ is invertible.*

The proof will follow from the following more general version.

Recall that if $f: A \to B$ is a morphism of modules for some ring, the *cokernel* of f is $Coker(f) := B/\text{Im}(f)$ and the *coimage* of f is $Coim(F)$ is $A/Ker(f)$. It is isomorphic to $\text{Im}(f)$.

The next result generalizes the result over fields. For now, we will be satisfied with the case of a PID. We mention a few points where linear algebra over a PID is different than over a field.

Recall that if $F \leq E$ be a containment of finite rank free modules over a PID. Then E/F is torsion if and only if $\text{rank}(E) = \text{rank}(F)$.

If M, N are finite rank free modules for a PID and their ranks are the same, a monomorphism $M \to N$ has torsion cokernel (over a field, the cokernel would be 0).

We use the term *corank* for a submodule A of the free module B to mean the rank (i.e., torsion free rank) of B/A. This is analogous to the term codimension for a containment of vector spaces.

Theorem 2.2.4. *Let R be a PID, and V a free R-module of rank n, $f: V \times V \to R$ a bilinear form. The following statements are equivalent:*

(1) *the left radical (or kernel) of f is 0;*

(2) *the right radical (or kernel) of f is 0;*

(3) *For any R-submodule W of V, the natural map $V \to Hom(W, R) : v \mapsto [w \mapsto f(v, w)]$ has torsion cokernel;*

(4) *For any R-submodule W of V, the natural map $V \to Hom(W, R) : v \mapsto [w \mapsto f(w, v)]$ has torsion cokernel;*

(5) *For any basis $v_i, \ldots, v_n$ of V, $\det(f(v_i, v_j))_{n \times n} \neq 0$.*

Proof. Let $n := \text{rank}(V)$. About (5), observe that the condition is independent of choice of basis.

(1) implies (3): Suppose W is a submodule of V such that the above map, S, does not have torsion cokernel. Let $m := \text{rank}(W)$. Then the image of S has rank at most $m - 1$ and the kernel of S has rank $d \geq 1 + n - m$. Define $r := n - d$ and let $p_1, \ldots, p_r$ span $\text{Im}(S)$. Then $Ker(p_i)$ has corank 1 in W for all i and $Ker(S) \cap Ker(p_1) \cap \cdots \cap Ker(p_r)$ has rank at least $1 + n - m - r = 1$. But $Ker(S) \cap Ker(p_1) \cap \cdots \cap Ker(p_r)$ is contained in the left radical of f, a contradiction to (1).

(3) implies (5): Suppose that $G := (f(v_i, v_j))$ has zero determinant. Let $v \neq 0$ be in its kernel. Then the column space of G has rank less than n. There exists a nonzero column vector $u = (c_1, \ldots, c_n)^t$ such that $Gu = 0$. Now take $v := \sum_{j=1}^{n} c_j v_j$. Then $f(v_i, v) = 0$ for all i, so v is in the left radical of f, a contradiction.

(5) implies (1): Suppose that v is in the left radical. Write $v := \sum_{i=1}^{n} c_i v_i$. Then the row vector $u = (c_1, \ldots, c_n)$ satisfies $uGy = 0$ for all column vectors y, so $uG = 0$. Since $\det(G) \neq 0$, $u = 0$.

The equivalences of (2), (4), (5) are proved similarly. □

Definition 2.2.5. *We say that the bilinear form f is nonsingular if any of the equivalent conditions of the previous theorem are satisfied.*

2.2.2 Extension of scalars

Let K be a field of characteristic 0 and L a free abelian group with a rational valued bilinear form. We may extend the given bilinear form $L \times L \to \mathbb{Q}$ to $V \times V \to K$ by use of the tensor product $V = L \otimes K$, as described below.

We have the containment $\mathbb{Q} \le K$ of rings and so we have a bilinear map $L \times L \to K$. This corresponds to a unique $\mathbb{Z}$-linear map $L \otimes_{\mathbb{Z}} L \to K$. We also have the bilinear map $K \times K \to K$ (multiplication in K) and so we get a K-linear map $K \otimes K \to K$. Combining these we get $(L \otimes_{\mathbb{Z}} L) \otimes_{\mathbb{Z}} (K \otimes_K K) \to K$.

By properties of tensor products, the left side is isomorphic to $(K \otimes_{\mathbb{Z}} L) \otimes_K (K \otimes_{\mathbb{Z}} L)$ and so we have a bilinear map. $(K \otimes_{\mathbb{Z}} L) \otimes_K (K \otimes_{\mathbb{Z}} L) \to K$. Then take $V := K \otimes_{\mathbb{Z}} L$, a K-vector space, together with the embedding $L \to V$ given by $x \to 1 \otimes x \in K \otimes_{\mathbb{Z}} L$.

2.2.3 Cyclicity of the values of a rational bilinear form

Let L be a rational lattice (2.0.9) with symmetric bilinear form $f : L \times L \to \mathbb{Q}$. Let $v_1, \ldots, v_n$ be a basis of L. Then $\text{Im}(f)$ is a subset of the additive subgroup of $\mathbb{Q}$ generated by the finite set $f(v_i, v_j), i, j = 1, \ldots, n$.

It turns out that this subgroup equals $\text{Im}(f)$. This follows from the following result.

Theorem 2.2.6. *Suppose that A and B are finitely generated free modules for the PID R and that $f : A \times B \to K$ is an R-bilinear map to K, the field of fractions of R. Then there exists $c \in K$ so that $Im(f) = cR$.*

Sketch of Proof. We may assume f is R-valued, and take $v_1, \ldots, v_m$ a basis of A and $w_1, \ldots, w_n$ a basis for B. Then $cR = \sum_{i,j} f(v_i, v_j)R$. Write

$$A = (a_{ij}) = \begin{pmatrix} f(v_1, w_1) & \cdots & f(v_1, w_n) \\ \vdots & & \vdots \\ f(v_n, w_1) & \cdots & f(v_n, w_n) \end{pmatrix}.$$

There exist two invertible integer matrices P and Q, so that

$$PAQ = D = \begin{pmatrix} d_1 & & \\ & d_2 & \\ & & \ddots \end{pmatrix},$$

and $d_1 \mid d_2 \mid d_3, \ldots$. Since cR is also spanned by the entries of D, we see that $c\mathbb{Z} = d_1\mathbb{Z}$. Clearly there exist $y \in A, z \in B$ so that $f(y, z) = d_1$. If so, then

$f(ky, z) = kd_1$, so we get all of d_1R in $\text{Im}(f)$. In fact, suppose $P = (p_{ij})$, $Q = (q_{ij})$, we take $y = \sum_{k=1}^{m} p_{1j}v_j$, $z = \sum_{k=1}^{n} q_{k1}w_k$, then

$$f(y, z) = \sum_{j,k} p_{1j} f(v_j, w_k) q_{k1} = d_1. \qquad \square$$

Exercise 2.2.7. $n = 2$, $f(v_1, v_1) = 30$, $f(v_1, v_2) = f(v_2, v_1) = 18$, $f(v_2, v_2) = 9$. Find u_1, u_2 so that $f(u_1, u_2) = 3$.

2.2.4 Gram matrix

Definition 2.2.8. *Let R be a commutative ring, M, N a pair of R-modules and suppose that $f : M \times N \to R$ is a bilinear form. Let $v_1, \ldots, v_p$ be a sequence of elements in M and $w_1, \ldots, w_q$ a sequence of elements in N. The Gram matrix for this pair of sequences is the $p \times q$ matrix $G := (f(v_i, v_j))$.*

Often, M and N are free modules and the sequences are bases for M, N, respectively. It is trivial to check that an R-linear map T from M to N is an isometry if an ordered set of generators and its image under T have the same Gram matrix.

We suppose that the two input sequences $\mathbf{v}, \mathbf{w}$ for the Gram matrix $G_{\mathbf{v},\mathbf{w}}$ are bases of the free modules M, N.

Consider a replacement of $\mathbf{v}$ by $\mathbf{v}'$, where $\mathbf{v} = (v_1, \ldots, v_p)$, $\mathbf{v}' = (v_1', \ldots, v_p')$, and these sequences are related by the matrix $A_{p \times p} := (a_{ij})$, where $v_i' = \sum_{k=1}^{p} a_{ik} v_k$. The (i, j) entry of $G_{\mathbf{v}',\mathbf{w}}$ is $f(v_i', w_j) = \sum_{k=1}^{p} a_{ik} f(v_k, w_j)$. In other words, $G_{\mathbf{v}',\mathbf{w}} = A G_{\mathbf{v},\mathbf{w}}$.

Similarly, we consider a replacement of $\mathbf{w}$ by $\mathbf{w}'$, where $\mathbf{w} = (w_1, \ldots, w_q)$, $\mathbf{w}' = (w_1', \ldots, w_q')$, and these sequences are related by $B_{q \times q} := (b_{ij})$, where $w_j' = \sum_{k=1}^{q} b_{jk} w_k$. The (i, j) entry of $G_{\mathbf{v},\mathbf{w}'}$ is $f(v_i, w_j') = \sum_{k=1}^{q} b_{jk} f(v_i, w_k)$. In other words, $G_{\mathbf{v},\mathbf{w}'} = G_{\mathbf{v},\mathbf{w}} B^t$.

If the input sequences $\mathbf{v}, \mathbf{w}$ have equal length and are in duality, i.e. $f(v_i, w_j) = \delta_{ij}$, then $G_{\mathbf{v},\mathbf{w}} = I$. If the form is not symmetric, we may not have $G_{\mathbf{w},\mathbf{v}} = I$.

We now suppose that M and N span the same K-vector space, where K is the field of fractions of R. Suppose also that the bases $\mathbf{x}, \mathbf{y}$ are in duality. (We are still not assuming that f is symmetric).

Let $A = (a_{ij})$ be a matrix as above which expresses $\mathbf{y}$ in terms of $\mathbf{x}$ (i.e., $y_j = \sum_{i=1}^{n} a_{ji} x_i$) and let $B = (b_{ij})$ be a matrix as above which expresses $\mathbf{x}$ in terms of $\mathbf{y}$ (i.e., $x_j = \sum_{i=1}^{n} b_{ji} y_i$). Direct substitution shows that $AB = I$: $y_j = \sum_{p}^{n} a_{jp} x_p = \sum_{p,q}^{n} a_{jp} b_{pq} y_q$.

We have

$$G_{\mathbf{x},\mathbf{y}} = AB = I,$$
$$G_{\mathbf{x},\mathbf{x}} = G_{\mathbf{x},\mathbf{y}} B^t = B^t,$$
$$G_{\mathbf{y},\mathbf{x}} = A G_{\mathbf{x},\mathbf{x}} = A B^t,$$
$$G_{\mathbf{y},\mathbf{y}} = G_{\mathbf{y},\mathbf{x}} A^t = A G_{\mathbf{x},\mathbf{x}} A^t = A B^t A^t = A(AB)^t = A.$$

We compute that $G_{\mathbf{x},\mathbf{x}} G_{\mathbf{y},\mathbf{y}} = B^t A = A^{-t} A$.

Now assume that f is a symmetric form. Then $I = G_{\mathbf{x},\mathbf{y}} = G_{\mathbf{y},\mathbf{x}} = AB^t$, whence $A = A^t$ and $B = B^t$ and $G_{\mathbf{x},\mathbf{x}}G_{\mathbf{y},\mathbf{y}} = I$. i.e., $G_{\mathbf{x},\mathbf{x}}$, $G_{\mathbf{y},\mathbf{y}}$ form an inverse pair of matrices.

Example 2.2.9. *Suppose that there is a bilinear form on $M \cong \mathbb{Z}^2$, which has rank 2 and basis v_1, v_2 with Gram matrix $G := \begin{pmatrix} 1 & 2 \\ 3 & 4 \end{pmatrix}$. Here the form is not symmetric.*

We may extend the form to $\mathbb{Q}^2$ by the rule $f((a,b)^t, (c,d)^t) = (a,b)G(c,d)^t$, taking $v_1 = (1,0)^t$, $v_2 = (0,1)^t$. Then the right dual basis is given by the columns of $G^{-1} = \begin{pmatrix} -2 & 1 \\ \frac{3}{2} & -\frac{1}{2} \end{pmatrix}$.

The left dual basis is given by the columns of $G^{-t} = \begin{pmatrix} -2 & \frac{3}{2} \\ 1 & -\frac{1}{2} \end{pmatrix}$.

Definition 2.2.10. *A lattice L is integral if the bilinear form takes integral values only, i.e., $(L, L) \leq \mathbb{Z}$.*

Definition 2.2.11. *Suppose L is an integral lattice with symmetric bilinear form $(\,,\,)$. The even sublattice is $L^{even} = \{x \in L \mid (x,x) \in 2\mathbb{Z}\}$. Call an integral lattice even if it equals its even sublattice and otherwise call it odd.*

Remark 2.2.12. *(1) Equation $(x+y, x+y) = (x,x) + 2(x,y) + (y,y)$ says that the map $\alpha : L \to \mathbb{Z}/2\mathbb{Z}, x \mapsto (x,x) \bmod 2$ is a group homomorphism, so L^{even} is the kernel of the map α, which is a sublattice of index 1 or 2.*

(2) Suppose L is a rational lattice, S a subset of L, which spans L over $\mathbb{Z}$. Then we have the following statements:

(a) L is integral if $(S, S) \subseteq \mathbb{Z}$;

(b) L is even if L is integral and $(x, x) \in 2\mathbb{Z}$ for all $x \in S$.

If L is a lattice with a symmetric $\mathbb{Q}$-valued bilinear form, we let V be the ambient rational vector space, i.e. $V = \mathbb{Q} \otimes L$ with L identified with $1 \otimes L$. We assume that the form is nondegenerate.

Definition 2.2.13. *We define the dual of L (or, the dual lattice of L) to be*

$$L^* = \{x \in V \mid (x, y) \in \mathbb{Z}, \text{ for all } y \in L\} = \{x \in \mathbb{R}^n \mid (x, L) \subseteq \mathbb{Z}\}.$$

It does not matter in this definition if we replace V by $V_K := K \otimes L$, where K is any field of characteristic 0. To prove this, let $x_1, \ldots, x_n$ be a basis of L and $y_1, \ldots, y_n$ the dual basis, which lies in V. Then if $u \in V_K$ satisfies $(u, x) \in \mathbb{Z}$, for all $x \in L$, then we define $a_i := (u, x_i) \in \mathbb{Z}$. By nondegenacy, $u = \sum_i a_i y_i$.

Consequently, the dual of L always lies in $\mathbb{Q} \otimes L$ and so there exists an integer $m > 0$ so that $mL^* \leq L$. We may choose m to also satisfy $mL \leq L^*$. Therefore $L \cap L^*$ has finite index in both L and L^*.

Suppose that $\mathbf{x}, \mathbf{y}$ are bases of V which are in duality, such that $\mathbf{x}$ is a $\mathbb{Z}$-basis of L. Suppose also that L is an integral lattice. Then $G_{\mathbf{x},\mathbf{x}}$ is an integer matrix. Since $G_{\mathbf{x},\mathbf{x}}G_{\mathbf{y},\mathbf{y}} = I$, the product of their determinants is 1. Therefore, the determinant formula for the inverse of a matrix implies that $G_{\mathbf{y},\mathbf{y}}$ is an integer matrix if and only if $\det(G_{\mathbf{x},\mathbf{x}}) = \pm 1$.

Therefore, for an integer lattices, the dual lattice is not integral except when the determinant of a Gram matrix is ± 1.

Let us suppose that the rank n lattice spans Euclidean space $\mathbb{R}^n$. The bilinear form on L is just the restriction of the usual inner product.

Suppose that $\mathbf{x}, \mathbf{y}$ are finite length sequences in $\mathbb{R}^n$. Then we may write $G_{\mathbf{x},\mathbf{y}}$ as $X^t Y$, where X is a matrix whose columns are the vectors of $\mathbf{x}$ and Y is a matrix whose columns are the vectors of $\mathbf{y}$. This is because an entry of a matrix product is an "inner product" of a row vector of the left matrix and a column vector of the right matrix.

Assuming $\mathbf{x}, \mathbf{y}$ are in duality, $X^t Y = I$. By matrix theory, $Y X^t = I$ and $X Y^t = I$.

We then have $G_{\mathbf{x},\mathbf{x}} G_{\mathbf{y},\mathbf{y}} = X^t X Y^t Y = X^t (X Y^t) Y = X^t I Y = X^t Y = I$, giving another proof that Gram matrices for a lattice and its dual are inverses.

Exercise 2.2.14. *Let G be any symmetric real matrix which is positive definite. Prove that there exists $w_1, \ldots, w_n \in \mathbb{R}^n$ whose Gram matrix is G. It follows that if L is a rank n positive definite rational lattice, it is realized as the $\mathbb{Z}$-span of a basis of $\mathbb{R}^n$.*

Exercise 2.2.15. *Let $G := \begin{pmatrix} 2 & 1 \\ 1 & 2 \end{pmatrix}$. If $X = \begin{pmatrix} \sqrt{2} & \frac{1}{\sqrt{2}} \\ 0 & \frac{\sqrt{3}}{\sqrt{2}} \end{pmatrix}$, then the columns of X span a rank 2 free lattice in $\mathbb{R}^2$ with $G = X^t X$ as a Gram matrix.*

Find a few other X (essentially different from the above) which satisfy $X^t X = G$ and sketch the picture around the origin of the lattices which they generate.

Show that if X satisfies $G = X^t X$, then X does not have rational entries. In other words, a corresponding lattice in $\mathbb{R}^2$, does not lie in $\mathbb{Q}^2$.

(Moral: we may need square roots to draw pictures of a rational lattice.)

2.3 Discriminant group

Analyses of integral lattices L very often make use of the quotient L^*/L.

Definition 2.3.1. *Let L be a nonsingular integral lattice. The quotient $\mathcal{D}(L) := L^*/L$ is called the discriminant group of lattice L.*

Let L be an integral lattice, which is nonsingular of rank n, then the Gram matrix G has integral entries. Its Smith normal form is

$$D := \begin{pmatrix} d_1 & & \\ & d_2 & \\ & & \ddots \end{pmatrix},$$

with $d_1 d_2 \cdots d_n \neq 0, d_i \mid d_{i+1}$ for $i = 1, \ldots, n-1$.

We study the dual L^* of L. Since $\mathbb{Z}$ is a PID and L is a free $\mathbb{Z}$-module of rank n, there exists a basis $u_1, \ldots, u_n$ of L^*, so that $v_1 = d_1 u_1, v_2 = d_2 u_2, \ldots, v_n = d_n u_n$ is a basis of L. Hence

$$L^*/L = \frac{\mathbb{Z}u_1 \oplus \mathbb{Z}u_2 \oplus \cdots \oplus \mathbb{Z}u_n}{\mathbb{Z}v_1 \oplus \mathbb{Z}v_2 \oplus \cdots \oplus \mathbb{Z}v_n} \cong \frac{\mathbb{Z}u_1 \oplus \mathbb{Z}u_2 \oplus \cdots \oplus \mathbb{Z}u_n}{\mathbb{Z}d_1 u_1 \oplus \mathbb{Z}d_2 u_2 \oplus \cdots \oplus d_n \mathbb{Z}u_n} \cong$$

$$(\mathbb{Z}u_1/\mathbb{Z}d_1 u_1) \oplus (\mathbb{Z}u_2/\mathbb{Z}d_2 u_2) \oplus \cdots \oplus (\mathbb{Z}u_n/\mathbb{Z}d_n u_n) \cong \mathbb{Z}_{d_1} \oplus \mathbb{Z}_{d_2} \oplus \cdots \oplus \mathbb{Z}_{d_n}.$$

Let $\mathbf{x}, \mathbf{y}$ be any bases of, respectively, L and L^*. Then $G_{\mathbf{x},\mathbf{y}}$ (in earlier notation) is an integral matrix of determinant ± 1. If we take such $\mathbf{y} = y_1, \ldots, y_n$ and positive integers $d_1, \ldots, d_n$ so that $d_1, \ldots, d_n$ is a basis of L. Then $G_{\mathbf{x},\mathbf{y}} D$ is integrally equivalent to the Gram matrix $G_{\mathbf{x},\mathbf{x}}$. At once we deduce that a Gram matrix for L has D as its Smith normal form.

The determinant of L is therefore $\pm \mid \mathcal{D}(L) \mid$.

Example 2.3.2. *Let L be of rank 2 with basis α, β, set $G_L = \begin{pmatrix} 2 & 1 \\ 1 & 2 \end{pmatrix}$. It has invariants 1 and 3, so the discriminant group $\mathcal{D}(L) = L^*/L \cong \mathbb{Z}_3$. Consider $\nu = \frac{1}{3}(\alpha + \beta) \notin L$, thus $(\alpha, \nu) = \frac{1}{3}(\alpha, \alpha + \beta) = \frac{1}{3}(2+1) = 1$, similarly we have $(\beta, \nu) = 1$. Therefore $L^* = L + \mathbb{Z}\nu = L \cup (L+\nu) \cup (L+2\nu)$.*

Theorem 2.3.3. *("Index-determinant formula") Let L be a rational lattice, and M a sublattice of L of finite index $|L:M|$. Then*

$$\det(L) \, |L:M|^2 = \det(M).$$

Proof. Choose a basis $x_1, \ldots, x_n$ for L, denote the Gram matrix of L by $G_L = (f(x_i, x_j))$, with positive integers $d_1, d_2, \ldots, d_n$, so that M has a basis $d_1 x_1, d_2 x_2, \ldots, d_n x_n$. While the Gram matrix of lattice M is $G_M = (f(d_i x_i, d_j x_j)) = D G_L D$, where

$$D = \begin{pmatrix} d_1 & & \\ & d_2 & \\ & & \ddots \end{pmatrix},$$

and $G_L = (f(x_i, x_j))$, thus $\det(G_M) = \det(D)^2 \cdot \det(G_L)$. $\square$

Example 2.3.4. *Let $L = \mathbb{Z}^n$ be a square lattice with basis $\{e_i\}$ and $(e_i, e_j) = \delta_{ij}$. The Gram matrix is*

$$G_L = \begin{pmatrix} 1 & & & \\ & 1 & & \\ & & \ddots & \\ & & & 1 \end{pmatrix}.$$

Let $M := \{(x_1, \ldots, x_n) \in \mathbb{Z}^n \mid \sum x_i \text{ is even}\}$.
We have homomorphisms

$$L \xrightarrow{\alpha} \mathbb{Z} \xrightarrow{\beta} \mathbb{Z}/2\mathbb{Z}$$

and

$$\gamma := \beta \circ \alpha : (x_1, \ldots, x_n) \longmapsto \sum x_i \longmapsto \sum x_i \pmod{2}.$$

18   2 Bilinear Forms, Quadratic Forms and Their Isometry Groups

Thus, we have $L/M = Domain(\gamma)/Ker(\gamma) \cong \text{Im}(\gamma) \cong \mathbb{Z}_2$. So M has index 2 in L, and $|M^* : M| = \det(M) = \det(L) \cdot |L : M|^2 = 1 \cdot 2^2 = 4$.

We now describe M^* and the isomorphism types of $\mathcal{D}(M) = M^*/M$. Note that $M \leq L \leq M^*$.

Theorem 2.3.5. *Let M be the D_n lattice. Then $\mathcal{D}(M) \cong \mathbb{Z}_4$ if and only if n is odd, and $\mathcal{D}(M) \cong \mathbb{Z}_2 \oplus \mathbb{Z}_2$ if and only if n is even.*

Proof. Let M be the standard model of the D_n-lattice, i.e., $M := \{(x_1, \ldots, x_n) \in \mathbb{Z}^n \mid \sum x_i \in 2\mathbb{Z}\}$ Set $\nu := (1, \ldots, 1)$. Then $(\nu, x) \in 2\mathbb{Z}$ by the definition of M, $(\frac{1}{2}\nu, x) \in \mathbb{Z}$, so $\frac{1}{2}\nu \in M^*$. Since $|M^* : L| = 2$ and $\frac{1}{2}\nu \notin L$, we have $M^* = L + \mathbb{Z} \cdot \frac{1}{2}\nu$. We conclude that $\mathcal{D}(M) \cong \mathbb{Z}_4$ if and only if n is odd, and $\mathcal{D}(M) \cong \mathbb{Z}_2 \oplus \mathbb{Z}_2$ if and only if n is even, since in M^*/M, $2(\frac{1}{2}\nu + M) = \nu + M$. $\square$

Notation 2.3.6. *We use the notation of (2.3.5) to describe some overlattices of D_n. As above, $\nu := (1, 1, \ldots, 1)$ and we let $\nu' := (-1, 1, \ldots, 1)$.*

Define lattices $HS_n^+ := D_n + \mathbb{Z}\frac{1}{2}\nu$ and $HS_n^- := D_n + \mathbb{Z}\frac{1}{2}\nu'$. (The "HS" is used since HS is spanned by the weights of a half-spin module for a Lie algebra of type D_n. These lattices are sometimes denoted $D_n^{\pm}$.) They are isometric, since a reflection at any standard basis vector interchanges them.

If n is odd, they both equal the dual of D_n, so have determinant $\frac{1}{4}$.

Now assume that n is even. Then they are distinct and contain D_n with index 2 (since n is even), so are unimodular. They are integral if and only if n is divisible by 4 and they are even integral if and only if n is divisible by 8.

We conclude this section with a general observation.

Proposition 2.3.7. *Let L be an integral lattice.*

(1) On $\mathcal{D}(L)$, there is a natural $\mathbb{Q}/\mathbb{Z}$-valued bilinear form. If L is even, the function $x \mapsto \frac{1}{2}(x,x) + \mathbb{Z}$ is constant on cosets of L, so gives a $\mathbb{Q}/\mathbb{Z}$-valued quadratic form whose associated bilinear form is the above bilinear form.

(2) The overlattice K of L in $\mathbb{Q} \otimes L$ is integral if and only if $K \leq L^*$ and the additive subgroup K/L is totally singular with respect to the form of (1).

Proof. The form is the well-defined bilinear function which pairs $x + L, y + L$ to $(x, y) + \mathbb{Z} \in \mathbb{Q}/\mathbb{Z}$. Verifications are formal. $\square$

2.4 Relations between a lattice and sublattices

The results of this section will be very helpful in computing invariants of lattices.

Theorem 2.4.1. *Let L be a nonsingular integral lattice and M a sublattice which is a direct summand of L (i.e., there exists a sublattice $B \leq L$ such that $L = M \oplus B$ as abelian groups). Then*

(1) the natural map $\psi : L^* \to M^*$ is onto;
(2) the cokernel of $\psi|_L$ is a quotient of the discriminant group $\mathcal{D}(L)$;
(3) define π to be the composition of ψ and the quotient map $M^* \to M^*/M$, then we have $Ker(\pi) = M + ann_{L^*}(M)$ and $Ker(\pi|_L) = M + ann_L(M)$;

(4) $Coker(\pi|_L)$ is a quotient of the cokernel $Coker(\psi|_L)$, so is a finite abelian group whose order divides the greatest common divisor $gcd(|\mathcal{D}(L)|, |\mathcal{D}(M)|)$. Therefore, $\pi|_L$ is surjective if these group orders are relatively prime.

Proof. (1) Suppose $x_1, \ldots, x_p$ is a basis of M, then extend it to be a basis $x_1, \ldots, x_p, \ldots, x_n$ of L, so that we get a basis of L^* by the dualization of it.

(2) Denote $\psi|_L$ by h, then $h(L) \cong L/Ker(h) \cong L/L \cap Ker(\psi) \cong L + Ker(\psi)/Ker(\psi)$, so we obtain that $Coker(h) = M^*/h(L) \cong \psi(L^*)/h(L) \cong \psi(L^*)/\psi(L+Ker(\psi)) \cong L^*/(L+Ker(\psi))$, which is a quotient of $L^*/L \cong \mathcal{D}(L)$.

(3) As we defined, the map π is the composition $L^* \xrightarrow{\psi} M^* \xrightarrow{\zeta} M^*/M$, so $Ker(\pi) = M + ann_{L^*}(M)$. Therefore $Ker(\pi|_L) = L \cap Ker(\pi) = M + L \cap ann_{L^*}(M) = M + ann_L(M)$.

(4) Note that $Coker(\pi|_L) = \mathcal{D}(M)/\text{Im}(\pi|_L)$ is a quotient of the finite abelian group $\mathcal{D}(M)$ whose order divides $|\mathcal{D}(M)|$, and $Coker(\pi|_L)$ is a quotient of $Coker(\psi|_L)$ whose order divides $|\mathcal{D}(L)|$ due to (2). □

Corollary 2.4.2. *If* $\det(L) = 1$, *then the restriction* $\psi|_L$ *of the natural map* $\psi: L^* \to M^*$ *to L is surjective. Furthermore, M and $ann_L(M)$ have isomorphic discriminant groups.*

2.5 Involutions on quadratic spaces

We study isometries of order 2 on quadratic spaces in general and in more detail on lattices. The basic example of reflection generalizes to RSSD maps.

Let R be a ring, (A, Q) a quadratic space over R. Take $0 \neq v \in A$ so that $0 \neq Q(v)$ divides every value of f, the associated symmetric bilinear form. This happens, for example, if $Q(v)$ is a unit in R.

Definition 2.5.1. *Define* $r_v \in End(A)$ *by the formula*

$$x \mapsto x - \frac{f(x,v)}{Q(v)} v.$$

This transformation is called the reflection on A at v.

Lemma 2.5.2. r_v *is an isometry of Q.*
Proof. Let $x \in A$. Then

$$Q(r_v(x)) = Q\left(x - \frac{f(x,v)}{Q(v)} v\right) = Q(x) + Q\left(-\frac{f(x,v)}{Q(v)} v\right) + f\left(x, \frac{f(x,v)}{Q(v)} v\right)$$
$$= Q(x) + \frac{f(x,v)^2}{Q(v)^2} Q(v) - \frac{f(x,v)}{Q(v)} f(x,v) = Q(x).$$

□

Definition 2.5.3. *An endomorphism of a vector space is called a transvection if it has the form $1 + u$, where u has rank 1 and $u^2 = 0$. There exists a basis so that the transvection has matrix*

$$\begin{pmatrix} 1 & 1 & 0 & \ldots & 0 & 0 \\ 0 & 1 & 0 & \ldots & 0 & 0 \\ 0 & 0 & 1 & \ldots & 0 & 0 \\ \vdots & \vdots & \vdots & & \vdots & \vdots \\ 0 & 0 & 0 & \ldots & 1 & 0 \\ 0 & 0 & 0 & \ldots & 0 & 1 \end{pmatrix}$$

The map r_v is called the *reflection* on A at v, though when R has characteristic 2, we usually call it the *transvection at v*. Note that r_v takes v to $-v$ and is the identity on $v^\perp$. Thus, we have a diagonal matrix if 2 is a unit in the commutative ring R.

When R has characteristic 2, r_v fixes v. The map r_v is the identity if and only if $(v, A) = 0$. When r_v is not the identity, $0 \neq (r_v - 1)A \leq Rv$ and so r_v is a nondiagonalizable transformation with minimum polynomial $(x-1)^2$. In fact, $r_v - 1$ has rank 1. When R is a field of characteristic 2, r_v has a Jordan canonical form which is a block diagonal sum of 1×1 blocks and a single Jordan block of size 2×2 in its Jordan canonical form.

2.6 Standard results on quadratic forms and reflections, II

We continue to state general results on quadratic forms.

Theorem 2.6.1. *Let F be a field, V a finite dimensional vector space over F and Q a nondegenerate quadratic form. Suppose the characteristic of the field F is not 2. Then every isometry is a product of reflections.*

Recall that we say a field F is perfect if all roots of the polynomial $x^p - a$ are in F for any $x \in F$, where p is the characteristic of F.

Theorem 2.6.2. *Let F be a field, V a finite dimensional vector space over F and Q a nondegenerate quadratic form. Suppose F is perfect with characteristic 2. Then every isometry is a product of transvections $\left(i.e., x \mapsto x - \frac{(x,v)}{Q(v)}v\right)$, except when F is the two-element field $\mathbb{F}_2$, $\dim(V) = 4$ and Q has a "maximal Witt index": there exists a subspace $U \leq V$ with $\dim(U) = 2$ and $Q|_U \equiv 0$ (i.e., U is totally singular of dimension 2).*

Theorem 2.6.3. *Let F be a perfect field of characteristic 2, V an n dimensional vector space over F and Q a nondegenerate quadratic form on V. Given $n = 2m$, there are, up to isometry, just two quadratic spaces of dimension n. They are distinguished by having Witt index $m - 1$ or m (the dimension of the maximal totally singular subspace).*

2.6.1 Involutions on lattices

On an integral lattice, L, the reflection at v leaves the integral lattice invariant if $\frac{2(x,v)}{(v,v)}v$ is in the lattice, for all $x \in L$. This is one of the simplest ways to create nontrivial isometries in a lattice. A sufficient condition for this to happen is $(v,v) = \pm 1$ or ± 2. The condition is not necessary, however.

2.6 Standard results on quadratic forms and reflections, II

Definition 2.6.4. *In an integral lattice L, a root is an element $v \in L$ with norm $(v, v) = 2$.*

Remark 2.6.5. *If v is a root in the integral lattice L, $r_v : L \to L$ is an isometry and it induces a map $\overline{r}_v : L/2L \to L/2L$. This map is the identity if and only if $(v, L) \leq 2\mathbb{Z}$, or equivalently, if $L = \mathbb{Z}v \perp \text{ann}_L(v)$. It is a transvection if and only if $(v, L) = \mathbb{Z}$, or equivalently, if and only if $\overline{(v, L)} = \overline{\mathbb{Z}}$.*

Now suppose that L is an even lattice.

We claim that $\overline{r}_v$ is a transvection which respects the quadratic form $x + 2L \mapsto \frac{1}{2}(x, x) \pmod 2$. Since $Q(\overline{v}) = 1 \pmod 2$, this is clear.

When do two roots give the same transvection on the lattice mod 2?

Lemma 2.6.6. *Let L be an even lattice and let v, w be roots such that $(L, v) = \mathbb{Z} = (L, w)$ (whence $\overline{r}_v$ and $\overline{r}_w$ are not the identity). Then v and w are proportional if and only if $\overline{r}_v = \overline{r}_w$.*

Proof. We assume that $\overline{r}_v = \overline{r}_w$. There exists $x \in L$ so that $r_v(x) = r_w(x) \neq x$, i.e., $x - (x, v)v \equiv x - (x, w)w \not\equiv x \pmod{2L}$. This means that (x, v) and (x, w) are odd. Then $\overline{v}$ and $\overline{w}$ are proportional. This means that there exists $y \in L$ so that $v - w = 2y$. Assuming v and w are not proportional, the triangle inequality implies that left side has length strictly less than $2\sqrt{2}$ and the right side has length at least $2\sqrt{2}$ since $y \neq 0$. This contradiction proves that v and w are proportional. $\square$

We treat the important topic of root lattices later in this book. Many lattices of interest, however, do not contain roots despite having large automorphism group. The following concept is a useful generalization of a reflection on a lattice.

Definition 2.6.7. *A sublattice M of an integral lattice L is RSSD (relatively semiselfdual) if and only if $2L \leq M + \text{ann}_L(M)$. The property that $2M^* \leq M$ is called SSD (semiselfdual). Associated to an RSSD sublattice is the orthogonal transformation t_M, defined on the ambient vector space, by the condition that it acts as -1 on M and as 1 on $\text{ann}_L(M)$. We call t_M an RSSD involution.*

The SSD property implies the RSSD property. The RSSD property is more general and is often more useful. For example, if M is RSSD in L and $M \leq J \leq L$, then M is RSSD in J, whence (by the next result) the involution t_M leaves both J and M invariant.

Proposition 2.6.8. *Suppose that the sublattice M of L is RSSD. The involution t_M is an orthogonal transformation which maps L onto L, i.e., $t_M \in O(L)$.*

Proof. We may assume that M is a direct summand of L. The action of t_M on the quotient group $\mathbb{Q}L/M \cong M^\perp \times (\mathbb{Q}M/M)$ fixes L/M because the RSSD hypothesis implies that $L \leq M^\perp + \frac{1}{2}M$. Therefore, t_M takes L into L. Since t_M is orthogonal, it takes L onto L. $\square$

The next result shows that isometries of order 2 on lattices and RSSD sublattices are essentially equivalent concepts.

Definition 2.6.9. *Suppose that t is an involution on the lattice L. For $\varepsilon = \pm$, define $L^\varepsilon := L^\varepsilon(t) := \{x \in L \mid tx = \varepsilon x\}$. These are the eigenlattices of t. The sum $Tel(L,t) := L^+(t) + L^-(t)$ is the total eigenlattice for t on L.*

Proposition 2.6.10. *Suppose that t is an involution on the lattice L. Then each eigenlattice L^ε is RSSD in L.*

Proof. Let $x \in L$. Then $2x = (x + tx) + (x - tx)$. Since $x + \varepsilon tx \in L^\varepsilon$ and L^+ and L^- are orthogonal, we are done. $\square$

Example 2.6.11. *An example of a SSD lattice is $\sqrt{2}\,U$, where U is a unimodular lattice. More examples include the family of D_n lattices for n even and the family of Barnes-Wall lattices. They are treated later in this book.*

Example 2.6.12. *Let the integral lattice L have basis $v_1, \ldots, v_4$ and Gram matrix*

$$G := \begin{pmatrix} 6 & 2 & 3 & 1 \\ 2 & 4 & 1 & 2 \\ 3 & 1 & * & * \\ 1 & 2 & * & * \end{pmatrix}.$$

Then $M := \mathrm{span}\{v_1, v_2\}$ is RSSD, so L has an isometry of order 2 which is -1 on M and 1 on $\mathrm{ann}(M)$. Note that the entries of the integer matrix G which do not involve M are irrelevant to existence of such an involution.

Lemma 2.6.13. *If the sublattice M is a direct summand of the integral lattice L and $(\det(L), \det(M)) = 1$, then SSD and RSSD are equivalent properties for M.*

Proof. It suffices to assume that M is RSSD in M and prove that it is SSD.

Let V be the ambient real vector space for L and define $A := \mathrm{ann}_V(M)$. The natural image of L^* in M^* is M^*. Since $(\det(L), \det(M)) = 1$, the natural image of L in $\mathcal{D}(M)$ is $\mathcal{D}(M)$, i.e., $L + A = M^* + A$. We have $2(L+A) = 2(M^* + A)$, or $2L + A = 2M^* + A = M \perp A$. The left side is contained in $M + A$, by the RSSD property. So, $2M^* \leq M + A$. If we intersect both sides with $\mathrm{ann}_V(A)$, we get $2M^* \leq M$. This is the SSD property. $\square$

Lemma 2.6.14. *Suppose that L is an integral lattice and $N \leq M \leq L$ and both M and N are RSSD in L. Assume that M is a direct summand of L. Then $\mathrm{ann}_M(N)$ is an RSSD sublattice of L.*

Proof. This is easy to see on the level of involutions. Let $t := t_M, u := t_N$ be the involutions associated to M, N. They are in $O(L)$ and they commute since u is the identity on $\mathrm{ann}_L(M)$, where t acts as the scalar 1, and since u leaves invariant $M = \mathrm{ann}_L(\mathrm{ann}_L(M))$, where t acts as the scalar -1. Therefore, $s := tu$ is an involution. Its negated sublattice $L^-(s)$ is RSSD (2.6.7), and this is $\mathrm{ann}_M(N)$. $\square$

Later, as we study particular lattices, we shall mention examples of RSSD sublattices and their associated involutions.

2.7 Scaled isometries: norm doublers and triplers

There are interesting and useful maps which carry a lattice to itself and multiply norms by a scale factor. We know of examples which come from isometries and which double and triple norms.

Definition 2.7.1. *A fourvolution is an orthogonal transformation whose square is -1.*

Lemma 2.7.2. *Suppose that f is a fourvolution on the lattice L. Then $f - 1$ doubles norms. Furthermore, $(f - 1)^2 = -2f$. so $(f - 1)^2 = 2L$.*

Proof. We calculate that $((f - 1)x, (f - 1)y) = (fx, fy) + (x, y) - (fx, y) - (x, fy)$. We have $(fx, fy) = (x, y)$ and $(fx, y) = (ffx, fy) = -(x, fy)$. The second statement is easy to prove. $\square$

Lemma 2.7.3. *Suppose that h is an order 3 isometry without eigenvalue 1. Then the map $h - 1$ triples norms. Also, $(h - 1)^2 = -3h$, so $(h - 1)^2 L = 3L$.*

Proof. The map h satisfies $h^2 + h + 1 = 0$. We calculate $((h - 1)x, (h - 1)y) = (hx, hy) + (x, y) - (hx, y) - (x, hy)$. We have $(hx, hy) = (x, y)$ and $(hx, y) + (x, hy) = (hx, y) + (hx, h^2 y) = (hx + h^2 x, y) = -(x, y)$. The second statement is easy to prove. $\square$

Endomorphisms which multiply norms by other primes do not seem to arise as in (2.7.2),(2.7.3).

3
General Results on Finite Groups and Invariant Lattices

Suppose L is a rational lattice with a positive definite bilinear form f. We will assume that $L \leq \mathbb{R}^n$ from now on.

3.1 Discreteness of rational lattices

Definition 3.1.1. *We say a subset $S \subseteq \mathbb{R}^n$ is discrete if for any element $v \in \mathbb{R}^n$, there is a positive real number $\epsilon > 0$, such that $B_\varepsilon(v) \cap S \subseteq \{v\}$. Here, $B_\varepsilon(v)$ means the open ball centered at v.*

Example 3.1.2. (1) *The set of integers $\mathbb{Z}$ is discrete in $\mathbb{R}$;*
(2) *For a real number $\alpha > 1$, the set $\{\alpha^{-i} \mid i = 0, 1, 2, \ldots\} \subseteq \mathbb{R}$ is not discrete (since $0 \in \mathbb{R}$ is a limit point).*

Theorem 3.1.3. *The rank n rational lattice L in $\mathbb{R}^n$ is discrete.*

Proof. We know that $(L, L) = \{(x, y) \mid x, y \in L\} = \mathbb{Z}c$, for some $c \in \mathbb{Q}$, and $c \neq 0$ if $n > 0$. Suppose that there are elements $x_1, x_2, \ldots$ in L so that the sequence x_i converges to $x^* \in \mathbb{R}^n$. Then, there is an integer m so that if $i > m$, $|x_i - x^*| < \sqrt{|c|}/2$. That means $(x_i - x_j, x_i - x_j) = |x_i - x_j|^2 \leq (|x_i - x^*| + |x_j - x^*|)^2 < |c|$ for all $i, j > m$. But $x_i - x_j \in L$, and if $x_i \neq x_j$, $0 \neq (x_i - x_j, x_i - x_j) \in \mathbb{Z}c$. We conclude that $x_i = x_j$ for all $i, j \geq m$. $\square$

3.2 Finiteness of the isometry group

Definition 3.2.1. *Suppose L is a positive definite rational lattice with finite rank. We define the isometry group of L to be*

$$O(L) := \{g \in O(\mathbb{R}^n) \mid g(L) = L\}.$$

Theorem 3.2.2. *Suppose L is a positive definite rational lattice with a finite rank. Then the isometry group $O(L)$ of L is a finite group.*

Proof. We observe that the orthogonal group on $\mathbb{R}^n$ preserves all spheres $\{x \in \mathbb{R}^n \mid (x,x)^{1/2} = r, 0 < r \in \mathbb{R}\}$. Now, L has a basis $v_1, \ldots, v_n$ which is also a basis of $\mathbb{R}^n$. So we can take a ball $B = B_r(0)$, so that all these elements $v_1, \ldots, v_n \in B$. We claim that the intersection $L \cap B$ is finite. Suppose not. Then there is a sequence of distinct points $x_1, x_2, \ldots$ in $L \cap B$ converging to a point $x^* \in B$ for the compactness of B. Since L is discrete, this sequence is eventually constant. The claim follows.

Note that the group $O(L)$ permutes the finite set $L \cap B$, so we have a group homomorphism $\alpha : O(L) \to Sym_{L \cap B}$ Let $K = Ker(\alpha) \leq GL(\mathbb{R}^n)$. This map is a monomorphism since an element of the kernel fixes each $v_1, \ldots, v_n$. $\square$

Remark 3.2.3. (1) *We remark that if L is a lattice and g is an orthogonal transformation which satisfies $g(L) \leq L$, then $g(L) = L$. The reason is that $g(L)$ and L have the same Gram matrix, so by the index-determinant formula (2.3.3), $|L : g(L)| = 1$.*

(2) An indefinite lattice may have finite or infinite automorphism group. The lattice with Gram matrix $\begin{pmatrix} 0 & 1 \\ 1 & 0 \end{pmatrix}$ is indefinite and has automorphism group which is dihedral of order 8 (hint: prove that there are just two rank 1 sublattices which are totally singular). The rank $2n$ lattice with Gram matrix $\begin{pmatrix} 0 & I_n \\ I_n & 0 \end{pmatrix}$ contains in its isometry group the set of matrices $\begin{pmatrix} I_n & A \\ 0 & I_n \end{pmatrix}$, where $A + A^t = 0$. This is an infinite set if $n \geq 2$.

3.3 Construction of a G-invariant bilinear form

Existence of lattices invariant under a rational representation of a finite group is easy to prove. Classifying them all is generally hard.

Theorem 3.3.1. *If $G \leq GL(n, \mathbb{Q})$ is a finite group, then $\mathbb{Q}^n$ contains a G-invariant free abelian group L of rank n. Also, there is a symmetric bilinear form $f : L \times L \to \mathbb{Z}$ which is positive definite and G-invariant.*

Proof. Let $v_1, \ldots, v_n$ be any basis of $\mathbb{Q}^n$. Then the set $\{gv_i \mid g \in G, i = 1, \ldots, n\}$ is a finite set of vectors in $\mathbb{Q}^n$. Let L be the $\mathbb{Z}$-span of this set. It is a finitely generated free abelian group so is isometric to $\mathbb{Z}^m$ for some integer m. Since $L \supseteq \mathbb{Z}v_1 \oplus \cdots \oplus \mathbb{Z}v_n$, we get $m \geq n$ by PID theory.

Let $u_1, \ldots, u_m$ be a basis for L. We claim that $u_1, \ldots, u_m$ are linearly independent in $\mathbb{Q}^n$. Suppose $\sum_{i=1}^m \frac{a_i}{b_i} u_i = 0$ in $\mathbb{Q}^n$, with $a_i, b_i \in \mathbb{Z}$ and $b_1, \ldots, b_m \neq 0$. Then $\sum_{i=1}^m a'_i u_i = 0$ where $a'_i = a_i \prod_{j \neq i} b_j \in \mathbb{Z}$. This is a linear dependence relation in L. Hence all coefficients a'_i are zero and all a_i are zero. That implies $m \leq n$, therefore $m = n$.

Next, we will find an f so that it is integral, positive definite and G-invariant. Take a bilinear form $h : L \times L \to \mathbb{Z}$ which is positive definite. For example, one could take a basis of L to be an orthogonal set. For $v, w \in \mathbb{Q}^n$, we define $f(v, w) := \sum_{g \in G} h(gv, gw)$. If $g' \in G$, then $f(g'v, g'w) = \sum_{g \in G} h(gg'v, gg'w) =$

$f(v, w)$, whence f is G-invariant. For the positive definite property, take any vector $0 \neq v \in \mathbb{Q}^n$. We have $f(v, v) = \sum_{g \in G} h(gv, gv) > 0$ since h is positive definite. $\square$

3.4 Semidirect products and wreath products

We review a few constructions from group theory. These will be relevant to the study of isometry groups of general lattices, which we begin in the next section.

Definition 3.4.1. *Let A, B be groups and let $\theta : B \to Aut(A)$ be a group homomorphism. The semidirect product $A \rtimes_\theta B$ is defined to the group with underlying set $\{(x, y) \mid x \in A, y \in B\}$ and group operation $(a, b)(a', b') = (a\, (\theta(b)(a')), bb')$*

Definition 3.4.2. *Let B be a group and suppose that B acts on sets X, Y. Let $F := Functions(X, Y)$ be the set of all functions from X to Y. The rule $B \times F \to F$, given by $x \mapsto bf(b^{-1}x)$, for $x \in X, f \in F, b \in B$, gives a left action of B on F. This action satisfies $b(f(x)) = (bf)(bx)$, for $x \in X, f \in F, b \in B$.*

Definition 3.4.3. *If we are given an action π of B on the set X and A is a group, we have an action of B on $Functions(X, A)$, where we take the action of B on A to be trivial. Note that $Functions(X, A)$ is a group with respect to pointwise multiplication of the function values. The semidirect product of B and $Functions(X, A)$ is called the wreath product of B and A with respect to π.*

When the set X is finite, the wreath product may be expressed with n-tuples because an n-tuple $(a_1, \ldots, a_n)$ may be thought of as a function with domain $\{1, 2, \ldots, n\}$ and codomain A. We write this out in the next definition.

Definition 3.4.4. *Let A, B be groups and $n \geq 0$. Let $\pi : B \to Sym_n$ be a homomorphism. Then the wreath product $A \wr_\pi B$ is the semidirect product $A^n \rtimes_\theta B$ of B, where $A^n := A \times A \times \cdots \times A$, and $\theta : B \to Aut(A^n)$ is given by $\theta(b)(a_1, a_2, \ldots, a_n) := (a_{\pi(b)^{-1}(1)}, a_{\pi(b)^{-1}(2)}, \ldots, a_{\pi(b)^{-1}(n)})$.*

A common notation for wreath product is $A \wr B$ (with π not specified), the usual meaning of which is that π is the regular representation of B, i.e., the action of B on B by left-multiplication. However, the context may suggest another interpretation of π.

Exercise 3.4.5. (1) $\mathbb{Z}_2 \wr \mathbb{Z}_2 \cong Dih_8$.

(2) *Find an example of a wreath product and a subgroup of it which is not a wreath product in a nontrivial way. (Any group is a wreath product in two trivial ways: $G \wr 1 \cong G \cong 1 \wr G$).*

(3) *Give examples of wreath products $A \wr_\pi B$ where the defining normal subgroup A^n is a characteristic subgroup and an example where A^n is not a characteristic subgroup.*

(4) *The group of all monomial matrices in $GL(n, \mathbb{C})$ is $\mathbb{C}^\times \wr Sym_n$ (the missing π would be the natural action of the group of permutation matrices on a set of n basis elements).*

Is the intersection of this monomial group with $SL(n, \mathbb{C})$ a nontrivial wreath product?

Lemma 3.4.6. *Let $W := (G_1 \times G_2)\langle t \rangle$ be a wreath product where $|t| = 2$ and $G_1^t = G_2$. Then every involution of W outside $G_1 \times G_2$ is conjugate to t by an element of G_1 (and the same for conjugacy by G_2).*

Proof. Suppose $(g_1, g_2) \in G_1 \times G_2$ so that $(g_1, g_2)t = u$ is an involution. Then $1 = u^2 = ((g_1, g_2)t)^2 = (g_1, g_2)t(g_1, g_2)t = (g_1, g_2)(g_2, g_1)t^2 = (g_1 g_2, g_2 g_1)t^2 = 1$. So g_1, g_2 are inverses. Furthermore, $(g_1, 1)t(g_1, 1)^{-1} = (g_1, 1)(1, g_1^{-1})t = (g_1, g_1^{-1})t = (g_1, g_2)t$. □

3.5 Orthogonal decomposition of lattices

Definition 3.5.1. *Suppose L is an integral lattice. Denote the root set by $\Phi(L) = \{x \in L \mid (x, x) = 2\}$. The root sublattice of the lattice L is the $\mathbb{Z}$-span of $\Phi(L)$. We call L a root lattice if L equals its root sublattice.*

Definition 3.5.2. *Suppose L is a positive definite integral lattice. A vector $x \in L$ is said to be decomposable if there are vectors $y, z \in L$ so that $x = y + z$ with $(y, z) = 0$ and $(y, y) > 0, (z, z) > 0$. If $x \neq 0$ is not decomposable, then x is said to be indecomposable. We call the nonzero lattice L indecomposable if there is an orthogonal decomposition $L = A \perp B$ for sublattices A and B implies $A = 0$ or $B = 0$.*

Theorem 3.5.3. *Suppose L is a nonzero positive definite integral lattice. Then there exist nonzero sublattices $L_1, \ldots, L_m$, such that*
(1) $L = L_1 \perp \cdots \perp L_m$; and
(2) L_i is orthogonally indecomposable for all $i = 1, \ldots, m$.
Furthermore, if $L = L'_1 \perp \cdots \perp L'_{m'}$ is another decomposition as in (1) and (2), then $m' = m$ and there exists a permutation π of $\{1, 2, \ldots, m\}$ such that $L_i = L'_{\pi(i)}$.

Proof. Write $I(L)$ the set of indecomposable vectors in L. Since L is integral and positive definite, every nonzero vector in L is a sum of indecomposables, i.e., $L = Span_{\mathbb{Z}}(I(L))$.

We define a relation $\sim$ on $I(L)$ so that $x \sim y$ if and only if $(x, y) \neq 0$. Such a relation $\sim$ generates an equivalence relation, i.e., any two indecomposable vectors are equivalent if and only if there exist a chain connecting them by the relation $\sim$. We partition $I(L)$ into equivalence classes. The number of such classes is less than the rank of L, since the orthogonal sets of nonzero vectors are linearly independent. Let $C_1, C_2, \ldots, C_n$ be such equivalence classes, and define $L_i := Span_{\mathbb{Z}}(C_i)$, for $i = 1, \ldots, n$.

We show that (a) $L = \sum_{i=1}^n L_i$ and (b) $L_i \cap (\sum_{j \neq i} L_j) = 0$ for $i = 1, \ldots, n$. The statement (a) is easy to see since L is spanned by $I(L)$. For (b), take any $v \in L_i \cap (\sum_{j \neq i} L_j)$. Note that for all i, $L_i \cap (\sum_{j \neq i} L_j) \subseteq \sum_{j \neq i} L_j \subseteq ann_L(L_i)$, so that $(v, v) = 0$. This implies that implies $v = 0$.

We prove that each L_i is orthogonally indecomposable. Suppose the decomposition $L_i = A \perp B$ is nontrivial. Take a vector $x \in C_i$. Since x is indecomposable, $x \in A$ or $x \in B$. This implies that $C_i = (C_i \cap A) \cup (C_i \cap B)$, a contradiction to the fact that C_i is an equivalence class.

3.5 Orthogonal decomposition of lattices

Suppose that M is a sublattice of L which is an orthogonal direct summand, i.e., there exists a sublattice $N \leq L$ such that $L = M \perp N$. We shall prove that M is a sum of subset of the L_i, i.e., $M = Span\{L_i \mid L_i \leq M\}$. In fact, if $x \in L$ is indecomposable then $x \in M$ or $x \in N$. So if $x, y \in C_i$, then $\{x,y\} \subseteq M$ or $\{x,y\} \subseteq N$, which implies that $L_i = (L_i \cap M) \perp (L_i \cap N)$, hence one of $L_i \cap M$, $L_i \cap N$ is zero. $\square$

In the following, we give some applications to the structure of the isometry group $O(L)$, where the lattice L is positive definite and integral.

Suppose the lattice has a decomposition $L = L_1 \perp \cdots \perp L_n$ with each L_i indecomposable. Take an isometry $g \in O(L)$. Then we have $L = g(L) = g(L_1) \perp \cdots \perp g(L_n)$, each $g(L_i)$ is orthogonally indecomposable. So we get a permutation of $\{L_1, \ldots, L_n\}$.

Suppose we have an isometry $\psi : L_i \to L_j$. For $x_i, y_i \in L_i, i = 1, \ldots, n$, we define $h : L \to L$ by $(x_1, \ldots, x_n) \mapsto (y_1, \ldots, y_n)$, $y_k = x_k$ if $k \neq i, j$; $y_i = \psi^{-1}(x_j)$; $y_j = \psi(x_i)$. This gives a transposition on the set $\{L_1, \ldots, L_n\}$. We partition the indices as follows, $\{L_{i_1}, \ldots, L_{i_r}\}$ are all L_i's of a fixed isometry type; $\{L_{j_1}, \ldots, L_{j_s}\}$ are all L_j's of a second, distinct isometry type; etc. So we get an embedding of groups $Sym_r \times Sym_s \times \cdots \hookrightarrow O(L)$. For the special case $L_1 \cong \cdots \cong L_n$, we have $O(L) \cong (O(L_1) \times \cdots \times O(L_n)) \rtimes Sym_n \cong O(L_1) \wr Sym_n$, a wreath product.

Example 3.5.4. *The isometry group of the lattice $L = \mathbb{Z}^n$. We have a canonical decomposition $\mathbb{Z}^n = \mathbb{Z}e_1 \perp \cdots \perp \mathbb{Z}e_n$. Since $O(\mathbb{Z}e_i) = \{\pm 1_{\mathbb{Z}e_i}\} \cong \mathbb{Z}_2$, then $O(\mathbb{Z}^n) \cong \mathbb{Z}_2 \wr Sym_n$, which has order $2^n \cdot n!$. Therefore $O(\mathbb{Z}^n) = \{DP \mid D$ is a diagonal matrix*

$$\begin{pmatrix} \pm 1 & & \\ & \ddots & \\ & & \pm 1 \end{pmatrix}_{n \times n},$$

P is a permutation matrix $\}$, with respect to the standard basis $e_1, \ldots, e_n$.

4
Root Lattices of Types A, D, E

4.1 Background from Lie theory

A root lattice is an integral lattice spanned by roots (vectors of norm 2). Since roots have even norm, a root lattice is an even lattice.

The terminology comes from the theory of root systems for finite dimensional Lie algebras over the complex numbers. The classification of such simple Lie algebras gives a bijection between isomorphism types of simple Lie algebras and indecomposable root systems, which may be classified as an independent combinatorial problem. We shall not require knowledge of Lie algebras in this book.

The meaning of "root" in the theory of finite dimensional complex Lie algebras allows different lengths. When unequal lengths occur in an orthogonally indecomposable root system, the ratio of lengths is $\sqrt{2}$ (for systems of type B_n, C_n and F_4) or $\sqrt{3}$ (for G_2). For background, see [11, 47].

We shall assume this well-known classification of root systems and use the root lattices of types A_n, D_n and E_6, E_7 and E_8 in this book. Thus, a root system in an integral lattice is an orthogonal direct sum of indecomposable root systems of types A_n, D_n and E_6, E_7 and E_8. Reference to other kinds of root system or usage of the term "root" shall be quite explicit.

The other types of systems, B_n, C_n and F_4 or G_2, do appear in our theory of integral lattices. For example, in the dual lattice of D_4, the vectors of norms 1 and 2 form a root system of type F_4.

A root lattice is orthogonally indecomposable if and only if its root system is orthogonally indecomposable. This follows from (3.5.3).

Vectors given an n-tuples (rows or columns) have the usual dot product as their pairing, unless specified otherwise. Finally, we let e_i be the usual n-tuple with 1 at coordinate i and 0s elsewhere. Sometimes, n is determined by the context.

4.2 Root lattices, their duals and their isometry groups

We begin with some preliminaries about reflections, then proceed to define the root lattices, type by type. The root system (in the above sense, from Lie theory) is just the set of roots which occur in the given root lattice.

Recall that an element v of an integral lattice is called a root if $(v, v) = 2$.

Notation 4.2.1. *Denote the set of roots in the integral lattice L by $\Phi(L)$. Given a root $v \in L$, the reflection at v is defined by $r_v : x \mapsto x - 2\frac{(x,v)}{(v,v)}v = x - (x,v)v$, which is in the isometry group $O(L)$ of the lattice L. Let $R(L) := R(\Phi)$ denote the group generated by the reflections $\langle r_v \mid v \in \Phi(L) \rangle$. This is also called the Weyl group of the root system or root lattice. (Warning: this comment applies only to root systems of types ADE since the definition of reflection is different for root systems of types BCFG).*

Theorem 4.2.2. *The group $R(L)$ generated by reflections at roots is a normal subgroup of the isometry group $O(L)$.*

Proof. We will show that if $g \in O(L)$, then $gr_v g^{-1} = r_{g(v)}$. Note that $g(v)$ is a root in L since g preserves lengths and L. For any $x \in L$, we have $(gr_v g^{-1})(x) = gr_v(g^{-1}x) = g(g^{-1}x - (g^{-1}x, v)v) = x - (g^{-1}x, v)gv$. Also, $r_{g(v)}(x) = x - (x, gv)gv$. Since $g \in O(L)$, it preserves the form, hence $(g^{-1}x, v) = (g(g^{-1}x), gv) = (x, gv)$ as we desired. □

Lemma 4.2.3. *(Products of two reflections) Suppose L is an integral positive definite lattice, and v, w are roots. Then we have*

(1) the value $(v, w) \in \{0, \pm 1, \pm 2\}$; and $(v, w) = \pm 2$ if and only if $w = \pm v$.

(2) the reflections r_v, r_w have order 2; and the product $r_v r_w$ has order 1 if and only if $w = \pm v$, order 2 if and only if $(v, w) = 0$, order 3 if and only if $(v, w) = \pm 1$.

Proof. (1) By the Cauchy-Schwartz inequality, $|(v, w)| \leq |v| \cdot |w|$, in which the equality holds if and only if v, w are linearly dependent. In our case, this inequality reads $|(v, w)| \leq \sqrt{2} \cdot \sqrt{2} = 2$.

(2) The first statement is easy to check. For the second one, observe that the reflection r_u acts as identity on the codimension one subspace $ann(u)$, since $r_u(x) = x - 2\frac{(x,u)}{(u,u)}u = x$ for any $x \in ann(u)$. So the group $\langle r_v, r_w \rangle$ acts as identity on the subspace $ann(v) \cap ann(w)$ which is of codimension less than 2. Obviously we have $(ann(v) \cap ann(w)) \perp span\{v, w\}$. It is sufficient to study r_v, r_w on the subspace $U = span\{v, w\}$.

The case that v, w are linearly dependent is easy. Assume that they are linearly independent. Let $s = r_v r_w$. Since $s(v) = r_v(v - (v,w)w) = v - (v,w)w - (v - (v,w)w, v)v = (-1 + (v,w)^2)v - (v,w)w$, and $s(w) = r_v(-w) = (v,w)v - w$, the linear transformation s has the matrix $T = \begin{pmatrix} -1 + (v,w)^2 & (v,w) \\ -(v,w) & -1 \end{pmatrix}$.

Case I: $(v, w) = 0$. Then $T = \begin{pmatrix} -1 & 0 \\ 0 & -1 \end{pmatrix}$, i.e., the reflections r_v, r_w act as negatives on U and commute with each other.

Case II: $c := (v, w) = \pm 1$. We have $T^2 = \begin{pmatrix} 0 & c \\ -c & -1 \end{pmatrix}^2 = \begin{pmatrix} -1 & -c \\ c & 0 \end{pmatrix}$, and the equation $T^2 + T + I = 0$ holds. So s satisfies $X^3 - 1 = (X^2 + X + 1)(X + 1) = 0$, whence s has order 3. $\square$

4.2.1 Definition of the A_n lattices

Take $L := \mathbb{Z}^{n+1} = \{(x_0, x_1, \ldots, x_n) \mid x_i \in \mathbb{Z}\}$, and $\nu = (1, 1, \ldots, 1) \in L$. Define $M =: \mathbb{Z}\nu$ and are $N := ann_L(M) = ann_L(\nu)$. Then both M and N are direct summands and annihilators of each other. Also, L contains $M \oplus N$ with finite index. Since $\det(L) = \det(I_{n+1}) = 1$, the maps $L \to M^*$ and $L \to N^*$ are surjections due to corollary 5.15, thus we obtain two short exact sequences:

$$0 \to M + N \to L \to M^*/M \to 0, \quad 0 \to M + N \to L \to N^*/N \to 0.$$

The Gram matrix of M is $G_M = \nu\nu^t = (n+1)_{1\times 1}$

As usual, we let e_i be the n-tuple with 1 at index i and 0s elsewhere.

Lemma 4.2.4. *The following sets are bases of N: (1) $e_1 - e_2, e_2 - e_3, \ldots, e_{n-1} - e_n$; (2) $e_i - e_j$, for all $j = 1, 2, \ldots, n$, $j \neq i$.*

For N, we choose a standard basis $\{e_i - e_{i+1} \mid i = 0, 1, \ldots, n-1\}$ of N, then

$$G_N = \begin{pmatrix} 2 & -1 & 0 & \cdots & 0 \\ -1 & 2 & -1 & \cdots & 0 \\ 0 & -1 & 2 & \cdots & 0 \\ \vdots & \vdots & \vdots & & \vdots \\ 0 & 0 & 0 & \cdots & 2 \end{pmatrix}_{n \times n}$$

which, by ( ), has determinant $\det(G_N) = \det(G_M) = n+1$.

Definition 4.2.5. *The above lattice N is called the A_n lattice.*

Let $x = (x_0, x_1, \ldots, x_n) \in N$ satisfy $(x, x) = \sum_{i=0}^n x_i^2 = 2$. Then $\pm x = (0, \ldots, 0, 1, 0, \ldots, 0, -1, 0, \ldots, 0)$, with 1 in the i-th position and -1 in the j-th, i.e., $\pm e_i \mp e_j$. So the roots correspond to the ordered pairs $(i, j), i \neq j, i, j = 0, 1, \ldots, n$. The number of such is $n(n+1)$.

Take a root $v = v_{ij} = (0, \ldots, 0, 1, 0, \ldots, 0, -1, 0, \ldots, 0)$, and define $r = r_v$, which gives an isometry of L. Let $e_i = (0, \ldots, 0, 1, 0, \ldots, 0)$ with 1 in the i-th position. Then, $v = e_i - e_j$ and so $r(e_i) = e_i - 2\frac{(e_i, v)}{(v, v)}v = e_j$. Similarly we have $r(e_j) = e_i$, and $r(e_k) = e_k, k \neq i, j$. So $r = r_{e_i - e_j}$ transposes e_i and e_j and fixes all $e_k, k \neq i, j$. Therefore we have a subgroup of $O(L)$ generated by $\{r_{e_i - e_j} \mid i \neq j, i, j = 0, \ldots, n\}$, acting as the symmetric group Sym_{n+1} on the basis $\{e_0, e_1, \ldots, e_n\}$ of L.

For any lattice L, the scalar map $-1 \in O(L)$. Clearly -1 does not fix the set $\{e_0, e_1, \ldots, e_n\}$, so $-1 \notin \langle r_{e_i - e_j} \mid i \neq j, i, j = 0, \ldots, n \rangle$. We have $\langle -1, r_{e_i - e_j} \mid i, j \in \{0, 1, \ldots, n\} \rangle \cong \langle -1 \rangle \times Sym_{n+1}$.

4.2.2 Definition of the D_n lattices

Let $L := \mathbb{Z}^n$ and define $M := \{(x_1, \ldots, x_n) \in \mathbb{Z}^n \mid \sum x_i \in 2\mathbb{Z}\}$.

Definition 4.2.6. *The sublattice M is called the D_n-lattice.*

Its roots are $\{\pm e_i \pm e_j \mid i \neq j\}$. There are $2n(n-1)$ roots in all. We choose the standard basis $\{e_1 - e_2, e_2 - e_3, \ldots, e_{n-1} - e_n, e_{n-1} + e_n\}$ of the lattice M. Then its Gram matrix is

$$G_M = \begin{pmatrix} 2 & -1 & \cdots & 0 & 0 & 0 \\ -1 & 2 & \cdots & 0 & 0 & 0 \\ \vdots & \vdots & & \vdots & \vdots & \vdots \\ 0 & 0 & \cdots & 2 & 0 & -1 \\ 0 & 0 & \cdots & 0 & 2 & 0 \\ 0 & 0 & \cdots & -1 & 0 & 2 \end{pmatrix}_{n \times n}$$

This has determinant $\det(G_M) = 4$ since M has index 2 in $\mathbb{Z}^n$.

Suppose $v = e_i + e_j$ and $i \neq j$. It is easily checked that $r(e_i) = -e_j, r(e_j) = -e_i$, and $r(e_k) = e_k, k \neq i, j$ where $i, j, k = 1, 2, \ldots, n$. Hence the group $\langle r_{e_i \pm e_j} \mid i \neq j, i, j = 1, \ldots, n\rangle$ permutes the set $\{\pm e_1, \pm e_2, \ldots, \pm e_n\}$.

The discriminant group of D_n is isomorphic to $\mathbb{Z}_2 \oplus \mathbb{Z}_2$ if n is even and to $\mathbb{Z}_4$ if n is odd. This was proved in (2.3.5).

4.2.3 Definition of the E_n lattices

The E_8 lattices is obtained from the standard model of the D_8 lattice as follows: $E_8 := D_8 + \frac{1}{2}\mathbb{Z}(1,1,1,1,1,1,1,1)$. We remark that the reflection at any of the e_i takes this lattice to $D_8 + \frac{1}{2}\mathbb{Z}(1,\ldots,1,-1,1,\ldots,1)$. Either $D_8 + \frac{1}{2}\mathbb{Z}(1,1,1,1,1,1,1,1)$ or $D_8 + \frac{1}{2}\mathbb{Z}(1,\ldots,1,-1,1,\ldots,1)$ could be taken as the standard model for E_8. It is customary to use $D_8 + \frac{1}{2}\mathbb{Z}(1,\ldots,1,1,1,\ldots,1)$ as the standard model for E_8.

Definition 4.2.7. *The E_8-lattice is $D_8 + \frac{1}{2}\mathbb{Z}(1,1,1,1,1,1,1,1)$.*
The E_7 lattices is the annihilator in E_8 of any A_1 sublattice.
The E_6 lattices is the annihilator in E_8 of any A_2 sublattice.

We shall analyze these lattices in detail later. The E_n lattices represent an exceptional branch in the classification proof of root systems.

4.2.4 Analysis of the A_n root lattices

We give two different arguments to determine the isometry group of A_n. The first involves a study of minimal vectors in the dual lattice.

Proposition 4.2.8. *(1) Suppose that $n \geq 2$. The minimal vectors in L^* are the $2(n+1)$ vectors obtained from $\pm\frac{1}{n+1}(1,1,1,\ldots,1,1,-n-1)$ by coordinate permutations.*
(2) Two vectors as in (1) are proportional or have inner product $\pm\frac{1}{n+1}$.

First proof of (4.2.8)

The displayed vectors are in the dual and have norms $\frac{n}{n+1}$. Now, take a vector $v \neq 0$ in the dual such that $(v,v) \leq \frac{n}{n+1}$. Then $v \notin L$. We may therefore write $v = \frac{1}{n+1}u$, where there exists an integer $c \in (0, n+1)$ such that u has p coordinates in $c + (n+1)\mathbb{Z}$ and q coordinates in $-c + (n+1)\mathbb{Z}$ and where $p + q = n + 1$. Since $(u, u) \leq n(n+1)$, no coordinate may have absolute value $n+1$ or greater. Therefore, every coordinate of u has absolute value c or $n + 1 - c$.

We can moreover say that no coordinate of u has absolute value $\frac{n+1}{2}$ modulo $(n+1)\mathbb{Z}$ (or else all coordinates have this property and we violate the inequality $(\frac{n+1}{2})^2 \leq n$ because $n \geq 2$). This implies in particular that c and $n+1-c$ are not congruent modulo $n+1$ and that no coordinate is congruent to its negative. Therefore, p and q are uniquely determined and the coordinates of absolute value c have common sign ε_1, that the coordinates of absolute value $n + 1 - c$ have a common sign ε_2 and that $\varepsilon_1 = -\varepsilon_2$. Therefore, either u has p coordinates equal to c and q coordinates equal to $c - n - 1$ or u has p coordinates equal to $-c$ and q coordinates equal to $n + 1 - c$.

From the two conditions $p + q = n + 1$ and $pc + q(c - n - 1) = 0$ (coordinate sum 0), we get a linear system with matrix $\begin{pmatrix} c & c-n-1 \\ 1 & 1 \end{pmatrix}$. Since this matrix has determinant $n + 1 \neq 0$, a solution is unique, so must be $p = n + 1 - c, q = c$. It follows that $(u, u) = (n + 1 - c)c^2 + c(n + 1 - c)^2 = c(n + 1 - c)(n + 1)$. As a function of c, the graph of (u, u) is a parabola opening downward. Its least value for $c \in [1, n]$ is at $c = 1$ and $c = n$. This proves the result. $\square$

Corollary 4.2.9. *When $n = 1$, $O(L) = \langle -1 \rangle$. For $n \geq 2$, $O(L) = \langle r_{e_i - e_j} \mid i \neq j, i, j = 0, \ldots, n \rangle \times \langle -1 \rangle \cong Sym_{n+1} \times \mathbb{Z}_2$.*

Proof. The case $n = 1$ is trivial, so we assume $n \geq 2$. We have a faithful representation of $O(L)$ on the set X of $2(n+1)$ minimal vectors of L^*. Let K be the kernel of the action of $O(L)$ on the set $X/\{\pm 1\}$. Let $h \in K$. We claim that h is a scalar map. Suppose not. Then there exists a pair $x, y \in X$ so that $hx = x, hy = -y$. This is impossible since h is orthogonal and $(x, y) \neq 0$ by the previous result. $\square$

Second proof of (4.2.8),(4.2.9)

Assume that $n \geq 2$. We know that $O(L)$ contains a subgroup isomorphic to $\langle \pm 1 \rangle \times Sym_{n+1}$.

In the previous section, we showed that the reflections in $R(\Phi)$ acts as transpositions in the symmetric group $Sym_{\{e_0, e_1, \ldots, e_n\}} \cong Sym_{n+1}$. The reflection $r_{e_i - e_j}$ interchanges e_i to e_j and leaves e_k fixed if $k \neq i, j$. The scalar map $-1 \in O(L)$ is not in $R(\Phi)$. So we have $\langle R(\Phi), -1 \rangle = R(\Phi) \times \langle -1 \rangle \leq O(A_n)$. (Recall that if H, K are subgroups of G with trivial commutator $[H, K] = 1$ and intersection $H \cap K = 1$, then $\langle H, K \rangle \cong H \times K$).

Notation 4.2.10. *For $\varepsilon = \pm 1, i = 0, 1, \ldots, n$, define $X_i^\varepsilon = \{\varepsilon(e_i - e_j) \mid j \neq i\}$. For ε fixed, the $n + 1$ sets X_i^ε partition the root set Φ. Obviously, $X_i^\varepsilon = -X_i^{-\varepsilon}$.*

Example 4.2.11. *Suppose $n = 2$, we have*

$$X_0^+ = \{(1,-1,0),(1,0,-1)\}, X_0^- = \{(-1,1,0),(-1,0,1)\};$$
$$X_1^+ = \{(0,1,-1),(-1,1,0)\}, X_1^- = \{(0,-1,1),(1,-1,0)\};$$
$$X_2^+ = \{(-1,0,1),(0,-1,1)\}, X_2^- = \{(1,0,-1),(0,1,-1)\}.$$

Define $X := \{Y \mid Y \subseteq \Phi, |Y| = n \text{ and } (a,b) = 1 \text{ for all pairs of distinct } a, b \in Y\}$. We will show that if $Y \in X$, then there exist ε and i so that $Y = X_i^\varepsilon$.

Consider two roots $a = e_i - e_j, b = e_k - e_l$. Then $(a,b) = 1$ if and only if $i = k$ or $j = l$ and $a \neq b$. Suppose $c = e_m - e_n$ is a root. Then, $(a,c) = 1$ and $(b,c) = 1$ require that the intersection of the set of subscripts be a single point $\{i,j\} \cap \{k,l\} \cap \{m,n\} = \{p\}$, and consequently $a, b, c = \pm\{e_p - e_{q_1}, e_p - e_{q_2}, e_p - e_{q_3}\}$, where $q_1 \in \{i,j\}, q_2 \in \{k,l\}$, and $q_3 \in \{m,n\}$. So $X = \{X_i^\varepsilon \mid \varepsilon = \pm 1 \text{ and } i = 0, 1, \ldots, n\}$, and the isometry group $O(L)$ permutes X and permutes the set of $n+1$ pairs $\overline{X} = \{X_i^+ \cup X_i^- \mid i = 0, 1, \ldots, n\}$ transitively. In fact, the map $\alpha : O(L) \to Sym_{\overline{X}}$ induced by the action is onto, since $R(\Phi) \subseteq O(L)$ acts as the symmetric group $Sym_{\overline{X}}$.

We conclude that $O(L) = K \cdot R(\Phi)$, where $K = Ker(\alpha : O(L) \to Sym_{\overline{X}})$. Observe that $-1 \in K$ exchanges X_i^ε and $X_i^{-\varepsilon}$. We will show $K = \{\pm 1\}$.

The set $X_i := X_i^+ \cup X_i^-$ defines the sublattice W_i as the span of all $a - b$ where $a, b \in X_i$ and $(a,b) = 1$. The calculation $(e_i - e_j) - (e_i - e_k) = e_k - e_j$ shows that W_i may be identified with the standard model of the A_{n-1}-lattice associated to the orthonormal basis $\{e_j \mid j \neq i\}$.

We start induction at $n = 2$. (For $n = 1$, the A_1 lattice is rank 1 so the full isometry group is only ± 1). In this case, the lattice is easy to picture and the statement may be verified directly: The roots span the space and form a hexagon, whose isometry group is the dihedral group $Dih_{12} \cong Dih_6 \times \{\pm 1\}$.

We now suppose that $n \geq 3$. Let $G := O(L)$, and denote by G_i the stabilizer of X_i in G. It is contained in the stabilizer of W_i in G. Note that $L \supseteq W_i \perp ann_L(W_i)$, $rank(L) = n$, $rank(W_i) = n - 1$ and $rank(ann_L(W_i)) = 1$. So, $ann_L(W_i) = \mathbb{Z}y$, for $y = (\sum_{j \neq i} e_j) - ne_i \in L$. Therefore the isometry group of the lattice $ann_L(W_i)$ is $\{\pm 1\}$. By induction, G_i acts on W_i as $Sym_n \times \{\pm 1\}$, whence the order $|G_i|$ is divisible by the number $2 \cdot n!$. Note that $|G| = (n+1)|G_i|$, and $\{-1\} \times R(W_i) \leq G_i$. We will prove $|G_i|$ is just $2 \cdot n!$. If so, then we are done by induction. It suffices to prove that the kernel of the action of G_i on W_i is the identity. It it is not the identity, it is generated by the reflection r_y.

We show that $r_y \notin G$. Take indices i, j such that $i \neq j$. We have

$$r_y(e_i) = e_i - 2\frac{(e_i, y)}{(y,y)}y = e_i + 2\frac{n}{n^2+n}y$$

while

$$r_y(e_j) = e_j - 2\frac{1}{n^2+n}y,$$

and so

$$r_y(e_i - e_j) = (e_i - e_j) + 2\frac{n+1}{n^2+n}y = (e_i - e_j) + 2\frac{1}{n}y \notin L.$$

Since $r_y \notin G, r_y \notin G_i$, and so we conclude that $|G_i| = 2 \cdot n!$, as required. $\square$

4.2.5 Analysis of the D_n root lattices

The standard model of the D_n lattice is the even sublattice of $L := \mathbb{Z}^n$. In this section, we study the isometry group of the D_n lattice, and prove the following result.

Theorem 4.2.12. *The isometry group of the D_n lattice contains the monomial group on the orthonormal basis.*

(1) For $n \neq 4$, this monomial group is the full group $O(D_n)$.

(2) When $n = 4$, this monomial group is a subgroup of index 3 in the full isometry group. The quotient of $O(D_4)$ by the subgroup generated by reflections at the roots is isomorphic to Sym_3.

Let $M := L^{even}$. We have an inclusion $O(L) \hookrightarrow O(M)$. We observe that if X is a lattice, its isometry group $O(X)$ leaves the dual X^* invariant. In fact, for $x \in X, y \in X^*$, the condition $(x, gy) \in \mathbb{Z}$ for $g \in O(X)$, is equivalent to $(g^{-1}x, g^{-1}gy) = (g^{-1}x, y) \in \mathbb{Z}$. The isometry group $O(M)$ acts on M and also on its dual M^*. Recall that M^*/M is isometric to $\mathbb{Z}_2 \times \mathbb{Z}_2$ if n is even while is isometric to $\mathbb{Z}_4$ if n is odd. (See Example 4.3.)

Let H be the subgroup of $O(M)$ which leaves invariant the set of vectors $\{\pm e_1, \ldots, \pm e_n\}$. Clearly, H is isomorphic to the natural monomial group on this double basis. We conclude that $O(M) = H$ if

(∗) *the only vectors in M^* which have norm 1 are $\{\pm e_1, \ldots, \pm e_n\}$.*

We now determine when (∗) is true.

Note that $M^* = L + \mathbb{Z} \cdot \frac{1}{2}\nu$, where $\nu = (1, 1, \ldots, 1) \in \mathbb{Z}^n$. Take any $x = (x_1, \ldots, x_n) \in L + \mathbb{Z} \cdot \frac{1}{2}\nu$. For all i, $x_i = a_i/2$ for an odd number a_i. Then $(x, x) = \frac{1}{4}(\sum_{i=1}^{n} a_i^2)$.

If n is odd, the norm $(x, x) = \frac{1}{4}(\sum_{i=1}^{n} a_i^2)$ cannot be equal to 1. Suppose n is even. If $n \geq 5$ then $(x, x) > 1$. If $n = 4$, the only solutions to the equation $\frac{1}{4}(\sum_{i=1}^{4} a_i^2) = 1$ are $(a_1, \ldots, a_4) = (\pm 1, \ldots, \pm 1)$. Suppose $n = 2$. The D_2 lattice has a decomposition $D_2 = \mathbb{Z}(e_1 - e_2) \perp \mathbb{Z}(e_1 + e_2)$. Since $Sym_2 \cong \mathbb{Z}_2$, we have $O(D_2) \cong \mathbb{Z}_2 \wr \mathbb{Z}_2 \cong Dih_8$, the dihedral group of order 8, by our discussion of orthogonal decompositions of lattices and wreath products.

Thus, (∗) is true and we have identified $O(M)$ if $n \neq 4$. We now assume that $n = 4$, and let $H \leq O(L)$ be the subgroup which fixes $\{\pm e_1, \ldots, \pm e_4\}$. From above discussion, we have $H \cong O(L)$, which has order $2^4 \cdot 4!$.

Definition 4.2.13. *A frame is a maximal subset in $\mathbb{R}^4$ consisting of unit vectors so that any two members are orthogonal or equal to negative.*

We shall classify the set U of unit vectors in M^* where $M = D_4$, and see what frames are contained in U. We determined that the members of U are $\pm e_1, \pm e_2, \pm e_3, \pm e_4$, and $(\pm \frac{1}{2}, \pm \frac{1}{2}, \pm \frac{1}{2}, \pm \frac{1}{2})$, a total of 24 unit vectors.

The only frame containing e_i is $F_1 := \{\pm e_1, \ldots, \pm e_4\}$.

The only frame containing $(\frac{1}{2}, \frac{1}{2}, \frac{1}{2}, \frac{1}{2})$ is

38   4 Root Lattices of Types A, D, E

$$F_2 = \left\{ \left(\frac{1}{2}, \frac{1}{2}, \frac{1}{2}, \frac{1}{2}\right), \left(\frac{1}{2}, \frac{1}{2}, -\frac{1}{2}, -\frac{1}{2}\right), \left(\frac{1}{2}, -\frac{1}{2}, -\frac{1}{2}, \frac{1}{2}\right), \left(\frac{1}{2}, -\frac{1}{2}, \frac{1}{2}, -\frac{1}{2}\right),\right.$$

$$\left.\left(-\frac{1}{2}, -\frac{1}{2}, -\frac{1}{2}, -\frac{1}{2}\right), \left(-\frac{1}{2}, -\frac{1}{2}, \frac{1}{2}, \frac{1}{2}\right), \left(-\frac{1}{2}, \frac{1}{2}, \frac{1}{2}, -\frac{1}{2}\right), \left(-\frac{1}{2}, \frac{1}{2}, -\frac{1}{2}, \frac{1}{2}\right)\right\};$$

Similarly we have another frame

$$F_3 = \left\{ \left(\frac{1}{2}, \frac{1}{2}, \frac{1}{2}, -\frac{1}{2}\right), \left(\frac{1}{2}, \frac{1}{2}, -\frac{1}{2}, \frac{1}{2}\right), \left(\frac{1}{2}, -\frac{1}{2}, \frac{1}{2}, \frac{1}{2}\right), \left(-\frac{1}{2}, \frac{1}{2}, \frac{1}{2}, \frac{1}{2}\right),\right.$$

$$\left.\left(-\frac{1}{2}, -\frac{1}{2}, -\frac{1}{2}, \frac{1}{2}\right), \left(-\frac{1}{2}, -\frac{1}{2}, \frac{1}{2}, -\frac{1}{2}\right), \left(-\frac{1}{2}, \frac{1}{2}, -\frac{1}{2}, -\frac{1}{2}\right), \left(\frac{1}{2}, -\frac{1}{2}, -\frac{1}{2}, -\frac{1}{2}\right)\right\}.$$

These three frames are pairwise disjoint. Observe that, if L_i is the $\mathbb{Z}$-span of F_i, then $L_i \cong \mathbb{Z}^4$. Take a linear transformation on $\mathbb{R}^4$ which takes F_i to F_j (use the fact that a linear transformation is determined by an assignment of basis elements). Call it $t_{ij} : L_i \to L_j$. This takes an orthonormal basis to an orthonormal basis, so is an orthogonal transformation. Therefore we have an isometry $t_{ij} : L_i^{even} \to L_j^{even}$. It is easy to check that for $i = 1, 2, 3$, $L_i^{even} = M$. Therefore, $t_{ij} \in O(M)$ and $O(M)$ acts transitively on the set $\{L_1, L_2, L_3\}$. So we have $|O(M) : H| = 3$. The order of the isometry group of the D_4 lattice is $|O(M)| = |O(M) : H| \cdot |H| = 3 \cdot 2^4 \cdot 4! = 2^7 \cdot 3^2$. This completes the proof of the theorem.

What does this group $O(D_4)$ look like? Here is one description.

Notation 4.2.14. Let $E_i, i = 1, 2, 3$ be the frames of roots, i.e., maximal sets of roots in Φ_{D_4}, such that any two are proportional or orthogonal. It is easy to see that such a set is given by a partition of index set $\{1, 2, 3, 4\}$ into a pair of 2-sets, say A and B. To this partition corresponds the set $\{\pm e_i \pm e_j \mid \{i, j\} = A, B\}$.

Proposition 4.2.15. Let M continue to denote the D_4-lattice. Let $F_i, i = 1, 2, 3$ be the frames of unit vectors in M^* Let $E_i, i = 1, 2, 3$ be the frames of roots in M. Let Q be the kernel of the natural homomorphism $\pi : O(M) \to Sym_{\{E_1, E_2, E_3\}} \times Sym_{\{F_1, F_2, F_3\}}$. Then Q is a nonabelian group of order 2^5 (in fact, is an extraspecial 2-group; see (7.2.1)). The normal subgroup Q is complemented in $O(M)$, for example, by the subgroup $S \times T$, where S is generated by the reflections at $(1, -1, 0, 0)$ and $(0, 1, -1, 0)$ and where T is generated by the reflections at $(\frac{1}{2}, \frac{1}{2}, \frac{1}{2}, \frac{1}{2})$ and $(0, 0, 0, 1)$.

Proof. We have $S \cong T \cong Sym_3$. It is straightforward to check that S acts trivially on the set of F_i and faithfully on the set of E_i and that T acts trivially on the set of E_i and faithfully on the set of F_i. Therefore π is onto and by group order considerations, Q has order 32. In fact one can check that Q is the subgroup $D_1 K$ of H, where D_1 is the set of diagonal matrices of determinant 1 and where K is the normal subgroup of order 4 in the group of permutation matrices (isomorphic to Sym_4). $\square$

4.2 Root lattices, their duals and their isometry groups

Remark 4.2.16. (1) *More structure of $O(D_n)$ will be discussed in the next section.*

(2) *The A_3 and D_3 lattices are isometric. This follows from the theory of root systems, which shows that a set of fundamental roots for each has the same Cartan matrix. A second proof is to show that the Gram matrix for the set of vectors $(1, -1, 0), (0, 1, -1), (1, 1, 0)$ equals that of A_3. The $\mathbb{Z}$-span of this set is a sublattice of D_3 of determinant 4, whence the $\mathbb{Z}$-span is the full D_3-lattice.*

(3) *The set of vectors in $(D_4)^*$ which have norm 1 or 2 forms a root system of type F_4. Therefore the isometry group of the D_4-lattice is the isometry group of the root system of type F_4. This observation is sometimes helpful.*

4.2.6 More on the isometry groups of type D_n

Suppose L is an integral lattice. Denote $\Phi := \Phi(L)$ the set of roots $\{x \in L \mid (x, x) = 2\}$ in L, and $R(\Phi)$ the subgroup of the isometry group $O(L)$ generated by the reflections $\{r_\alpha \mid \alpha \in \Phi\}$.

We now analyze the group $R(\Phi)$ more in case Φ is of type D_n. Let L be the D_n lattice.

A natural guess would be that $R(\Phi)$ is the monomial group, denoted H in the previous section. It is easy to disprove this statement with an observation.

Example 4.2.17. *Let $L = \mathbb{Z}^n$, $v = (1, 0, 0, \ldots, 0) \in L$. Then the reflection $r_v : (x_1, x_2, \ldots, x_n) \mapsto (-x_1, x_2, \ldots, x_n)$ takes L to itself. However, $r_v \in O(L)$ but $r_v \notin R(\Phi)$ (because the space of negated vectors does not contain a root).*

Notation 4.2.18. *Denote by $OMon(n, \mathbb{Q})$ the group of orthogonal $n \times n$ rational monomial matrices, i.e. those degree n matrices which have the form pd, where p is a permutation matrix and d is a diagonal matrix with diagonal entries from $\{\pm 1\}$. This group may be written DP, where D is the set of such diagonal matrices and P is the set of permutation matrices, Define $D_1 := \{d \in D \mid \det(d) = 1\}$.*

Theorem 4.2.19. *We write matrices with respect to the orthonormal basis $\{e_1, \ldots, e_n\}$. Let Φ have type D_n. The group $R(\Phi)$ may be identified with $D_1 P$. Here, the product $D_1 P$ is a semidirect product.*

We prove this theorem in a sequence of steps.

First, we have some discussion about projections and the reflections on the Euclidean space $V = \mathbb{R}^n$. Suppose $W \leq V$ is a subspace, and $P_W : V \to W$ is the orthogonal projection to the subspace W. Let $w_1, \ldots, w_m$ be any basis of W and $w^1, \ldots, w^m$ is the corresponding dual basis with $(w_i, w^j) = \delta_{ij}$.

We have, for any $x \in V$, $P_W(x) = \sum_{i=1}^m (x, w_i) w^i$. The normal component of x is $N_W(x) = x - P_W(x) = P_{W^\perp}(x)$. Thus the reflection across W equals $R_W(x) = P_W(x) - N_W(x) = 2P_W(x) - x$.

Lemma 4.2.20. *If U is a subspace of Euclidean space, we write P_U for the orthogonal projection to U. If $W = W_1 \perp \cdots \perp W_p$ is an orthogonal direct sum of subspaces W_i, then projection $P_W = \sum_{i=1}^p P_{W_i}$. Consequently, the reflection across W satisfies $R_W(x) = 2(\sum_{i=1}^p P_{W_i}(x)) - x$.*

Notation 4.2.21. *Suppose $e_1, \ldots, e_n$ is an orthonormal basis. Define the orthogonal transformation ε_i by sending e_i to $-e_i$ and fixing e_j if $j \neq i$. If $A \subseteq \{1, 2, \ldots, n\}$, define $\varepsilon_A := \prod_{i \in A} \varepsilon_i$. We may write $\varepsilon_{ij\ldots}$ for $\varepsilon_{\{i,j,\ldots\}}$.*

We have the following identity

Lemma 4.2.22. $r_{e_i - e_j} r_{e_i + e_j} = r_{e_i} r_{e_j} = \varepsilon_{ij}$, *where $i \neq j$.*

Corollary 4.2.23. $O(L) = RD \geq R(\Phi) \geq D_1 P$.

Proof. The first containment follows from (4.2.12)(1). Recall that the group of permutation matrices, P, is just the reflection group for the A_{n-1} sublattice consisting of all elements of L with coordinate sum 0. So, $P \leq R(\Phi)$. The corollary follows from the lemma since the ε_{ij} generate D_1. □

Remark 4.2.24. *Let G be a group which acts on the abelian group A. If $B \leq A$ is a G-submodule, then G acts on B and also on the quotient module A/B, where $g \in G$ sends $a + B$ to $ga + B$, for all $a \in A$.*

We shall study the action of the isometry group $O(L)$ on the dual L^* and on the discriminant group $\mathcal{D}L = L^*/L$.

Lemma 4.2.25. *Let L be a root lattice, and Φ the set of roots in L. Then the group $R(\Phi)$ acts trivially on the discriminant group L^*/L.*

Proof. For a $\alpha \in \Phi \subseteq L$, we have a reflection $r = r_\alpha$. For $v \in L^*$, we have $r(v) = v - 2\frac{(v,\alpha)}{(\alpha,\alpha)} = v - (v,\alpha)\alpha \equiv v \pmod{L}$. □

Lemma 4.2.26. *Let L be a root lattice of type D_n. Let $u \in L^*$ be a unit vector, and let $v \in L^*$ satisfy $(u, v) \in \frac{1}{2} + \mathbb{Z}$. Then $r_u(v) \not\equiv v \pmod{L}$, so r_u acts nontrivially on the discriminant group L^*/L.*

Proof. Since $u \in L^*$ and u is a unit vector, r_u leaves invariant L and L^*. We have $r_u(v) = v - 2\frac{(v,u)}{(u,u)}u = v - mu$, where m is an odd integer. Observe that $(mu, mu) = m^2$ is odd so that $mu \notin L$, an even lattice. Consequently, $r_u(v) \not\equiv v \pmod{L}$. □

Corollary 4.2.27. *For L a root lattice of type $D_n, n \neq 4$, $R(\Phi) \neq DP$. Consequently, $R(\Phi) = D_1 P$.*

Proof. Let $u = (1, 0, \ldots, 0) \in L^*$, and $v = (\frac{1}{2}, \ldots, \frac{1}{2}) \in L^*$. Then $r_u \notin R(\Phi)$. Since $D_1 P$ has index 2 in $H = DP$, $R(\Phi) = D_1 P$ follows. □

We had actually observed earlier that $R(\Phi) \neq DP$, so this Corollary is an alternate proof. The main theorem of this section follows. Note that the diagonal group D_1 acts with determinant 1, but that $R(\Phi)$ has elements which act with determinant -1. In particular, a natural complement P to D_1 acts with elements of determinants both 1 and -1.

We can think of $R(\Phi) = D_1 P$ as the kernel of a homomorphism $\psi : DP \to \{\pm 1\}$, where ψ is the product of the determinant and the sign map for $P \cong Sym_n$ lifted to DP by the quotient map $DP \to P$.

4.2.7 Analysis of the E_n root lattices

Recall that the D_n lattice has dual $\mathbb{Z}^n + \mathbb{Z} \cdot \frac{1}{2}\nu$, where $\nu = (1,\ldots,1) \in \mathbb{Z}^n$. The vector $\frac{1}{2}\nu$ has norm $\frac{1}{4}n$. So the lattice $D_n + \mathbb{Z} \cdot \frac{1}{2}\nu$ is an integral lattice if and only if 4 divides n, and is an even lattice if and only if $\frac{1}{4}n \in 2\mathbb{Z}$, i.e., $n \in 8\mathbb{Z}$.

The E_8 lattice is defined to be the lattice $D_8 + \mathbb{Z} \cdot \frac{1}{2}\nu$, which is an even lattice with determinant 1. In fact, $|E_8 : D_8| = 2$, and we have calculated that $\det(D_n) = 4$ (2.3.5). The result $\det(E_8) = 1$ follows the formula $\det(D_8) = |E_8 : D_8|^2 \det(E_8)$.

This lattice is special in many ways. Besides being associated to the largest rank exceptional root system, it gives a dense sphere packing in dimension 8. Later, after we study the Hermite and Minkowski bounds, we shall prove an important characterization of it (5.2.1).

Recall that the discriminant group $\mathcal{D}D_n$ of the D_n lattice is isomorphic to $\mathbb{Z}_4$ if n is odd, and is isometric to $\mathbb{Z}_2 \oplus \mathbb{Z}_2$ if n is even (2.3.5). If L is a lattice satisfying $D_n \lneqq L \lneqq D_n^*$, then $|L : D_n| = 2$ and $\det(L) = 1$.

Take n even and let M be the standard model of the D_n lattice. Its dual is $M^* = \mathbb{Z}^n + \mathbb{Z} \cdot \frac{1}{2}\nu$, where $\nu = (1,\ldots,1) \in \mathbb{Z}^n$ is the standard model for D_n. Define $M^+ := M + \mathbb{Z} \cdot \frac{1}{2}\nu$, $M^- := M + \mathbb{Z} \cdot \frac{1}{2}\nu'$, where $\nu' = (1,\ldots,1,-1) \in \mathbb{Z}^n$. Then the three lattices strictly between M and M^* are $\mathbb{Z}^n, M^+$ and M^-. Note that M^+ and M^- are interchanged by the orthogonal transformation ε_n, so they are isometric. We have shown that, $M^\pm$ is integral if $n \in 4\mathbb{Z}$, and is even if $n \in 8\mathbb{Z}$.

Remark 4.2.28. *In some literature, the lattice $M^\pm$ is denoted by $D_n^\pm$ or by $HS_n^\pm$. The latter notation refers to the two half-spin lattices (named after the weights of a half-spin module for a Lie algebra of type D).*

Notation 4.2.29. *Let $L := M_8^+$, the standard E_8-lattice. Define $L_n := \{x \in L \mid (x,x) = n\}$. Thus, L_2 is the set of roots.*

Lemma 4.2.30. *There are 240 norm 2 vectors in L.*

Proof. The root set is the disjoint union $\Phi_L = \Phi_M \cup \Phi_{L-M}$. For any $x = (x_1,\ldots,x_8) \in \Phi_{L-M}$, $|x_i| \geq \frac{1}{2}$ and $(x,x) \geq 8 \cdot \frac{1}{4}$, so the number of $x = (\pm\frac{1}{2},\ldots,\pm\frac{1}{2})$ which have evenly many minus signs is 2^7. We have $\Phi(D_n) = \{\pm e_i \pm e_j \mid i \neq j\}$, a set of size $\binom{n}{2} \cdot 2^2 = 2n(n-1)$. So there are $2^7 + \binom{8}{2} \cdot 2^2 = 240$ roots in L. □

Lemma 4.2.31. *There are 2160 norm 4 vectors in L.*

Proof. A norm 4 vector x has the one of the following forms:
  (a) $(\pm 1, \pm 1, \pm 1, \pm 1, 0, 0, 0, 0)$, accounting for $\binom{8}{4} \cdot 2^4 = 1120$ vectors; or
  (b) $(\pm 2, 0, 0, 0, 0, 0, 0, 0)$, accounting for $\binom{8}{1} \cdot 2 = 16$ vectors; or
  (c) x has coefficients in $\frac{1}{2} + \mathbb{Z}$.

In the latter case, $(x,x) = 4$ implies that not every coefficient has absolute value $\frac{1}{2}$ and there exists just one coefficient with absolute value greater than $\frac{1}{2}$, i.e., up to signs and permutations of coefficients, $x = (\frac{1}{2},\frac{1}{2},\frac{1}{2},\frac{1}{2},\frac{1}{2},\frac{1}{2},\frac{1}{2},\frac{3}{2})$. The number of these is $\binom{8}{1} \cdot 2^7 = 1024$ since the $\pm\frac{3}{2}$ may be located at any coordinate and just seven of the signs are free, due to integrality of $(x, \frac{1}{2}\nu)$. Therefore we have

$1120 + 16 + 1024 = 2160$ norm 4 vectors in all. □

Next, we prove some transitivity theorems.

Notation 4.2.32. $\mathcal{A} := \{(X, Y) \mid X \cong E_8, Y \cong D_8, Y \leq X\}$.

We shall prove that the isometry group $O(\mathbb{R}^n)$ has only one orbit on the set of such pairs. This will help analyze the isometry group $O(X)$ of the lattice X.

Consider the E_8 lattice L. Set $S := L_0 \cup L_2 \cup L_4$, which has cardinality $|S| = 1 + 240 + 2160 = 2401 = 7^4$ in all. We look at $L/2L \cong \mathbb{Z}_2^8$ and ask when $a, a' \in S$ can be in the same coset of $2L$. Note that $|L/2L| = 2^8 = 256$.

Suppose $a, a' \in S, a \neq \pm a'$, and $a + 2L = a' + 2L$. Then $a - a' = 2b$ for some $0 \neq b \in L$. We have $p = (a - a', a - a') = (2b, 2b) = 4(b, b) \geq 8$. Also $a + a' = 2c$ for some $0 \neq c \in L$. Similarly we have $q = (a + a', a + a') = (2c, 2c) = 4(c, c) \geq 8$. The summation gives $p + q = (a, a) - 2(a, a') + (a', a') + (a, a) + 2(a, a') + (a', a') = 2(a, a) + 2(a', a') \geq 16$, that is $(a, a) + (a', a') \geq 8$. So $(a, a) = (a', a') = 4$, all inequalities are equalities, and $p = q$, which implies $(a, a') = 0$. The intersection $(a + 2L) \cap S$ has at most 16 elements. Therefore the number of cosets of $2L$ in L represented by S is at least $1 + \frac{240}{2} + \frac{2160}{16} = 256$, which is the order $|L/2L|$. Therefore, we have accounted for all cosets.

We have proved

Lemma 4.2.33. *Let $L = E_8$, Set $S := L_0 \cup L_2 \cup L_4$. Define equivalence in S by congruence modulo $2L$. The equivalence classes are just $\{\pm x\}$ except for classes which contain vectors of norm 4, in which case the equivalence classes are double orthogonal bases, i.e. 16-sets of norm 4 vectors, two of which are proportional or orthogonal.*

Notation 4.2.34. *A frame in L is a set of 16 norm 4 vectors in L which lie in the same coset of $2L$. Let F be a frame, say $a, a' \in F, a' \neq \pm a$, then $a - a' = 2b$ has norm 8, so b has norm 2, so that b is a root. We define a lattice $L(a)$ to be the $\mathbb{Z}$-span of $\frac{1}{2}F$, a set of 16 unit vectors. Such a lattice $L(a)$ is isometric to $\mathbb{Z}^8$.*

Remark 4.2.35. *An interesting (and useful) result is the classification of more general frames of norm 4 vectors in E_8, i.e., 16-sets of norm 4 vectors, two of which are proportional or orthogonal. Two vectors in such a frame need not be congruent modulo $2E_8$. There are four orbits of frames under the action of $O(E_8)$. This was proved first by Conway and Sloane [22]. A different proof is in [40].*

Example 4.2.36. *Take $a = (2, 0, 0, 0, 0, 0, 0, 0)$. The frame F (4.2.34) containing a is the set of 16 vectors of shape $\pm e_i = (0, \ldots, 0, \pm 2, 0, \ldots, 0)$. Note that the difference of two nonproportional frame elements is a norm 8 vector $2(\pm e_i \pm e_j) \in 2M \subseteq 2L$. Therefore, $L(a) = \mathbb{Z}e_1 + \cdots + \mathbb{Z}e_8$ (4.2.34).*

Notice $L(a) \cap L = M$. Since $a' - a'' \in 2L$ for all $a', a'' \in F$, it follows that $L(a) \cap L$ is the even sublattice of the $\mathbb{Z}$-span of $\frac{1}{2}F$, which is isometric to D_8.

Let a, b be two norm 4 vectors in L. There is an isometry $g \in O(\mathbb{R}^8)$ which takes $L(a)$ to $L(b)$. Such a g takes $L(a)^{even}$ to $L(b)^{even}$. These two lattices are isometric to D_8. The lattice L contains $L(b)^{even}$, so does $g(L)$, since the isometry g takes $L(a)^{even} \subseteq L$ to $L(b)^{even} = gL(a)^{even} \subseteq g(L)$.

Our discussion of overlattices of D_n showed that there is an $h \in O(L(b)^{even})$ which takes $g(L)$ to L. So, we may assume $g(L) = L$. We conclude that $O(L)$ acts transitively on frames.

Suppose $Y \leq L$ and $Y \cong D_8$. We claim that if $c \in Y^*$ is a unit vector, then $2c$ is a norm 4 vector of L. This is so because $Y^*/Y \cong \mathbb{Z}_2 \oplus \mathbb{Z}_2$ implies $2Y^* \subseteq Y \subseteq L$. We claim that $Y \leq L(c) \cong \mathbb{Z}^8$. This follows from the form of the standard model for $Y = D_8$, in which we have $c = (\pm 1, 0, 0, 0, 0, 0, 0, 0)$ and $Y \leq L(c)$.

We conclude that $O(L)$ is transitive on the set $\{Y \leq L \mid Y \cong D_8\}$. If $Y \cong D_8$, $O(Y)$ is transitive on the set of two overlattices isometric to E_8.

The transitivity results of the preceding discussion imply the following.

Proposition 4.2.37. *The orthogonal group $O(\mathbb{R}^8)$ is transitive on the set of pairs $\mathcal{A} = \{(X, Y) \mid X \cong E_8, Y \cong D_8, Y \leq X\}$.*

Corollary 4.2.38. *Let $X \cong E_8$ and let F be a frame in X. Then $Stab_{O(X)}(F)$ contains a group isometric to R, the reflection group of type D_8, acting faithfully on F. With respect to the double basis F, R is a group of monomial matrices, inducing Sym_8 on the 8-set $F/\{\pm 1\}$.*

Proof. This follows once we observe that $Y := \{x \in X \mid (x, F) \subseteq 2\mathbb{Z}\}$ is a lattice isometric to D_8. $\square$

Theorem 4.2.39. *The order of $O(L)$ is $2^{14} \cdot 3^5 \cdot 5^2 \cdot 7$.*

Proof. Consider the orthogonal group $G := O(\mathbb{R}^8)$, we define the stabilizers $G_{(X,Y)} := Stab_G((X,Y))$, $G_X := Stab_G(X) = O(X)$ and $G_Y := Stab_G(Y) = O(Y) \cong \mathbb{Z}_2 \wr Sym_8$. Define $\mathcal{A}_Y := \{X \subseteq \mathbb{R}^8 \mid X \cong E_8, X \geq Y\}$ (e.g. $\mathcal{A}_{D_8} = \{M^+, M^-\}$ is a two-element set), and $\mathcal{A}_X := \{Y \subseteq \mathbb{R}^8 \mid Y \cong D_8, Y \leq X\}$. The latter set is in bijection with the $\frac{2160}{16} = 135$ frames of norm 4 vectors. Since $|G_X : G_{(X,Y)}| = 135$ and $|G_Y| = 2^8 \cdot 8!$, we obtain $|G_{(X,Y)}| = \frac{|G_Y|}{|X:Y|} = \frac{1}{2} \cdot 2^8 \cdot 8!$ and $|O(X)| = |G_X| = 2^{14} \cdot 3^5 \cdot 5^2 \cdot 7$. $\square$

Lemma 4.2.40. *The action of $O(L)$ on $L/2L$ induces the full orthogonal group $O^+(8,2)$ and the kernel is $\langle -1 \rangle$.*

Proof. This follows from the fact that the 240 roots give the full set of 120 nonsingular cosets of $2E_8$ in E_8. These give all transvections. By (2.1.3), these transvections generate the full orthogonal group. From (2.1.7), $|O^+(8,2)| = 2^{13} 3^5 5 \cdot 7$. It follows from (4.2.39) that the kernel has order at most 2. Since -1 is in the kernel, the kernel has order exactly 2. $\square$

Corollary 4.2.41. *The commutator subgroup $O(L)'$ acts transitively on the set of norm 4 vectors.*

Proof. First we prove that $O(L)$ acts transitively. A norm 4 vector gives a singular point in $L/2L$. By Witt's theorem (2.1.1), any two cosets containing norm 4 vectors are in the same orbit under $O(L)$. We may therefore assume that our norm 4 vector is a member of the standard frame of 16 vectors of shape $(\pm 2^1 0^7)$. These vectors lie in the standard model of the D_8 lattice, which lies in L. The reflection group associated to the roots in D_8 acts transitively on such vectors and is contained in $O(L)$. Transitivity of $O(L)$ follows. Given a norm 4 vector, there exists a root orthogonal to it (for example, in the norm 4 vectors in (4.2.36), we may take a root of the form $e_i + e_j$). The reflection associated to that root fixes the norm 4 vector. Therefore, $O(L)'$ acts transitively. $\square$

Corollary 4.2.42. $O(L)$ *acts transitively on the set of sublattices isometric to* $\sqrt{2}L$.

Proof. Let M be such a sublattice. Since $\mathcal{D}(M) \cong 2^8$, $L \leq \frac{1}{2}M$ and so $2L \leq M \leq L$. By determinant considerations, $|L{:}M| = 2^4$. Since $M = 2M^*$, $M/2L$ is a maximal totally singular subspace of $L/2L$. The set of such forms an orbit under the orthogonal group of $L/2L$, by Witt's theorem. We finish by quoting (4.2.40). $\square$

Lemma 4.2.43. $O(L) \not\cong O^+(8,2) \times \mathbb{Z}_2$.

Proof. This follows because we can show that some $g \in O(L)$ has the property $g^2 = -1$. If we take $p = (12)(34)(56)(78) \in Sym_8$ and ε_A, where $A = \{1,3,5,7\}$, then

$$g = \varepsilon_A p = \begin{pmatrix} -1 & & & & \\ & 1 & & & \\ & & \ddots & & \\ & & & -1 & \\ & & & & 1 \end{pmatrix} \cdot \begin{pmatrix} 0 & 1 & & & \\ 1 & 0 & & & \\ & & \ddots & & \\ & & & 0 & 1 \\ & & & 1 & 0 \end{pmatrix} = \begin{pmatrix} 0 & -1 & & & \\ 1 & 0 & & & \\ & & \ddots & & \\ & & & 0 & -1 \\ & & & 1 & 0 \end{pmatrix}$$

satisfies $g^2 = -I_8$. $\square$

It is worth recording the following general result, which gives an alternate proof that the kernel of the homomorphism in (4.2.40) is just $\{\pm 1\}$. Note that the E_8 lattice is indecomposable.

Lemma 4.2.44. *Suppose that t is an involution on the free abelian group M such that t acts trivially on $M/2M$. Then M is the orthogonal direct sum of the two eigenlattices for t: $M = \{x \in M \mid tx = x\} \perp \{x \in M \mid tx = -x\}$. Consequently, if M is indecomposable, then the kernel of the action of $O(M)$ on $M/2M$ is $\{\pm 1\}$.*

Proof. The second statement follows from the first, which in turn follows from the discussion in [40] of involutions acting on free abelian groups. Here, we give a direct proof.

We may write $t = 1 + 2E$, where $E \in Hom(M, M)$ and $E(M) \leq 2M$. Since $t^2 = 1$, $1 + 4(E + E^2)$, or $E(1+E) = 0$. For $x \in M$, we have $x = 1x = (1+E)x - Ex$. We have $t(1+E)x = (1+2E)(1+E)x = (1+3E+2E^2)x = (1+E)x$ and $tEx = (1+2E)Ex = (E+2E^2)x = -Ex$. $\square$

Notation 4.2.45. *Denote*

$$L_2^0 := \{(\pm 1, \pm 1, 0, 0, 0, 0, 0, 0)\}$$

and

4.2 Root lattices, their duals and their isometry groups

$$L_2^1 := \left\{ (a_1, a_2, a_3, a_4, a_5, a_6, a_7, a_8) \mid a_i \in \left\{ \frac{1}{2}, -\frac{1}{2} \right\} \text{ and } \sum_i a_i \in 2\mathbb{Z} \right\}.$$

Then $|L_2^0| = \binom{8}{2} \cdot 2^2 = 112$, $|L_2^1| = 2^8/2 = 128$ and $L_2 = L_2^0 \cup L_2^1$.

Lemma 4.2.46. *Take $Y \leq X$, where $Y \cong D_8$ and $X \cong E_8$. We take each to be standard models on a common orthonormal basis $e_1, \ldots, e_8$.*
(1) *$O(Y)$ acts transitively on both L_2^0 and L_2^1.*
(2) *$O(X)$ acts transitively on L_2.*

Proof. (1) The orbit $Dv = \{d(v) \mid d \in D\}$ is one-to-one corresponding to the cosets of $Stab_D(v)$ in D. If all $v_i \neq 0$, then $Stab_D(v) = \{\varepsilon_A \mid A = \emptyset\} = 1$, and $|D(v)| = 2^8$. For $v \in L_2^1$, $Stab_{D_1}(v) = 1$. Then we have $|D_1(v)| = 128$, and L_2^1 is an orbit for D_1. For elements in L_2^0, the orbit of $e_1 + e_2$ contains all $e_i + e_j, i \neq j$, and so all $\pm e_i \pm e_j$. Therefore L_2^0 is an orbit for $D_1 P$.

(2) To prove $O(X)$ is transitive on L_2, we show that there exists an isometry $g \in O(X)$ so that $g(L_2^0) \cap L_2^1 \neq \emptyset$. In fact, we have a reflection $g = r_{(\frac{1}{2}, \ldots, \frac{1}{2})}$ and a vector $v = e_1 + e_2 \in L_2^0$, so that $g(v) = (\frac{1}{2}, \frac{1}{2}, -\frac{1}{2}, -\frac{1}{2}, -\frac{1}{2}, -\frac{1}{2}, -\frac{1}{2}, -\frac{1}{2}) \in L_2^1$. $\square$

Definition 4.2.47. *The E_7-lattice is the lattice $ann_X(v)$, where X is an E_8 lattice and v is any root in X. By the previous result, the isometry type is independent of choice of root.*

Theorem 4.2.48. *The isometry group of $ann_X(v)$ is just the group generated by reflections of the roots in it. This is the same as the group induced by the stabilizer in $O(X)$ of $ann_X(v)$.*

Proof. Let $U = ann_X(v)$, a direct summand of X. Since $\det(E_8) = 1$, and $\det(\mathbb{Z}v) = (v, v) = 2$, we get $\det(E_7) = \det(U) = 2$, and $|X : \mathbb{Z}v \perp U| = 2$.

Define $G := O(X)$. We have $Stab_G(U) \leq G$ and $Stab_G(U) = Stab_G(U^\perp) = Stab_G(\mathbb{Z}v)$. The generators of the cyclic group $\mathbb{Z}v$ are just roots $\pm v$ in $\mathbb{Z}v$. So $Stab_G(U)$ has index $\frac{1}{2} \cdot 240 = 120$ in G, whence $|Stab_G(U)| = |G|/|G : Stab_G(U)| = \frac{2^{14} \cdot 3^5 \cdot 5^2 \cdot 7}{2^3 \cdot 3 \cdot 5} = 2^{11} \cdot 3^4 \cdot 5 \cdot 7$.

Note that $1 \neq r_v \in Stab_G(U)$ since $r_v(u) = u$ for $u \in U$. The restriction gives a homomorphism $\alpha : Stab_G(U) \to O(U)$ with $r_v \in Ker(\alpha)$. The kernel $Ker(\alpha)$ acts on U trivially, and acts on $\mathbb{Z}v$ as ± 1. It follows that $Ker(\alpha) = \langle r_v \rangle \cong \mathbb{Z}_2$, which implies that $|O(U)|$ is divisible by $2^{10} \cdot 3^4 \cdot 5 \cdot 7$. We will show that $|O(U)| = 2^{10} \cdot 3^4 \cdot 5 \cdot 7$.

The lattice X is the unique lattice which lies between $\mathbb{Z}v + U$ and $\mathbb{Z}v + U^*$, which contains U with index 2 and is unequal to $U + \mathbb{Z}\frac{1}{2}v$ and $U^* + \mathbb{Z}v$. Evidently, it is the unique lattice between $\mathbb{Z}v + U$ and $\mathbb{Z}v + U^*$ which is integral.

The action of $O(U)$ on U extends to an action on the ambient vector space by making $O(U)$ act trivially on the span of v. This action preserves $\mathbb{Z}v + U$ and $\mathbb{Z}v + U^*$ and so, by the preceding discussion, preserves X. It follows that $O(U)$ is realized as the restriction of $Stab_G(U) \cap Stab_G(v)$. The restriction map here is an isomorphism onto. We conclude that $|O(U)| = 2^{10} \cdot 3^4 \cdot 5 \cdot 7$.

Note that the natural pair of restriction maps gives isomorphisms $Stab_G(U) \cong O(U) \times O(\mathbb{Z}v) \cong O(U) \times \mathbb{Z}_2$. $\square$

Lemma 4.2.49. *There is one orbit for the action of $O(L)$ on A_2-sublattices of L.*

Proof. The action of $O(L)$ on $L/2L$ is that of the full orthogonal group. By Witt's theorem, there is just one orbit on nonsingular 2-dimensional subspaces without singular vectors. The image of any A_2-sublattice in $L/2L$ is such a subspace. Furthermore, if $L \geq J \geq 2L$ and $J/2L$ is such a subspace, the set of roots in J is exactly a root system of type A_2 since two roots of L which are congruent modulo $2L$ are proportional. □

Definition 4.2.50. *The E_6-lattice is the lattice $ann_L(N)$, where N is isometric to an A_2-lattice. By the previous result, the isometry type is independent of the choice of A_2-sublattice.*

For the E_6-lattice, we determine the isometry group like we did for E_7. Let X be an E_8-lattice.

Theorem 4.2.51. *The E_6 lattice has isometry group equal to $\langle -1 \rangle \times T$, where T is the group generated by reflections at the roots in the E_6-lattice.*

Proof. We choose a particular E_6-lattice to be $J := ann_X(U)$, where $U \leq X$ and $U \cong A_2$, e.g.

$$U := Span\left\{\left(\frac{1}{2}, \frac{1}{2}, \frac{1}{2}, \frac{1}{2}, \frac{1}{2}, \frac{1}{2}, \frac{1}{2}, \frac{1}{2}\right), \left(-\frac{1}{2}, -\frac{1}{2}, -\frac{1}{2}, -\frac{1}{2}, -\frac{1}{2}, -\frac{1}{2}, \frac{1}{2}, \frac{1}{2}\right)\right\}.$$

Thus $U \perp J$ has index 3 in L. The stabilizer $S := Stab_{O(L)}(U)$ of $(U+2L)/2L$ in $O(L)$ induces the direct product of orthogonal groups $O^-(2,2) \times O^-(6,2)$ on $L/2L$. This direct product has order $2 \cdot 3 \times 2^6 3^4 5$ and so S has order twice as much (because the kernel of the action of $O(X)$ on $X/2X$ is $\{\pm 1\}$), i.e. $2^8 3^5 5$.

As we did in the E_7-program, we show that $O(J)$ extends to an action of L. This will prove that $O(J)$ is a quotient group of S. We have $\mathcal{D}U \cong \mathcal{D}J \cong \mathbb{Z}_3$. If $g \in O(J)$ acts trivially on $\mathcal{D}J$, we may extend g to an isometry of the ambient vector space by making it act trivially on U. If $g \in O(J)$ acts by -1 on $\mathcal{D}J$, then we may extend g to an isometry of the ambient vector space by making g act as -1 on U. This extension of the action of $O(J)$ has the property that it fixes $U+J$, its dual and all lattices between $U+J$ and its dual (since an element of $O(J)$ acts as a scalar, 1 or -1 on $\mathcal{D}U+J$). □

Remark 4.2.52. *The group $O(J)$ occurs in studies of the configuration of 27 lines on cubic surfaces.*

Remark 4.2.53. *For the case of E_7 and E_6, we have $O(E_7) \cong \mathbb{Z}_2 \times Sp(6,2)$, and $O(E_6) \cong O(5,3) \cong \mathbb{Z}_2 \times \Omega(5,3)$. We have $Sp(6,2) \cong O(7,2)$ (in general, $Sp(2n,2) \cong O(2n+1,2)$).*

These points may be verified in the case of E_7 by showing that sufficiently many roots are available to realize the transvections which generate these respective groups. Note that -1_{E_7} is a product of reflections.

Let $X \cong E_8$ and let $J \cong E_6, U \cong A_2$ be as above. The number of J in X is the number or pairs x, y of roots such that $(x, y) = -1$, modulo suitable equivalence. For such a pair, there are 240 choices for x and, given x, there are 56 possible y. Since $O(J) \cong Dih_{12}$, the number of sublattices of X isomorphic to J is $240 \cdot 56/12 = 1120 = 2^5 5 \cdot 7$. Therefore, $O(X, J) := Stab_{O(X)}(J)$, the stabilizer in $O(X)$ of J, has order $2^{14} 3^5 5^2 \cdot 7 / 2^5 5 \cdot 7 = 2^9 3^5 5$.

There is a natural embedding of $O(X, J)$ in $O(J) \times O(U)$. The action of an element of $O(J)$ on $\mathcal{D}J \cong \mathbb{Z}_3$ is by a scalar ± 1. Furthermore, there are elements which act as -1 (in fact, -1_J has this property). Analogous statements hold for $O(U)$. We now consider the subgroup of $O(J) \times O(U)$ which preserves the overlattice X. The natural maps of X to $\mathcal{D}J$ and $\mathcal{D}U$ are onto. Therefore, an element of $O(J) \times O(U)$ which stabilizes X acts as the same scalar on $\mathcal{D}J$ and $\mathcal{D}U$. It follows that $O(X, J)$ has index 2 in $O(J) \times O(U)$ and contains neither direct summand . Furthermore, it contains the reflection groups associated to each summand (since they act trivially on the discriminant group).

We conclude that $O(U)$ has order $2^9 3^5 5/6 = 2^7 3^4 5$. Obviously, it contains -1_U. Since -1_U acts nontrivially on $\mathcal{D}U$, we have a direct product decomposition $O(U) = \langle -1_U \rangle \times K$, where K is the kernel of the action of $O(U)$ on $\mathcal{D}U$.

The group K has order $2^6 3^4 5$. Its action on the nonsingular quadratic space $U/3U^* \cong \mathbb{F}_3^5$ embeds it as an index 2 subgroup of $O^+(5, 3)$, namely the subgroup S generated by reflections at vectors which have norm 2 mod 3 (these are just the cosets of roots mod $3U^*$). If we take four pairwise orthogonal norm 2 mod 3 elements, say v_1, v_2, v_3, v_4, their common annihilator is spanned by a norm 4 (mod 3) vector, say w. The product of the reflections at the v_i is minus the reflection at w. Thus $O^+(5,3)$ is $S\langle -1 \rangle = S \times \langle -1 \rangle$. We omit the details, which involve work with the finite orthogonal groups.

5

Hermite and Minkowski Functions

We introduce two useful functions here, due to Hermite and Minkowski, respectively. They have somewhat different natures. Both tell us that for a rational lattice of given rank and determinant, there is a nonzero vector whose norm is not too large. This can be quite useful for getting structure.

Definition 5.0.1. *For integers $n \geq 1$ and real numbers $d \geq 0$, we define:*
Function of Hermite: $H(n,d) := (\frac{4}{3})^{\frac{n-1}{2}} \cdot d^{\frac{1}{n}}$.
Function of Minkowski: $M(n,d) = \frac{4}{\pi}\Gamma(1+\frac{n}{2})^{\frac{2}{n}} \cdot d^{\frac{1}{n}}$.

Here Γ denotes the Gamma function, *i.e.*, for $z \in \mathbb{C}$ and $\text{Re}(z) > -1$, $\Gamma(z+1) := \int_0^\infty t^z e^{-t} dt$; it satisfies $\Gamma(z+1) = z\Gamma(z)$, $\Gamma(k+1) = k!$ for an integer $k > 0$ and $\Gamma(\frac{1}{2}) = \sqrt{\pi}$.

Definition 5.0.2. *Suppose L is a rational lattice, we define*

$$\mu(L) := \min\{|(x,x)| \mid x \in L, x \neq 0\}.$$

Here, the symmetric bilinear form $(\,,\,)$ is not assumed to be positive definite, so the value (x,x) could be zero or negative.

Theorem 5.0.3. *(Hermite) If L is a rational lattice of rank n, then $\mu(L) \leqslant H(n, |\det(L)|)$.*

We shall give a proof of this theorem, which goes by induction on the dimension. We discuss some linear algebra issues which arise.

Remark 5.0.4. *The image of a lattice under an orthogonal projection may or may not be a lattice. We discuss how these outcomes may happen.*

(1) Assume L is a rational lattice and S is a subset of L. We assume that $span_\mathbb{Q}(S)$ is nonsingular. This implies in particular that the natural map of S to $Hom(S, \mathbb{Z})$ has finite cokernel.

Let P be the orthogonal projection of L into $span_\mathbb{Q}(S)$.

We claim that $P(L)$ is a rational lattice.

(1.a) First, we observe that $P(L)$ is a free abelian group;

(1.b) The sublattice $span_{\mathbb{Z}}(S) + ann_L(S)$ has finite index in L. Call the index $k > 0$. Consider the orthogonal projection $P : L \to W$. We have $Ker(P) \supseteq ann_L(S)$ and $P(span_{\mathbb{Z}}(S)) = span_{\mathbb{Z}}(S)$. So $P(span_{\mathbb{Z}}(S) + ann_L(S)) = P(span_{\mathbb{Z}}(S))$, and this has finite index in $P(L)$ (in fact, this index divides k). Now we have $(P(L), P(L)) \cdot k^2 \subseteq (span_{\mathbb{Z}}(S), span_{\mathbb{Z}}(S)) \subseteq (L, L) \subseteq \mathbb{Q}$. Therefore, $P(L)$ is rational.

(2) We give an example to show that a projection of L need not be discrete. Let $u := (\alpha, \beta) \in \mathbb{R}^2$ be a unit vector so that α/β is irrational. Let P be the projection to $\mathbb{Q}u$. Then, we compute $P(1, 0) = \frac{((1,0), u)}{(u,u)} u = \alpha u$ and similarly $P(0, 1) = \beta u$, so that $P(\mathbb{Z}^2) = \mathbb{Z}\alpha^2 + \mathbb{Z}\beta^2 + \mathbb{Z}\alpha\beta$ is not rational, and is not even discrete.

Proof of Hermite's Theorem. (This proof follows [58].)

Suppose that there is a vector $0 \neq x \in L$, so that $(x, x) = 0$ i.e. $\mu(L) = 0$. Then we are done. So, we may assume that $\mu(L) > 0$. This means if $M \neq 0$ is any sublattice of L, then $\det(M) \neq 0$.

Suppose $n = 1$. The inequality (*) says $\mu(L) \leq 1 \cdot |\det(L)|$. Since $L = \mathbb{Z}v$ and $0 \neq (v, v) = \det(L)$, then $|(kv, kv)| = k^2|(v, v)| \geq (v, v)$ if $0 \neq k \in \mathbb{Z}$. So $\mu(L) = |(v, v)| = |\det(L)|$, and so the inequality (*) holds for $n = 1$.

Suppose $n \geq 2$. Let $0 \neq v \in L$ be a vector such that $|(v, v)| = \mu(L) = m > 0$. Let $V := span_{\mathbb{Q}}(L)$ and $W := ann_V(v)$, a subspace of dimension $n - 1$. Let P be the projection from L onto W. Then $L' = P(L)$ is a rational lattice, by the comments in (1), above. Suppose $m' = \mu(L')$ and $x' \in L'$ is a vector such that $|(x', x')| = m' > 0$. There exists an $x \in L$ such that $P(x) = x'$. Thus $P(x + \mathbb{Z}v) = x'$. We have $x - x' \in \mathbb{Q}v$. There is a scalar $r \in \mathbb{Q}$ so that $x - x' = rv$. We may replace x by an element of $x + \mathbb{Z}v$ to assume that $|r| \leq \frac{1}{2}$. Since $(v, x') = 0$, $m \leq |(x, x)| = |(x' + rv, x' + rv)| \leq |(x', x')| + r^2|(v, v)| \leq m' + \frac{1}{4}m$. That is, $m \leq \frac{4}{3}m'$.

Let $d = \det(L)$ and $d' = \det(L')$. We shall prove that $d = d'm$. Let $x_1, \ldots, x_{n-1}, x_n = v$ be a basis of L. Then $P(x_1), \ldots, P(x_{n-1})$ is a basis of L'. Recall that $P(x) = x - \frac{(x,v)}{(v,v)}v$. We have $d = \det(G)$, where $G = ((x_i, x_j)) = (g_{nn})$ is the Gram matrix of L. Let $H = (h_{ij})$, $h_{ij} = (P(x_i), x_j)$ for $i \neq n$ and $h_{nj} = (v, x_j)$ for $j = 1, 2, \ldots, n$. So H has the form

$$\begin{pmatrix} & & & 0 \\ & G' & & \vdots \\ & & & 0 \\ * & \cdots & * & g_{nn} \end{pmatrix}.$$

We get H from G by subtracting multiples of the last row from the other rows. By determinant theory, $\det(H) = \det(G)$. Note that H is in trianglular block form. So $\det(H) = d' \cdot g_{nn}$, where $d' := \det(G')$ and $g_{nn} = (v, v) = m$. This proves that $d = d'm$.

By induction, $m' \leq (\frac{4}{3})^{\frac{(n-1)-1}{2}} d'^{\frac{1}{n-1}}$. We have $m \leq \frac{4}{3}m' \leq (\frac{4}{3})^{\frac{n}{2}} (\frac{d}{m})^{\frac{1}{n-1}}$ and $m^{n-1} \leq (\frac{4}{3})^{\frac{n(n-1)}{2}} (\frac{d}{m})$. Therefore $m^n \leq (\frac{4}{3})^{\frac{n(n-1)}{2}} d$. Taking n-th roots, we get the inequality (*). $\square$

We present a general finiteness result.

Theorem 5.0.5. *For given integers $n > 0$ and $d \neq 0$, there are finitely many isometry types of integral lattices of rank n and determinant d.*

Proof. If $n = 1$, a lattice with determinant d must be $\mathbb{Z}x$ with $(x, x) = d$. Suppose $n \geqslant 2$. Then the lattice L contains a nonzero vector x with norm bounded by the Hermite bound $H(n, d)$.

Case 1. $(x, x) \neq 0$.

Then $L \supseteq \mathbb{Z}x \perp A$, where $A = ann_L(x)$. The index $k = |L : \mathbb{Z}x \perp A|$ divides (x, x) since the natural map $L \to (\mathbb{Z}x)^*$ carries $A \perp \mathbb{Z}x$ to a subgroup of index $|(x, x)|$ in $(\mathbb{Z}x)^*$.

Note that $\det(L) \cdot |L : \mathbb{Z}x \perp A|^2 = \det(\mathbb{Z}x \perp A)$, or $\det(L) \cdot k^2 = (x, x) \cdot \det(A)$. Thus $\det(A)$ has only finitely many possible integer values. By induction on rank, the number of isometry types possible for $\mathbb{Z}x \perp A$ is bounded. Between $\mathbb{Z}x \perp A$ and its dual, there are just finitely many lattices. One of these is L.

Case 2. If $0 \neq x \in L$ satisfies the Hermite bound, then $(x, x) = 0$.

We follow the idea of Case 1, but we replace $\mathbb{Z}x$ by a rank 2 nonsingular sublattice. We may assume that our x is a primitive lattice vector which satisfies $(x, x) = 0$. Since $\det(L) = d$, the image of L in $(\mathbb{Z}x)^*$ under the natural map has index e, some integer dividing e. Let $y \in L$ satisfy $(x, y) = e$. Then y is a primitive vector. Define $M := span\{x, y\}$, a sublattice of L which is a direct summand as an abelian group (because of the way y was chosen). We may replace y by a vector of the form $y - kx$ to assume that $(y, y) = f \in [-e, e]$. So M has a Gram matrix of the form $\begin{pmatrix} 0 & e \\ e & f \end{pmatrix}$, and so the number of possible isometry types for M is bounded (by $2e$ in fact). In any case, M is nonsingular of determinant $-e^2$. We have $ann_L(M) \cap M = 0$ and $M \perp ann_L(M)$ has index dividing e^2 and so determinant dividing de^4. Therefore, $\det(ann_L(M))$ is a divisor of de^2. We now finish as in Case 1. □

5.1 Small ranks and small determinants

We shall use the Hermite bound in to classify certain positive definite lattices with small determinant. We suggest that the reader study the first few results in this section to understand the methods. The later ones are more technical and could be postponed. The characterizations of the lattices E_6, E_7 and E_8 given in the next section are quite important.

For the case $n = 2$, there is a well-known, complete classification of integral quadratic forms. The positive definite case is stated below. There seems to be no general analogue of this definitive result for general dimensions (but see Section 10.1, Chapter 15 [21]).

Theorem 5.1.1. *Suppose that L is a positive definite rank 2 integral lattice of determinant d. Then there exists a basis so that the Gram matrix has a reduced form as below:*
$$\begin{pmatrix} a & b \\ b & c \end{pmatrix},$$
where

5 Hermite and Minkowski Functions

$$-a < 2b \leq a \leq c, \text{ with } b \geq 0 \text{ if } a = c.$$

Moreover, a positive definite form is properly equivalent (i.e., by a change of basis of determinant 1) to exactly one such reduced form.

Proof. See, for example [21], p. 358. □

Reduced forms may be enumerated as follows. Such a reduced form satisfies $b^2 \leq d/3$, so one may factor $d + b^2 = ac$ in all possible ways consistent with the reduction criteria: $-a < 2b \leq a \leq c$, with $b \geq 0$ if $a = c$.

Remark 5.1.2. *Sometimes, an integral lattice decomposes orthogonally, in which case applying induction to the orthogonal direct summands may be helpful.*

If L is integral and has a unit vector u, then we have an orthogonal decomposition $L = \mathbb{Z}u \perp ann_L(u)$, since $x = (x,u)u + (x - (x,u)u)$ for any $x \in L$. More generally, if L contains a unimodular sublattice U, then $L = U \perp ann_L(U)$ (reason: the natural maps $L \to Hom(U, \mathbb{Z})$ and $U \to Hom(U, \mathbb{Z})$ are both onto; so if $x \in L$, there exists $y \in U$ so that $x - y \in ann_L(U)$).

5.1.1 Table for the Minkowski and Hermite functions

We display a table of values of the Minkowski and Hermite functions for ranks up to 26 and determinants up to 10. The text of the Maple program used to create it is given at the beginning.

Table 5.1. Values of Hermite function ($H(n,x)$ is the (n,d)-entry)

	$d=1$	$d=2$	$d=3$	$d=4$	$d=5$	$d=6$	$d=7$	$d=8$	$d=9$	$d=10$
$n=1$	1	2	3	4	5	6	7	8	9	10
$n=2$	1.1547	1.6330	2.0000	2.3094	2.5820	2.8284	3.0550	3.2660	3.4641	3.6515
$n=3$	1.3333	1.6799	1.9230	2.1165	2.2800	2.4228	2.5506	2.6667	2.7734	2.8726
$n=4$	1.5396	1.8309	2.0262	2.1773	2.3022	2.4096	2.5043	2.5893	2.6667	2.7378
$n=5$	1.7778	2.0421	2.2146	2.3458	2.4529	2.5440	2.6236	2.6946	2.7589	2.8176
$n=6$	2.0528	2.3042	2.4653	2.5864	2.6844	2.7672	2.8392	2.9031	2.9606	3.0131
$n=7$	2.3704	2.6171	2.7732	2.8895	2.9831	3.0618	3.1300	3.1903	3.2444	3.2936
$n=8$	2.7371	2.9848	3.1400	3.2549	3.3470	3.4242	3.4908	3.5495	3.6022	3.6499
$n=9$	3.1605	3.4135	3.5708	3.6868	3.7794	3.8567	3.9233	3.9820	4.0344	4.0819
$n=10$	3.6494	3.9114	4.0732	4.1921	4.2867	4.3656	4.4334	4.4930	4.5462	4.5944
$n=11$	4.2140	4.4881	4.6566	4.7800	4.8780	4.9594	5.0294	5.0909	5.1457	5.1952
$n=12$	4.8659	5.1552	5.3324	5.4618	5.5643	5.6495	5.7225	5.7866	5.8436	5.8952
$n=13$	5.6187	5.9264	6.1141	6.2510	6.3591	6.4490	6.5259	6.5933	6.6533	6.7074
$n=14$	6.4879	6.8172	7.0175	7.1632	7.2783	7.3737	7.4554	7.5268	7.5904	7.6477
$n=15$	7.4915	7.8458	8.0608	8.2169	8.3400	8.4421	8.5293	8.6055	8.6733	8.7345
$n=16$	8.6505	9.0335	9.2653	9.4335	9.5659	9.6756	9.7692	9.8511	9.9238	9.9894
$n=17$	9.9887	10.4045	10.6556	10.8374	10.9806	11.0990	11.2001	11.2885	11.3669	11.4376
$n=18$	11.5340	11.9868	12.2599	12.4574	12.6127	12.7412	12.8508	12.9464	13.0314	13.1079
$n=19$	13.3183	13.8132	14.1111	14.3264	14.4956	14.6353	14.7545	14.8587	14.9511	15.0342
$n=20$	15.3786	15.9209	16.2471	16.4824	16.6674	16.8199	16.9502	17.0637	17.1644	17.2551
$n=21$	17.7577	18.3536	18.7115	18.9695	19.1721	19.3394	19.4818	19.6061	19.7164	19.8155

Continued

	$d=1$	$d=2$	$d=3$	$d=4$	$d=5$	$d=6$	$d=7$	$d=8$	$d=9$	$d=10$
$n=22$	20.5048	21.1612	21.5548	21.8385	22.0612	22.2447	22.4011	22.5375	22.6584	22.7674
$n=23$	23.6770	24.4015	24.8353	25.1481	25.3931	25.5953	25.7674	25.9173	26.0504	26.1702
$n=24$	27.3398	28.1409	28.6204	28.9655	29.2361	29.4589	29.6489	29.8143	29.9608	30.0926
$n=25$	31.5693	32.4567	32.9874	33.3694	33.6683	33.9149	34.1248	34.3073	34.4696	34.6151
$n=26$	36.4531	37.4380	38.0265	38.4494	38.7810	39.0536	39.2859	39.4881	39.6675	39.8287

Table 5.2. Values of Minkowski function ($M(n,d)$ is the (n,d)-entry)

	$d=1$	$d=2$	$d=3$	$d=4$	$d=5$	$d=6$	$d=7$	$d=8$	$d=9$	$d=10$
$n=1$	1	2	3	4	5	6	7	8	9	10
$n=2$	1.2732	1.8006	2.2053	2.5465	2.8471	3.1188	3.3687	3.6013	3.8197	4.0264
$n=3$	1.5393	1.9395	2.2201	2.4435	2.6322	2.7972	2.9447	3.0787	3.2020	3.3164
$n=4$	1.8006	2.1413	2.3698	2.5465	2.6926	2.8181	2.9289	3.0283	3.1188	3.2020
$n=5$	2.0585	2.3645	2.5643	2.7161	2.8401	2.9456	3.0378	3.1200	3.1944	3.2624
$n=6$	2.3136	2.5970	2.7785	2.9150	3.0254	3.1188	3.2000	3.2720	3.3369	3.3959
$n=7$	2.5667	2.8339	3.0029	3.1289	3.2302	3.3155	3.3893	3.4546	3.5132	3.5665
$n=8$	2.8181	3.0732	3.2330	3.3514	3.4461	3.5256	3.5942	3.6547	3.7089	3.7581
$n=9$	3.0682	3.3138	3.4665	3.5791	3.6689	3.7440	3.8087	3.8656	3.9166	3.9627
$n=10$	3.3170	3.5552	3.7022	3.8103	3.8962	3.9679	4.0295	4.0837	4.1321	4.1759
$n=11$	3.5649	3.7967	3.9393	4.0437	4.1265	4.1955	4.2547	4.3067	4.3530	4.3949
$n=12$	3.8118	4.0385	4.1773	4.2786	4.3589	4.4257	4.4829	4.5330	4.5778	4.6181
$n=13$	4.0580	4.2803	4.4159	4.5147	4.5928	4.6577	4.7133	4.7619	4.8053	4.8444
$n=14$	4.3036	4.5220	4.6548	4.7515	4.8279	4.8911	4.9453	4.9927	5.0349	5.0729
$n=15$	4.5485	4.7636	4.8942	4.9889	5.0637	5.1257	5.1786	5.2249	5.2661	5.3032
$n=16$	4.7929	5.0051	5.1336	5.2267	5.3001	5.3609	5.4128	5.4581	5.4984	5.5348
$n=17$	5.0368	5.2465	5.3731	5.4648	5.5370	5.5967	5.6477	5.6922	5.7318	5.7674
$n=18$	5.2804	5.4877	5.6127	5.7031	5.7742	5.8330	5.8832	5.9270	5.9659	6.0009
$n=19$	5.5234	5.7286	5.8522	5.9415	6.0117	6.0696	6.1190	6.1622	6.2006	6.2350
$n=20$	5.7662	5.9695	6.0918	6.1800	6.2494	6.3066	6.3554	6.3980	6.4357	6.4698
$n=21$	6.0085	6.2102	6.3313	6.4186	6.4871	6.5437	6.5919	6.6340	6.6713	6.7048
$n=22$	6.2506	6.4507	6.5707	6.6572	6.7251	6.7810	6.8287	6.8703	6.9072	6.9403
$n=23$	6.4924	6.6911	6.8101	6.8958	6.9630	7.0185	7.0657	7.1068	7.1432	7.1761
$n=24$	6.7340	6.9313	7.0494	7.1344	7.2010	7.2559	7.3027	7.3434	7.3796	7.4120
$n=25$	6.9753	7.1714	7.2886	7.3730	7.4391	7.4936	7.5399	7.5803	7.6161	7.6483
$n=26$	7.2163	7.4113	7.5278	7.6116	7.6772	7.7311	7.7771	7.8172	7.8526	7.8846

5.1.2 Classifications of small rank, small determinant lattices

The first result could be treated by the normal form of the theory of binary quadratic forms. Instead, we use the Hermite function.

Definition 5.1.3. *A lattice is rectangular if it has a basis whose Gram matrix is diagonal. A lattice is square if it is diagonal and all diagonal entries are equal.*

Lemma 5.1.4. *Let J be a rank 2 integral lattice. If $\det(J) \in \{1,2,3,4,5,6\}$, then J contains a vector of norm 1 or 2. If $\det(J) \in \{1,2\}$, J is rectangular. If J is even, $J \cong A_1 \perp A_1$ or A_2.*

54   5 Hermite and Minkowski Functions

Proof. The first two statements follows from values of the Hermite function. Suppose that J is even. Then J has a root, say u. Then $ann_J(u)$ has determinant $\frac{1}{2}\det(J)$ or $2\det(J)$. If $ann_J(u) = \frac{1}{2}\det(J)$, then J is an orthogonal direct sum $\mathbb{Z}u \perp \mathbb{Z}v$, for some vector $v \in J$. For $\det(J)$ to be at most 6 and J to be even, $(v,v) = 2$ and J is the lattice $A_1 \perp A_1$. Now assume that $ann_J(u)$ has determinant $2\det(J)$, an integer at most 12. Let v be a basis for $ann_J(u)$. Then $\frac{1}{2}(u+v) \in J$ and so $\frac{1}{4}(2+(v,v)) \in 2\mathbb{Z}$ since J is assumed to be even. Therefore, $2+(v,v) \in 8\mathbb{Z}$. Since $(v,v) \leq 12$, $(v,v) = 6$. Therefore, $\frac{1}{2}(u+v)$ is a root and we get $J \cong A_2$. □

Lemma 5.1.5. *Suppose that X is an integral lattice which has rank $m \geq 1$ and there exists a lattice W, so that $X \leq W \leq X^*$ and $W/X \cong 2^r$, for some integer $r \geq 1$. Suppose further that every nontrivial coset of X in W contains a vector with noninteger norm. Then $r = 1$.*

Proof. Note that if $u + X$ is a nontrivial coset of X in W, then $(u, u) \in \frac{1}{2} + \mathbb{Z}$.

Let $\phi : X \to Y$ be an isometry of lattices, extended linearly to a map between duals. Let Z be the lattice between $X \perp Y$ and $W \perp \phi(W)$ which is diagonal with respect to ϕ, i.e., is generated by $X \perp Y$ and all vectors of the form $(x, x\phi)$, for $x \in W$.

Then Z is an integral lattice. In any integral lattice, the even sublattice has index 1 or 2. Therefore, $r = 1$ since the nontrivial cosets of $X \perp Y$ in Z are odd. □

Lemma 5.1.6. *Let J be a rank 3 integral lattice. If $\det(J) \in \{1, 2, 3\}$, then J is rectangular or J is isometric to $\mathbb{Z} \perp A_2$. If $\det(J) = 4$, J is rectangular or is isometric to A_3.*

Proof. If J contains a unit vector, J is orthogonally decomposable and we are done by (5.1.4). Now use the Hermite function: $H(3,2) = 1.67989473\ldots$, $H(3,3) = 1.92299942\ldots$ and $H(3,4) = 2.11653473\ldots$. We therefore get an orthogonal decomposition unless possibly $\det(J) = 4$ and J contains no unit vector. Assume that this is so.

If $\mathcal{D}J$ is cyclic, the lattice $K = J + 2J^*$ which is strictly between J and J^* is integral and unimodular, so is isomorphic to $\mathbb{Z}^3$. So, J has index 2 in $\mathbb{Z}^3$, and the result is easy to check. If $\mathcal{D}J \cong 2 \times 2$, we are done by a similar argument provided a nontrivial coset of J in J^* contains a vector of integral norm. If this fails to happen, we quote (5.1.5) to get a contradiction. □

5.2 Uniqueness of the lattices E_6, E_7 and E_8

The first result characterizes positive definite unimodular lattices of rank not more than 8. In particular, it characterizes the E_8-lattice as the only even unimodular lattice rank at most 8. The proof is recent and relatively elementary. See [39], which contains a historical review. Our uniqueness proofs for E_7 and E_6 follow from uniqueness for E_8.

The E_8 lattice is important in many areas of mathematics. It is the smallest rank integral unimodular lattice which contains no unit vectors. Its roots form a root system for the largest of the exceptional finite simple complex Lie algebras. It is part of the series of Barnes-Wall lattices in dimensions a power of 2 and $\sqrt{2}E_8$ appears prominently within the important Leech lattice in dimension 24. Moreover, most even unimodular lattices in rank up to 24 contain sublattices isometric to $\sqrt{2}E_8$. The sphere packing associated to the E_8 lattice has been known for a long time to be extremely dense. Recently, Cohn and Kumar [60] proved that in 8 dimensions, the E_8 lattice packing is at least as dense as any packing [60], including non-lattice packings.

Theorem 5.2.1. *Let L be a positive definite integral lattice of rank no more than 8 and determinant 1. Then $L \cong \mathbb{Z}^n$ or $L \cong E_8$.*

Proof. Suppose L has a unit vector u. Then there exist an orthogonal decomposition $L = \mathbb{Z}u \perp L_1$ with $\det(L) = \det(L_1)$. Hence $L \cong \mathbb{Z}^n$ by induction. So we may assume that there are no unit vectors in L.

We calculate that $H(n,1) < 3$ for $n \leqslant 8$. Therefore, there is a vector $x \in L$ such that $(x,x) = 2$. Let $A := ann_L(x)$, which is of rank 7. Then $A \perp \mathbb{Z}x$ has index 1 or 2 in L. But it could not have index 1 since $\det(L) = \det(A) \cdot (x,x) \in 2\mathbb{Z}$. So, $\det(A \perp \mathbb{Z}x) = |L : A \perp \mathbb{Z}x|^2 \cdot \det(L) = 2^2 \cdot \det(L) = 4$, or $\det(A) = 2$. Since $H(n,2) < 3$ for $n \leqslant 7$, there is a vector $y \in A$ such that $(y,y) = 2$. Consider $z := \frac{1}{2}(x+y)$, which is a unit vector hence not in L. Since $2z = x+y \in L$, the lattice $L + \mathbb{Z}z$ contains L with index 2.

Let $L' := \{v \in L | (v,z) \in \mathbb{Z}\}$. We claim that $|L : L'| = 2$. Indeed, for any $v \in L$, $(v, 2z) \in \mathbb{Z}$ since $2z \in L$ and L is integral. So $(v, z) \in \frac{1}{2}\mathbb{Z}$. Since $L + \mathbb{Z}z$ contains L with index 2, it has determinant $\frac{1}{4}$, so is not integral. Therefore, $(z, L) = \frac{1}{2}\mathbb{Z}$. The sublattice L' is just the kernel of the map $u \mapsto (u,z) + \mathbb{Z} \in \frac{1}{2}\mathbb{Z}/\mathbb{Z}$, $u \in L$.

Observe that $2z = x+y \in L'$ since $(2z,z) = (x+y, \frac{1}{2}(x+y)) = \frac{1}{2}(x+y, x+y) = 2$. So $|L' + \mathbb{Z}z : L'| = 2$. We can replace L by $M := L' + \mathbb{Z}z$. By induction $M \cong \mathbb{Z}^n$.

Suppose $M = \mathbb{Z}e_1 + \cdots + \mathbb{Z}e_n$ where the e_i form an orthonormal basis. Note that $M \cap L = Ker(\phi)$ for some group homomorphism $\phi : M \to \mathbb{Z}_2$. Since $\phi(e_i) \in \mathbb{Z}_2$ and no e_i is in $Ker(\phi)$, then $\phi(e_i) = 1$ for all i. So $Ker(\phi)$ is the usual D_n lattice.

We have analyzed overlattices of the D_n lattice which contain it with index 2. So $n \neq 4, 8$ implies $L \cong \mathbb{Z}^n$. For the case $n = 8$, we get $L \cong E_8$ or $L \cong \mathbb{Z}_8$, and for the case $n = 4$, we get $L \cong \mathbb{Z}^4$. $\square$

We now proceed to characterizations of E_7, E_6 and related lattices. Our proofs make use of the previous characterization of $\mathbb{Z}^n$ and E_8.

Theorem 5.2.2. *Let L be an integral lattice of rank 7 and determinant 2. Then L is rectangular with Gram matrix of the form*

$$\begin{pmatrix} 1 & & & \\ & \ddots & & \\ & & 1 & \\ & & & 2 \end{pmatrix}$$

or $L \cong E_7$.

Proof. Let L be a lattice as above. Take $v \in L^* - L$. Then $(v,v) \notin \mathbb{Z}$ (or else the lattice $L + \mathbb{Z}v \cong L^*$ is integral while $\det(L^*) = \frac{1}{2}$). Since $L^*/L \cong \mathbb{Z}_2$, $2v \in L$. So $(v, 2v) \in \mathbb{Z}$ and $(v,v) \equiv \frac{1}{2} \pmod{\mathbb{Z}}$.

Define a new lattice $M := 2 \cdot \mathbb{Z}w$, where $w \in L^* - L$ and $(w,w) = \frac{1}{2}$. So, $M \cong A_1$. Define $N := (L \perp M) + \mathbb{Z}(v + w)$, which is an integral lattice of rank 8 (and a sublattice of $L \perp \mathbb{Z}w$). Note that $2(v+w) = 2v + 2w \in L \perp M$ and $\det(L \perp M) = \det(L) \cdot \det(M) = 4$, it follows $\det(N) = \det(L \perp M)|N : L \perp M|^{-2} = 4 \cdot 2^{-2} = 1$. So N is one of the two lattices $\mathbb{Z}^8$ and E_8.

Suppose that $N \cong \mathbb{Z}^8$. Let $M = \mathbb{Z}u$ for some root $u \in N$. Then u has one of the forms $(\pm 1, \pm 1, 0, 0, 0, 0, 0, 0)$, and $L = \operatorname{ann}_N(u)$. Say $u = ae_i + be_j$ for $a = \pm 1, b = \pm 1$. Then $L = \operatorname{span}(\{e_k | k \neq i, j\} \cup \{ae_i - be_j\})$ is rectangular with Gram matrix of the form

$$\begin{pmatrix} 1 & & & \\ & \ddots & & \\ & & 1 & \\ & & & 2 \end{pmatrix}.$$

Any two root vectors in $N \cong \mathbb{Z}^8$ are in the same orbit under the isometry group $O(N) \cong \mathbb{Z}_2 \wr \operatorname{Sym}_8$, so the isometry type of L in this case is determined.

Suppose that $N \cong E_8$. Since $L \leqslant N$ is an integral sublattice of rank 7 and determinant 2, L is a direct summand of N. Indeed, if S is an integral lattice of rank 7 containing L, then $\det(S) = \det(L)|S : L|^{-2} \in \mathbb{Z}$, which implies $|S : L| = 1$. Since $\det(N) = 1$, then $L \perp \operatorname{ann}_N(L)$ has index 2 in N. We obtain that $\operatorname{ann}_N(L)$ is a lattice of rank 1 and determinant 2. So $\operatorname{ann}_N(L) = \mathbb{Z}y$ for some root $y \in N$. The set of roots in E_8 form an orbit under the isometry group $O(N)$. We deduce that L is determined up to isometry. Therefore, L is the standard E_7-lattice. $\square$

Theorem 5.2.3. *Let L be an integral lattice of rank 6 and determinant 3. Then L is rectangular with Gram matrix of the form*

$$\begin{pmatrix} 1 & & & \\ & \ddots & & \\ & & 1 & \\ & & & 3 \end{pmatrix},$$

or $L \cong A_2 \perp \mathbb{Z}^4$, or $L \cong E_6$.

Proof. Let L be as in the hypothesis, then $L^*/L \cong \mathbb{Z}_3$. Take an element $u \in L^* - L$. Then $3u \in L$. So $(3u, u) \in \mathbb{Z}$, or $(u,u) = \frac{a}{3}$ for some $0 \neq a \in \mathbb{Z}$. If $a \equiv 0 \pmod{3}$, then L^* is integral. But $\det(L^*) = (\det(L))^{-1} = \frac{1}{3}$. So $a \equiv 1$ or $2 \pmod 3$.

First, we suppose that $a \equiv 2 \pmod 3$. We form a lattice $L \perp M$, where $M = \mathbb{Z} \cdot 3w$ and $(w,w) = \frac{1}{3}$. Then form a lattice $N = L \perp M + \mathbb{Z}(u + w)$. We have $\det(L \perp M) = \det(L) \cdot \det(M) = 3 \cdot 3 = 9$. Observe that $3(u + w) = 3u + 3w \in L \perp M$ and $u + w \notin L \perp M$. So $|N : L \perp M| = 3$, and $\det(N) = \det(L \perp M) \cdot |N : L \perp M|^{-2} = 9 \cdot 3^{-2} = 1$. So N is an integral lattice of rank 7 with determinant 1. That is, $N \cong \mathbb{Z}^7$. We have $L = \operatorname{ann}_N(x)$, where $x = 3w$. Since $(x,x) = 3$, x has form $(\pm 1^3, 0^7)$. We may apply an isometry to assume $x = (1, 1, 1, 0, 0, 0, 0)$. So

$L = ann_N(x) \cong A \perp B$, where $A = \{(y_1, y_2, y_3, 0, 0, 0, 0) \mid y_1 + y_2 + y_3 = 0\} \cong A_2$ and $B = \{(0,0,0,*,*,*,*)\} \cong \mathbb{Z}^4$.

Now, we suppose that $a \equiv 1 \pmod 3$. We form a lattice $L \perp M$, where $M \cong A_2$. We have $\det(L \perp M) = 9$, since $\det(A_2) = 3$. We want to find a vector w so that $N = L \perp M + \mathbb{Z}w$ is and integral lattice of rank 8 with determinant 1. Since $M \cong A_2$, there is a basis r, s of M so that its Gram matrix is $\begin{pmatrix} 2 & -1 \\ -1 & 2 \end{pmatrix}$. Take $y = \frac{1}{3}(r - s)$. We have $(r, y) = 1$ and $(s, y) = -1$. So $y \in M^*$ and $(y, y) = \frac{2}{3}$. If $x = y + z$ with $z \in M$, then $(x, x) = (y, y) + 2(y, z) + (z, z) \equiv (y, y) \pmod{2\mathbb{Z}}$. Therefore, elements of $M^* \setminus M$ have norms which are in $\frac{2}{3} + 2\mathbb{Z}$.

Now we define $w := u + y \in L^* \perp M^*$. We calculate that $(w, w) = (u, u) + (y, y) \equiv \frac{a}{3} + \frac{2}{3} \pmod{\mathbb{Z}} \equiv \frac{1}{3} + \frac{2}{3} \pmod{\mathbb{Z}} \equiv 0 \pmod{\mathbb{Z}}$. Thus N is integral as desired. By the previous theorem, $N \cong \mathbb{Z}^8$ or $N \cong E_8$.

We observe that $L = ann_N(M)$. If $N \cong \mathbb{Z}^8$, then the roots are of the forms $(\pm 1, \pm 1, 0, 0, 0, 0, 0, 0)$. Up to isometry of N, we may assume that $M = span\{(1, -1, 0, 0, 0, 0, 0, 0), (0, 1, -1, 0, 0, 0, 0, 0)\}$. So $ann_N(M)$ is spanned by $(1, 1, 1, 0, 0, 0, 0, 0)$ and all vectors of the form $(0, 0, 0, *, *, *, *, *)$, which has the Gram matrix

$$\begin{pmatrix} 1 & & & \\ & \ddots & & \\ & & 1 & \\ & & & 3 \end{pmatrix}.$$

Now assume that $N \cong E_8$. By Lemma (4.2.49), the isometry group $O(E_8)$ acts transitively on the set of A_2-sublattices of E_8. Then L is isometric to $ann_N(M)$. □

5.3 More small ranks and small determinants

We gather more characterizations of the sort treated in earlier sections. These proofs illustrate more techniques. These results are not necessary for the rest of the book, so could be skipped first time through. Many of these results are taken from [44].

Lemma 5.3.1. *Suppose that X is an integral lattice which has rank 4 and determinant 4. Then X embeds with index 2 in $\mathbb{Z}^4$. If X is odd, X is isometric to one of $2\mathbb{Z} \perp \mathbb{Z}^3, A_1 \perp A_1 \perp \mathbb{Z}^2, A_3 \perp \mathbb{Z}$. If X is even, $X \cong D_4$.*

Proof. Clearly, if X embeds with index 2 in $\mathbb{Z}^4$, X may be thought of as the annihilator mod 2 of a vector $w \in \mathbb{Z}$ of the form $(1, \ldots, 1, 0, \ldots, 0)$. The isometry types for X correspond to the cases where the weight of w is 1, 2, 3 and 4. It therefore suffices to demonstrate such an embedding.

First, assume that $\mathcal{D}X$ is cyclic. Then $X + 2X^*$ is an integral lattice (since $(2x, 2y) = (4x, y)$, for $x, y \in X^*$) and is unimodular, since it contains X with index 2. Then the classification of unimodular integral lattices of small rank implies $X + 2X^* \cong \mathbb{Z}^4$, and the conclusion is clear.

Now, assume that $\mathcal{D}X$ is elementary abelian. By (5.1.5), there is a nontrivial coset $u + X$ of X in X^* for which (u, u) is an integer. Therefore, the lattice $X' := X + \mathbb{Z}u$ is integral and unimodular. By the classification of unimodular integral lattices, $X' \cong \mathbb{Z}^4$. $\square$

Lemma 5.3.2. *If M is an even integral lattice of determinant 5 and rank 4, then $M \cong A_4$.*

Proof. Let $u \in M^*$ so that $u + M$ generates $\mathcal{D}M$. Then $(u, u) = \frac{k}{5}$, where k is an integer. Since $5u \in M$, k is an even integer. Since $H(4, \frac{1}{5}) = 1.029593054\ldots$, a minimum norm vector in M^* does not lie in M, since M is an even lattice. We may assume that u achieves this minimum norm. Thus, $k \in \{2, 4\}$.

Suppose that $k = 4$. Then we may form $M \perp \mathbb{Z}5v$, where $(v, v) = \frac{1}{5}$. Define $w := u + v$. Thus, $P := M + \mathbb{Z}w$ is a unimodular integral lattice. By the classification, $P \cong \mathbb{Z}^5$, so we identify P with $\mathbb{Z}^5$. Then $M = ann_P(y)$ for some norm 5 vector y. The only possibilities for such $y \in P$ are $(2, 1, 0, 0, 0)$, $(1, 1, 1, 1, 1)$, up to monomial transformations. Since M is even, the latter possibility must hold and we get $M \cong A_4$.

Suppose that $k = 2$. We let Q be the rank 2 lattice with Gram matrix $\begin{pmatrix} 3 & 2 \\ 2 & 3 \end{pmatrix}$. So, $\det(Q) = 5$ and there is a generator $v \in Q^*$ for Q^* modulo Q which has norm $\frac{3}{5}$. We then form $M \perp Q$ and define $w := u + w$. Then $P := M + Q + \mathbb{Z}w$ is an integral lattice of rank 6 and determinant 1. By the classification, $P \cong \mathbb{Z}^6$ In P, M is the annihilator of a pair of norm 3 vector, say y and z. Each corresponds in $\mathbb{Z}^6$ to some vector of shape $(1, 1, 1, 0, 0, 0)$, up to monomial transformation. Since M is even, the 6-tuples representing y and z must have supports which are disjoint 3-sets. However, since $(y, z) = 2$, by the Gram matrix, we have a contradiction. $\square$

Notation 5.3.3. *We denote by $\mathcal{M}(4, 25)$ an even integral lattice of rank 4 and determinant 25. By (5.3.4), it is unique.*

Lemma 5.3.4. (1) *There exists a unique, up to isometry, rank 4 even integral lattice whose discriminant group has order 25.*

(2) *It is isometric to a glueing of the orthogonal direct sum $A_2 \perp \sqrt{5}A_2$ by a glue vector of the shape $u + v$, where u is in the dual of the first summand and $(u, u) = \frac{2}{3}$, and where v is in the dual of the second summand and has norm $\frac{10}{3}$.*

(3) *The set of roots forms a system of type A_2; in particular, the lattice does not contain a pair of orthogonal roots.*

(4) *The isometry group is isomorphic to $Sym_3 \times Sym_3 \times 2$, where the first factor acts as the Weyl group on the first summand in (3) and trivially on the second, the second factor acts as the Weyl group on the second summand of (3) and trivially on the first, and where the third direct factor acts as -1 on the lattice.*

(5) *The isometry group acts transitively on (a) the six roots; (b) the 18 norm 4 vectors; (c) ordered pairs of norm 2 and norm 4 vectors which are orthogonal; (d) length 4 sequences of orthogonal vectors whose norms are 2, 4, 10, 20.*

(6) *An orthogonal direct sum of two such embeds as a sublattice of index 5^2 in E_8.*

5.3 More small ranks and small determinants

Proof. The construction of (2) shows that such a lattice exists and it is easy to deduce (3), (4), and (5).

We now prove (1). Suppose that L is such a lattice. We observe that if the discriminant group were cyclic of order 25, the unique lattice strictly between L and its dual would be even and unimodular. Since L has rank 4, this is impossible. Therefore, the discriminant group has shape 5^2.

Since $H(4, 25) = 3.44265186\ldots$, L contains a root, say u. Define $N := ann_L(u)$. Since $H(3, 50) = 4.91204199\ldots$, N contains a norm 2 or 4 element, say v.

Define $R := \mathbb{Z}u \perp \mathbb{Z}v$ and $P := ann_L(R)$, a sublattice of rank 2 and determinant $2(v,v) \cdot 25$. Also, the Sylow 5-group of $\mathcal{D}P$ has exponent 5. Then $P \cong \sqrt{5}J$, where J is an even, integral lattice of rank 2 and $\det(J) = 2(v,v)$. Since $\det(J)$ is even and the rank of the natural bilinear form on $J/2J$ is even, it follows that $J = \sqrt{2}K$, for an integral, positive definite lattice K. We have $\det(K) = \frac{1}{2}(v,v) \in \{1,2\}$ and so K is rectangular. Also, $P \cong \sqrt{10}K$. So, P has rectangular basis w, x whose norm sequence is $10, 5(v,v)$

Suppose that $(v, v) = 2$. Also, $L/(R \perp P) \cong 2 \times 2$. A nontrivial coset of $R \perp P$ contains an element of the form $\frac{1}{2}y + \frac{1}{2}z$, where $y \in span\{u, v\}$ and $z \in span\{w, x\}$. We may furthermore arrange for $y = au + bv$, $z = cw + dx$, where $a, b, c, d \in \{0, 1\}$. For the norm of $\frac{1}{2}y + \frac{1}{2}z$ to be an even integer, we need $a = b = c = d = 0$ or $a = b = c = d = 1$. This is incompatible with $L/(R \perp P) \cong 2 \times 2$. Therefore, $(v, v) = 4$.

We have $L/(R \perp P) \cong 5^2$. Therefore, $\frac{1}{5}w$ and $\frac{1}{5}x$ are in L^* but are not in L.

Form the orthogonal sum $L \perp \mathbb{Z}y$, where $(y, y) = 5$. Define $w := \frac{1}{5}x + \frac{1}{5}y$. Then $(w, w) = 1$. Also, $Q := L + \mathbb{Z}w$ has rank 5, is integral and contains $L \perp \mathbb{Z}y$ with index 5, so has determinant 5. Since w is a unit vector, $S := ann_Q(w)$ has rank 4 and determinant 5, so $S \cong A_4$. Therefore, $Q = \mathbb{Z}w \perp S$ and $L = ann_Q(y)$ for some $y \in Q$ of norm 5, where $y = e + f$, $e \in \mathbb{Z}w$, $f \in S$. Since S has no vectors of odd norm, $e \neq 0$ has odd norm. Since $(y, y) = 5$ and since (e, e) is a perfect square, $(e, e) = 1$ and $(f, f) = 4$. Since $O(A_4)$ acts transitively on norm 4 vectors of A_4, f is uniquely determined up to the action of $O(S)$. Therefore, the isometry type of L is uniquely determined.

It remains to prove (6). For one proof, use (5.3.5). Here is a second proof. We may form an orthogonal direct sum of two such lattices and extend upwards by certain glue vectors.

Let M_1 and M_2 be two mutually orthogonal copies of L. Let u, v, w, x be the orthogonal elements of M_1 of norm $2, 4, 10, 20$ as defined in the proof of (5). Let u', v', w', x' be the corresponding elements in M_2. Set

$$\gamma = \frac{1}{5}(w + x + x') \quad \text{and} \quad \gamma' = \frac{1}{5}(x + w' + x').$$

Their norm are both 2. By computing the Gram matrix, it is easy to show that $E = span_\mathbb{Z}\{M_1, M_2, \gamma, \gamma'\}$ is integral and has determinant 1. Thus, E is even and so $E \cong E_8$. □

In the next result, we are thinking of the $\mathbb{Q}/\mathbb{Z}$-valued form of (2.3.7), but interpreted as $\mathbb{F}_p$-valued, since the set of values in $\mathbb{Q}/\mathbb{Z}$ form a group of order p.

Lemma 5.3.5. *Let p be a prime which is $1 \pmod 4$. Suppose that M, M' are lattices such that $\mathcal{D}M$ and $\mathcal{D}M'$ are elementary abelian p-groups which are isometric as quadratic spaces over $\mathbb{F}_p$. Let ψ be such an isometry and let $c \in \mathbb{F}_p$ be a square root of -1. Then the overlattice N of $M \perp M'$ spanned by the "diagonal cosets" $\{\alpha + c\alpha\psi | \alpha \in \mathcal{D}M\}$ is unimodular. Also, N is even if M and M' are even.*

Proof. The hypotheses imply that N contains $M + M'$ with index $|\det(M)|$, so is unimodular. It is integral since the space of diagonal cosets so indicated forms a maximal totally singular subspace of the quadratic space $\mathcal{D}M \perp \mathcal{D}M'$. The last sentence follows since $|N : M \perp M'|$ is odd. $\square$

Lemma 5.3.6. *An even rank 4 lattice with discriminant group which is elementary abelian of order 125 is isometric to $\sqrt{5}A_4^*$.*

Proof. Suppose that L is such a lattice. Then $\det(\sqrt{5}L^*) = 5$. We may apply the result (5.3.2) to get $\sqrt{5}L^* \cong A_4$. $\square$

Lemma 5.3.7. *An even integral lattice of rank 4 and determinant 9 is isometric to A_2^2.*

Proof. Let M be such a lattice. Since $H(4,9) = 2.66666666\ldots$ and $H(3,18) = 3.494321858\ldots$, M contains an orthogonal pair of roots, u, v. Define $P := \mathbb{Z}u \perp \mathbb{Z}v$. The natural map $M \to \mathcal{D}P$ is onto since $(\det M, \det P) = 1$. Therefore, $Q := \text{ann}_M(P)$ has determinant 36 and the image of M in $\mathcal{D}Q$ is 2×2. Therefore, Q has a rectangular basis w, x, each of norm 6 or with respective norms 2, 18.

We prove that 2, 18 does not occur. Suppose that it does. Then there is a sublattice N isometric to A_1^3. Since there are no even integer norm vectors in $N^* \setminus N$, N is a direct summand of M. By coprimeness, the natural map of M to $\mathcal{D}N \cong 2^3$ is onto. Then the natural map of M to $\mathcal{D}\mathbb{Z}x$ has image isomorphic to 2^3. Since $\mathcal{D}\mathbb{Z}x$ is cyclic, we have a contradiction.

Since $M/(P \perp Q) \cong 2 \times 2$ and M is even, it is easy to see that M is one of $M_1 := span\{P, Q, \frac{1}{2}(u+w), \frac{1}{2}(v+x)\}$ or $M_2 := span\{P, Q, \frac{1}{2}(u+x), \frac{1}{2}(v+w)\}$. These two overlattices are isometric by the isometry defined by $u \mapsto u, v \mapsto v, w \mapsto x, x \mapsto w$. It is easy to see directly that they are isometric to A_2^2. For example, $M_1 = span\{u, x, \frac{1}{2}(u+x)\} \perp span\{v, w, \frac{1}{2}(v+w)\}$. $\square$

6

Constructions of Lattices by Use of Codes

To build and analyze a lattice, one often starts with a finite index sublattice which is relatively easy to understand. Then one considers how it may be enlarged to the full lattice. In may cases, coset representatives for the overlattice suggest an error correcting code, defined over some finite ring. We introduce elementary coding theory to deal with these situations.

We use the term *glue vector* to mean a coset representative of a rather simple form. This "definition" is not rigorous, but is natural to use in certain situations. A criterion could be smallest norm in the coset or vector with smallest coordinates. A coset may contain several candidates (or none) for the simplest.

6.1 Definitions and basic results

Definition 6.1.1. *Let K be a commutative ring, and $n \geq 1$ an integer, $v = (v_1, \ldots, v_n) \in K^n$. The weight of v (also called the Hamming weight) is the number of i so that $v_i \neq 0$. The distance between v and $w = (w_1, \ldots, w_n)$ is the number of i so that $v_i \neq w_i$ (this is just the weight of $v - w$).*

Definition 6.1.2. *A code $\mathcal{C}$ is a subset of K^n. If $\mathcal{C}$ is a K-submodule, it is called a linear code.*

If $\mathcal{C}$ is linear, then $v, w \in \mathcal{C}$ implies $v - w \in \mathcal{C}$. So, the minimum distance between pairs of distinct elements v, w in $\mathcal{C}$ with $v \neq w$ is just $\min\{wt(u) \mid u \in \mathcal{C}, u \neq 0\}$.

Definition 6.1.3. *Let K be a commutative ring, and $x = (x_1, \ldots, x_n)$, $y = (y_1, \ldots, y_n) \in K^n$. Define a symmetric bilinear form $x \cdot y = \sum_{i=1}^{n} x_i y_i \in K$. If $S \subseteq K^n$ is a subset, define $ann(S) = S^\perp = \{x \in K^n \mid x \cdot y = 0 \text{ for all } y \in S\}$. We call a linear code $\mathcal{C}$ self-orthogonal if $\mathcal{C} = \mathcal{C}^\perp$.*

Definition 6.1.4. *Let $Mon(n, K)$ be the group of all the $n \times n$ invertible monomial matrices with coefficients in the commutative ring K. It has normal subgroup $Diag(n, K)$ of all invertible diagonal matrices. This normal subgroup is complemented by $Perm(n, K)$, the group of all permutation matrices.*

62   6 Constructions of Lattices by Use of Codes

If $g \in Mon(n, K)$ and $v \in K^n$, then $wt(gv) = wt(v)$. It is not generally true that a monomial transformation preserved the natural dot product.

Definition 6.1.5. *Let K be a commutative ring and $n \geq 1$. We say that two codes $\mathcal{C}, \mathcal{C}'$ in K^n are equivalent if there exists $g \in Mon(n, K)$ so that g takes $\mathcal{C}$ to $\mathcal{C}'$. An equivalence class is just an orbit of $Mon(n, K)$ in its natural action on subsets of K^n. The group of the code $\mathcal{C} \subseteq K^n$ is the subgroup of $Mon(n, K)$ which takes $\mathcal{C}$ onto itself, i.e., it is the stabilizer in $Mon(n, K)$ of $\mathcal{C}$ in this natural action.*

For certain studies, one enlarges the group $Mon(n, K)$ to its semidirect product with Γ, a subgroup of $Aut(K)$ acting coordinatewise on K^n, to get a group $Mon^*(n, K)$ of semilinear transformations. One may then consider equivalence defined by action of $Mon^*(n, K)$.

Lemma 6.1.6. *Let V be a vector space over a field F with dimension n, and let g be a nonsingular bilinear form on V. Suppose $W \leqslant V$ is a subspace and totally singular, i.e. $g(W, W) = 0$. Then $\dim(W) \leqslant \frac{1}{2}n$.*

Proof. Let $h: V \to W^* = Hom(W, F)$ be the map $v \in V \mapsto [w \mapsto g(v, w)]$. Then h is onto. Let $m = \dim(W)$. So $m = \mathrm{rank}(h)$ and $m + nullity(h) = n$. Since the nullity of h is the dimension of the kernel $Ker(h) \geqslant W$, we have $nullity(h) \geqslant m$. Hence $2m \leqslant n$. □

Corollary 6.1.7. *The equality $2m = n$ holds if and only if W is its annihilator with respect to the form g.*

Notation 6.1.8. *Let K be a field. A linear code $\mathcal{C} \subseteq K^n$ has parameters $[n, k, d]$, where n is the number of coordinates, k is the dimension of $\mathcal{C}$ and d is the minimal weight in the code $\mathcal{C}$. In case $K = \mathbb{F}_q$ is a finite field of q elements, we sometimes write the parameters $[n, k, d]_q$.*

Definition 6.1.9. *Let K be a field and $n \geq 1$ an integer. The affine general linear group $AGL(n, K)$ is the semidirect product TG, where T is the group of translations on the set K^n and G acts on K as $GL(n, K)$. So, T is a normal, abelian subgroup.*

6.1.1 A construction of the E_8-lattice with the binary $[8, 4, 4]$ code

First, we introduce the famous Hamming binary code in dimension 8, then show how it is used to construct the E_8-lattice.

Notation 6.1.10. *Let $\mathcal{H}$ be the span of the four vectors $(1, 1, 1, 1, 0, 0, 0, 0)$, $(1, 1, 0, 0, 1, 1, 0, 0)$, $(1, 0, 1, 0, 1, 0, 1, 0)$, $(1, 1, 1, 1, 1, 1, 1, 1) \in \mathbb{F}_2^8$.*

These vectors are linearly independent. By checking dot products of spanning vectors, we verify that $\mathcal{H} \subseteq \mathcal{H}^\perp$. Since $\dim(\mathcal{H}) = 4$ and $\dim(\mathcal{H}^\perp) = 8 - \dim(\mathcal{H}) = 4$, we have $\mathcal{H} = \mathcal{H}^\perp$. A look at the 16 vectors of $\mathcal{H}$ shows that it has minimum weight 4 and so is a $[8, 4, 4]_2$ code (see (6.1.8)). Alternatively, one can note that $\mathcal{H}$ is doubly even (weights are divisible by 4) follows from $\mathcal{H} = \mathcal{H}^\perp$ and the fact that a spanning set consists of doubly even binary vectors.

6.1 Definitions and basic results

Any two codes with parameters $[8, 4, 4]_2$ are equivalent, *i.e.*, there is a permutation matrix P which takes one to the other. We leave the verification as an exercise (or look ahead to (6.2.2)).

The weight distribution in $\mathcal{H}$ is: 1 vector of weight 0; 14 are of weight 4; 1 is of weight 8 (we use the abbreviation $0^1 4^{14} 8^1$ for the enumeration of weights). Moreover, the group of this code is isomorphic to the affine general linear group (6.1.9) $AGL(3,2)$ on $\mathbb{F}_2^3$, which is a semidirect product $T : G$ of $G \cong GL(3,2)$ and T is the group of translations on $\mathbb{F}_2^3$. We shall prove these properties in the next section. Let us assume them for the moment.

We now use this code to give a new construction of the E_8-lattice. Let $\Omega = \{1, 2, \ldots, 8\}$ be an index set, and let $\{\alpha_1, \ldots, \alpha_8\}$ be a set in $\mathbb{R}^8$ such that $(\alpha_i, \alpha_j) = 2\delta_{ij}$, for $i, j \in \Omega$. Take $Q := \mathbb{Z}\alpha_1 \perp \cdots \perp \mathbb{Z}\alpha_8 \cong \mathbb{Z}^8$. For a subset $A \subseteq \Omega$, define $\alpha_A := \sum_{i \in A} \alpha_i$. We consider $\frac{1}{2}\alpha_A$, where A is in the code $\mathcal{H}$ from the above example. Such an α_A corresponds to the code word in $\mathbb{F}_2^8$ which has coordinates 1 at indices in A. Define $R := Q + \sum_{A \in \mathcal{H}} \mathbb{Z} \cdot \frac{1}{2}\alpha_A$. We shall prove that R is isometric to the E_8-lattice.

(1) The lattice R is integral: It suffices to show that $(x, y) \in \mathbb{Z}$ for x, y form a generating set. We already have $(\alpha_i, \alpha_j) = 2\delta_{ij} \in \mathbb{Z}$ and it is trivial to check that $(\alpha_i, \frac{1}{2}\alpha_A) = 0$ if $i \notin A$ and $(\alpha_i, \frac{1}{2}\alpha_A) = 1$ if $i \in A$. We have $(\frac{1}{2}\alpha_A, \frac{1}{2}\alpha_B) = \frac{1}{4}(\alpha_{A \cap B}, \alpha_{A \cap B}) = \frac{1}{4} \cdot 2 \cdot |A \cap B|$. Observe that $\mathcal{H} = \mathcal{H}^\perp$ implies the cardinality $|A \cap B|$ is even for $A, B \in \mathcal{H}$.

(2) The lattice R is even: This follows because R is integral and all $\alpha_i, \frac{1}{2}\alpha_A$ have even norms. Indeed, $(\frac{1}{2}\alpha_A, \frac{1}{2}\alpha_A) = \frac{1}{2}|A \cap A| \in \{0, 2, 4\}$.

(3) We have an isomorphism $R/Q \cong \mathcal{H}$ of abelian groups: Since $\mathcal{H}$ is a 4 dimensional vector space over $\mathbb{F}_2$, $|\mathcal{H}| = 16$. Define a map $h : R \to \mathcal{H}$ by sending $Q + \frac{1}{2}\alpha_A$ to A. To show that h is well-defined, it is sufficient to show that $Q + \frac{1}{2}\alpha_A = Q + \frac{1}{2}\alpha_B$ implies $A = B$ for $A, B \in \mathcal{H}$, or, equivalently that $\frac{1}{2}\alpha_A - \frac{1}{2}\alpha_B \in Q = \sum \mathbb{Z}\alpha_i$. Since the left side has integer coordinates, $A = B$. Since $A + B = (A \cup B) - (A \cap B)$, we have $\frac{1}{2}\alpha_A + \frac{1}{2}\alpha_B = \frac{1}{2}\alpha_{A+B} + \alpha_{A \cup B}$. Therefore, h is a homomorphism. It is onto and one-to-one.

(4) The lattice R is unimodular: the index-determinant theorem (2.3.3) shows that $\det(R) = \det(Q) \cdot |R : Q|^{-2} = 2^8 \cdot 16^{-2} = 1$.

(5) The lattice R is isometric to the E_8-lattice: This follows from properties (1), (2) and (4) and the characterization of the E_8-lattice (5.2.1).

(6) The isometry group $O(R)$ contains a group of the form EL, where E is the group of diagonal matrices

$$\begin{pmatrix} \pm 1 & & \\ & \ddots & \\ & & \pm 1 \end{pmatrix}$$

with respect to the basis $\alpha_1, \ldots, \alpha_8$. So, E is isomorphic to $\mathbb{Z}_2^8$, and L is the subgroup of the group of permutation matrices on the index set Ω which fixes the code $\mathcal{H}$. In more detail, $L \cong AGL(3, 2)$, and it has order $|L| = 2^3 \cdot 2^3 \cdot 3 \cdot 7 = 2^6 \cdot 3 \cdot 7$. Therefore, the group EL has order $2^8 \cdot 2^6 \cdot 3 \cdot 7 = 2^{14} \cdot 3 \cdot 7$. It is a subgroup of the

64   6 Constructions of Lattices by Use of Codes

full isometry group, which has order $|O(R)| = 2^{14} \cdot 3^5 \cdot 5^2 \cdot 7$. We shall give proofs later.

6.1.2 A construction of the E_8-lattice with the ternary $[4, 2, 3]$ code

We now work with ternary codes (i.e., codes over $\mathbb{F}_3$) and give another existence proof for the E_8-lattice.

Lemma 6.1.11. *Let Q be the A_2-lattice and let h be an isometry of order 3. Then $h-1$ triples norms and carries Q^* onto Q. There is $x \in Q^* \setminus Q$ so that $(x,x) = \frac{2}{3}$. Any element of $Q^* \setminus Q$ has norm in $\frac{2}{3} + 2\mathbb{Z}$.*

Proof. (2.7.2). $\square$

Definition 6.1.12. *We define the tetracode $\mathcal{T} := \{(s, a, a+s, a+2s) \mid s, a \in \mathbb{F}_3\} = \mathrm{span}_{\mathbb{F}_3}\{(0,1,1,1),(1,0,1,2)\}$. We can easily check that $\mathcal{T}$ is self-orthogonal and has minimum weight 3.*

Observe that for any $0 \neq c \in \mathbb{F}_3$, we have $c^2 = 1$. If $v = (c_1, c_2, c_3, c_4)$ and $(v, v) = 0$, then $(v, v) = wt(v) \pmod 3$. If $v \in \mathcal{T}$, $wt(v)$ is therefore divisible by 3.

We now give a lattice construction using $\mathcal{T}$.

Take $Q := Q_1 \perp Q_2 \perp Q_3 \perp Q_4$, where $Q_i \cong A_2$ for all i. There is a vector in $Q_i^* - Q_i$ so that it has norm congruent to $\frac{2}{3} \pmod{2\mathbb{Z}}$ for each i (6.1.11). Note that $Q_i^*/Q_i \cong \mathbb{Z}_3$, we have $Q^*/Q \cong \mathbb{F}_3^4$. We may consider $\mathcal{T}$ embedded in this copy of $\mathbb{F}_3^4$. Take vectors $u, v \in Q^*$ corresponding to $(0,1,1,1)$ and $(1,0,1,2)$ respectively. Note that $(u, u) \equiv 0 + \frac{2}{3} + \frac{2}{3} + \frac{2}{3} \pmod{2\mathbb{Z}} \equiv 2 \pmod{2\mathbb{Z}}$. Since $3u \in Q$ and Q is even, then $(3u, 3u) \in 2\mathbb{Z}$. This gives a second proof that $(u, u) \in 2\mathbb{Z}$.

Lemma 6.1.13. *$S := Q + \mathbb{Z}u + \mathbb{Z}v$ is isometric to the E_8 lattice.*

Proof. Above discussion shows that S is an even, integral lattice. We have $S/Q \cong \mathcal{T}$, which has order 9. Since $\det(Q) = 3^4$, $\det(S) = 1$ (2.3.3). By the characterization (5.2.1), S is isometric to the E_8 lattice. $\square$

6.2 The proofs

We now give the proofs of the coding results used in the two previous subsections.

6.2.1 About power sets, boolean sums and quadratic forms

We start with a review of the power set of a finite set, and associated linear algebra.

Let $\Omega = \{1, 2, \ldots, n\}$ be an index set. Then $\mathcal{P}(\Omega) \cong \mathbb{F}_2^n$, where $\mathcal{P}(\Omega)$ is the power set of Ω. For any $A, B \in \mathcal{P}(\Omega)$, the *Boolean sum of A and B* is $A + B := (A - B) \cup (B - A) = A \cup B - A \cap B$. The Boolean sum defines an abelian group structure on $\mathcal{P}(\Omega)$. (Furthermore, the abelian group $\mathcal{P}(\Omega)$ will be a commutative ring when equipped with the product $AB := A \cap B$). The 0-element of this additive group is $\emptyset$ and the identity for the ring structure is Ω. These elements are often written 0 and 1 (but see remarks below about the all-1 vector).

We have $|A \cup B| = |A - B| + |B - A| + |A \cap B|$, thus $|A + B| = |A \cup B| - |A \cap B| = |A| + |B| - 2|A \cap B|$. We have a group homomorphism $\psi : \mathcal{P}(\Omega) \to \mathbb{F}_2$ such that $\psi(A) = |A| (\mathrm{mod}\, 2)$. The kernel is the space $\mathcal{PE}(\Omega)$ of even sets in Ω. On the space $\mathcal{PE}(\Omega)$, there is a quadratic form $q : A \mapsto \frac{1}{2}|A| (\mathrm{mod}\, 2)$ so that $q(A + B) + q(A) + q(B) = (\frac{1}{2}|A| + \frac{1}{2}|B| - |A \cap B|) + \frac{1}{2}|A| + \frac{1}{2}|B| \equiv |A \cap B| (\mathrm{mod}\, 2)$, where A, B are even sets in $\mathcal{P}(\Omega)$.

The function $f(A, B) := q(A + B) + q(A) + q(B) = |A \cap B| (\mathrm{mod}\, 2)$ is bilinear. Indeed, $f(A, B_2 + B_2) = |A \cap (B_2 + B_2)| = |A \cap ((B_1 - B_2) \cup (B_2 - B_1))| = |A \cap (B_1 - B_2)| + |A \cap (B_2 - B_1)| \equiv |A \cap (B_1 - B_2)| + |A \cap (B_1 \cap B_2)| + |A \cap (B_2 - B_1)| + |A \cap (B_1 \cap B_2)| (\mathrm{mod}\, 2) \equiv |A \cap B_1| + |A \cap B_2| (\mathrm{mod}\, 2) = f(A, B_1) + f(A, B_2)$. Thus, $(\mathcal{PE}(\Omega), q)$ is a quadratic space.

The annihilator $ann_{\mathcal{P}(\Omega)}(\mathcal{PE}(\Omega))$ is the set of all $C \in \mathcal{P}(\Omega)$ so that $|C \cap A|$ is even when $|A|$ is even. We claim that $C = \emptyset$ or Ω. If not, take $A = \{x, y\}$ such that $x \in C$ and $y \notin C$, then $|A \cap C|$ is odd. So $Rad(\mathcal{PE}(\Omega), q) = \{\emptyset\}$ if n is odd, otherwise $\{\emptyset, \Omega\}$. Note that $q(\Omega) = \frac{1}{2}n (\mathrm{mod}\, 2)$ if n is even.

Here, the monomial group is just the group of permutation matrices, isomorphic to Sym_n. Two codes $\mathcal{C}_1, \mathcal{C}_2 \subseteq K^n$ are equivalent if there is a permutation matrix $P \in Mon(n, K)$ such that $P(\mathcal{C}_1) = \mathcal{C}_2$.

Note that this representation of Sym_n fixes Ω, which corresponds to the all-1 vector $\mathbf{1} = (1, \ldots, 1) \in \mathbb{F}_2^n$. The vector $\mathbf{1}$ annihilates every even code. So if $\mathcal{C} = \mathcal{C}^\perp$, then $\mathbf{1} \in \mathcal{C}$.

Proposition 6.2.1. *The group $GL(n, q)$ of $n \times n$ invertible matrices over the field $\mathbb{F}_q$ has order $\prod_{i=0}^{n-1}(q^n - q^i) = q^{n(n-1)/2} \cdot \prod_{j=1}^{n}(q^j - 1)$.*

Proof. Fix a basis $v_1, \ldots, v_n$ of $V \cong \mathbb{F}_q^n$. Let $\mathcal{B}$ be the set of all bases. Then there is a one-to-one correspondence between $GL(n, q)$ and $\mathcal{B}$, so that $g \in GL(n, q)$ corresponds to $\{gv_1, \ldots, gv_n\} \in \mathcal{B}$. But the cardinality of the set $\mathcal{B}$ is $(q^n - 1)(q^n - q) \cdots (q^n - q^{n-1})$. $\square$

The group $SL(n, q)$ is the kernel of the determinant map $\det : GL(n, q) \to \mathbb{F}_q^\times$. So it has index $q - 1$ in $GL(n, q)$. We obtain that the order of this group is $|SL(n, q)| = q^{n(n-1)/2} \cdot \prod_{j=2}^{n}(q^j - 1)$.

For example, we have orders $|GL(3, 2)| = 168$, $|GL(2, 3)| = 48$ and $|SL(2, 3)| = 24$.

6.2.2 Uniqueness of the binary [8, 4, 4] code

Theorem 6.2.2. *There is, up to equivalence, a unique $[8, 4, 4]_2$ even self-orthogonal code. Its weight distribution is $0^1 4^{14} 8^1$. The automorphism group is isomorphic to the affine general linear group $AGL(3, 2) \cong \mathbb{F}_2^3 : GL(3, 2)$ (6.1.9).*

Proof. Let $\mathcal{C}$ be such a code and $\Omega = \{1, 2, \ldots, 8\}$ be an index set. Since $\mathcal{C} = \mathcal{C}^\perp$ is an even code, we have $\mathbf{1} \in \mathcal{C}$. Let $v = (v_1, \ldots, v_8) \in \mathcal{C}$, $v \neq 0, \mathbf{1}$ and $supp(v) = \{i \in \Omega | v_i \neq 0\}$. Then $\mathbf{1} + v \in \mathcal{C}$ is supported by $\Omega - supp(v)$. Since $\mathcal{C}$ has minimum weight 4, we have the cardinalities $|supp(v)| \geq 4$ and $|\Omega - supp(v)| \geq 4$. So $supp(v)$ is a 4-element set. Define $\mathcal{A} := \{(x_1, x_2, x_3)|$ the weight of $x_i \in \mathbb{F}_2^8$ is 4 for $i = 1, 2, 3$ and $span\{\mathbf{1}, x_1, x_2, x_3\}$ is an $[8, 4, 4]_2$ code$\}$. Let us count the cardinality $|\mathcal{A}|$.

Since we are considering uniqueness up to equivalence, we may use the natural action of Sym_8 on coordinates.

Note that x_1 may be one of $\binom{8}{4}$ vectors of the form $(1,1,1,1,0,0,0,0)$. We may, by using symmetry, assume $x_1 = (1,1,1,1,0,0,0,0)$. Since $x_2 \neq x_1$ or its complement $x_1 + \mathbf{1}$, then $supp(x_1) \cap supp(x_2) \subseteq \Omega$ is a 2-element set. Suppose x_2 has weight 4 with a ones in the first 4 coordinates and $4-a$ ones in the other 4 positions. Then $x_1 + x_2$ has weight $2(4-a)$, and this number must be 4. Thus $a = 2$. So x_2 may be one of $\binom{4}{2}^2$ vectors. We may use symmetries fixing the first four coordinates to assume now that x_1 is as above and $x_2 = (0,0,1,1,1,1,0,0)$. By taking intersections of supports of x_1, x_2 and their complements (i.e., the $\mathbf{1}+x_i$), we partition Ω into four sets. The intersections of $Supp(x_3)$ with these four sets have cardinalities $(1^a, 1^b, 1^c, 1^d)$. Then we have a linear system $a+b = 2$, $b+c = 2$, $a+c = 2$ and $a+b+c+d = 4$, which has the unique solution $a = 1, b = 1, c = 1, d = 1$. So x_3 may be one of $\binom{2}{1}^4 = 16$ vectors. Therefore, we have $|\mathcal{A}| = \binom{8}{2}\binom{4}{2}^2 \cdot 2^4 = 8! = |Sym_8|$. Hence the symmetric group Sym_8 is transitive on the set $\mathcal{A}$. In fact, the action is *regular*, i.e., a transitive action in which the stabilizer of a point is the identity.

When do two elements $(x_1, x_2, x_3), (y_1, y_2, y_3) \in \mathcal{A}$ give the same code $span\{\mathbf{1}, x_1, x_2, x_3\}$. The number of such sequences in a given $[8,4,4]_2$ code $\mathcal{C}$ is $2(2^3 - 1) \cdot 2(2^3 - 2) \cdot 2(2^3 - 2^2) = 2^6 \cdot 3 \cdot 7$, which is the order of the affine group $|AGL(3,2)| = |T| \cdot |GL(3,2)|$ (6.1.9).

To complete the proof, we will exhibit a code in $\mathbb{F}_2^8$, which is $[8,4,4]_2$ and has $AGL(3,2)$ in its automorphism group. This follows from the next section, where we introduce Reed-Muller codes. We identify our index set Ω with $\mathbb{F}_2^3$. From the next section, we get the group $AGL(3,2)$ acting on the power set $\mathcal{P}(\Omega)$, stabilizing the code $RM(1,3)$, which a 4-dimensional code with weights 0,4,8. This code is equivalent to our $\mathcal{C}$. We conclude that $Aut(\mathcal{C}) \cong AGL(3,2)$ for any $\mathcal{C}$ which is $[8,4,4]_2$. □

6.2.3 Reed-Muller codes

Recall the affine geometries on $\mathbb{F}_2^d$.

Definition 6.2.3. Let $d \geq 0$ and $k \leq d$. The Reed-Muller code $RM(k,d)$, is the linear subspace of the power set $\mathcal{P}(\mathbb{F}_2^d)$ spanned by the set of affine subspaces of codimension k.

These codes are stable under the action of the affine group $AGL(d,2)$.

Trivially we have $RM(k,0) = 0$ for $k \leq -1$. The code $RM(0,d)$ is spanned by the vector $\mathbf{1}$, so is 1-dimensional.

Lemma 6.2.4. *The code $RM(1,d)$ consists of 0, the universe $\Omega = \mathbb{F}_2^d$ and $2(2^d - 1)$ affine subspaces of dimension $d-1$. Its weight distribution is $0^1 (2^{d-1})^{2(2^d-1)} (2^d)^1$.*

Proof. The code $RM(1,d)$ is spanned by all $(d-1)$-dimensional subspaces. There are $(2^d - 1)$ linear subspaces of dimension $d - 1$. So there are $2(2^d - 1)$ affine subspaces of dimension $d - 1$. There are $2^d - 1$ nonzero linear functionals. Any two linear functionals α, β with $Ker(\alpha) = Ker(\beta)$ are equal. Take two affine hyperplanes H_1, H_2. If $H_1 = H_2$, then $H_1 + H_2 = 0$. If $H_1 + \Omega = H_2$, then $H_1 + H_2 = \Omega$. Suppose now $H_1 \neq H_2, H_2 + \Omega$. We shall prove that $H_1 + H_2$ is an affine hyperplane. We can consider H_1, H_2 as the solution sets to linear equations $\alpha x = a, \beta x = b$, respectively, where $a, b \in \mathbb{F}_2$ and $\alpha, \beta \in \mathbb{F}_2^d$ with $\alpha \neq \beta$. Then $H_1 + H_2 = \{x \in \mathbb{F}_2^d \mid x \text{ satisfies just one of these equations}\}$, which is the solution set to the equation $(\alpha + \beta)x = a + b + 1$, so is an affine hyperplane. We conclude that $\dim(RM(1, d)) = 1 + d$. $\square$

Later we will discuss the codes $RM(2, d)$.

6.2.4 Uniqueness of the tetracode

Let $\mathcal{T} := \{s, a, a+s, a+2s | a, s \in \mathbb{F}_3\} \subseteq \mathbb{F}_3^4$. It is a $[4, 2, 3]_3$ code. The following result applies to $\mathcal{T}$.

Theorem 6.2.5. *Up to the action of the monoidal group $Mon(4, 3) \cong \mathbb{Z}_2 \wr Sym_4$, there is a unique $[4, 2, 3]_3$ code. Its automorphism group is isomorphic to the general linear group $GL(2, 3)$.*

Proof. We can follow the ideas for uniqueness of the binary $[8, 4, 4]_2$ code. Define $\mathcal{A} := \{(x, y) \mid wt(x) = wt(y) = 3; x, y \text{ independent and orthogonal}\}$. Such a pair (x, y) generates a $[4, 2, 3]_3$ code and conversely the basis of any $[4, 2, 3]_3$ code is such a pair.

Using symmetry, we may assume $x = (0, 1, 1, 1)$. By independence, $Supp(y) \neq \{2, 3, 4\}$. By symmetry, we may assume that $Supp(y) = \{1, 3, 4\}$. Write $y = (p, 0, q, r)$. Orthogonality implies that $q + r = 0$ and $wt(y) = 3$ implies that $p \neq 0$. Using permutations which fix index 1, we may assume that $y = (p, 0, 1, 2)$. Finally, using a diagonal transformation which fixes x, we may assume that $y = (1, 0, 1, 2)$. This completes the proof of uniqueness. Our argument really shows that $Mon(4, 3)$ acts transitively on $\mathcal{A}$.

We count $|\mathcal{A}|$. The number of weight 3 vectors is $\binom{4}{3} 2^3 = 32$. Given a weight 3 vector, x, the number of weight 3 vectors y which are orthogonal to it is $\binom{3}{2} 2^2 = 12$. Therefore $|\mathcal{A}| = 2^7 3$, which is also the order of $Mon(4, 3)$. Therefore the action of $Mon(4, 3)$ on $\mathcal{A}$ is regular.

A consequence is that the stabilizer of a particular $[4, 2, 3]$ code acts regularly on the bases for that code. Therefore, the group of the code is isomorphic to $GL(2, 3)$. $\square$

6.2.5 The automorphism group of the tetracode

We give some more details on $Aut(\mathcal{T})$. Note that $GL(2, 3)$ acts on $\mathbb{F}_3^2$ but $GL(2, 3)$ does not contain the symmetric group Sym_4 on 4 letters. This follows from the next result.

Lemma 6.2.6. *Let K be a field of characteristic not 2. Then the only involution in $SL(2, K)$ is -1.*

68   6 Constructions of Lattices by Use of Codes

Proof. Let $t \in SL(2, K)$ have order 2. Then it satisfies the equation $x^2 - 1 = 0$. So t has eigenvalues ± 1 and it is conjugate to some diagonal matrix $\begin{pmatrix} a & \\ & b \end{pmatrix} \in SL(2, K)$ for $a, b \in \{\pm 1\}$. Since $\det(t) = 1$, $ab = 1$. We obtain that $\begin{pmatrix} a & \\ & b \end{pmatrix} = \begin{pmatrix} 1 & \\ & 1 \end{pmatrix}$ or $\begin{pmatrix} -1 & \\ & -1 \end{pmatrix}$. But t has order 2, which implies $\begin{pmatrix} a & \\ & b \end{pmatrix} = \begin{pmatrix} -1 & \\ & -1 \end{pmatrix}$. $\square$

Recall that the symmetric group Sym_4 has a normal subgroup of order 4, namely, $\{1,(12)(34),(13)(24),(14)(23)\}$; it is the kernel of the action of Sym_4 on the unordered partitions of the set $\{1,2,3,4\}$ of type 2^2. The next result is an analogue for $GL(2,3)$.

Lemma 6.2.7. *The set* $U = \{x \in GL(2,3) \mid x^2 = -1\}$ *is*

$$\left\{ \pm \begin{pmatrix} 0 & 1 \\ -1 & 0 \end{pmatrix}, \pm \begin{pmatrix} 1 & 1 \\ 1 & -1 \end{pmatrix}, \pm \begin{pmatrix} 1 & -1 \\ 1 & -1 \end{pmatrix} \right\}.$$

The lemma can be obtained by direct calculations. Then $Q := U \cup \{\pm 1\}$ is a normal subgroup of $GL(2,3)$. In fact, it is isomorphic to the quaternion group $Quat_8$ of order 8. Furthermore, if H is any subgroup of $GL(2,3)$ which is isomorphic to the symmetric group Sym_3, then $H \cap Q = 1$ and $GL(2,3) = QH$, a semidirect product.

Lemma 6.2.8. *The elements*

$$\sigma = \begin{pmatrix} -1 & 0 & 0 & 0 \\ 0 & 0 & 1 & 0 \\ 0 & 1 & 0 & 0 \\ 0 & 0 & 0 & 1 \end{pmatrix}, \tau = \begin{pmatrix} -1 & 0 & 0 & 0 \\ 0 & 0 & 0 & 1 \\ 0 & 0 & 1 & 0 \\ 0 & 1 & 0 & 0 \end{pmatrix} \text{ and } \rho = \begin{pmatrix} 0 & 0 & -1 & 0 \\ 0 & 0 & 0 & 1 \\ 1 & 0 & 0 & 0 \\ 0 & -1 & 0 & 0 \end{pmatrix}$$

are in the automorphism group $Aut(\mathcal{T})$.

Proof. Direct calculations. $\square$

We shall prove that these elements generate a group isomorphic to the group $GL(2,3)$.

Finally, we prove

Theorem 6.2.9. $\langle \sigma, \tau, \rho \rangle \cong GL(2,3)$.

Proof. Since $\tau\sigma\tau = 1$ and $\sigma^3 = \tau^2 = 1$, we have an isomorphism $\langle \sigma, \tau \rangle \cong Sym_3$. Note that $\rho^2 = -1$. let $H = \langle \sigma, \tau, \rho \rangle < Aut(\mathcal{T})$. We have a natural homomorphism $\psi : H \to GL(\mathcal{T}) \cong GL(2,3)$ by restriction. Also, $\psi(\langle \sigma, \tau \rangle) \cong Sym_3$ is a subgroup of $GL(2,3)$, and $\psi(\rho)^2 = -1$. So $\psi(\rho)$ is in the set $U = \{x \in GL(2,3) | x^2 = -1\}$ of the previous lemma. Direct calculation shows that if $u \in U$ and $w \in GL(2,3)$ has order 3, then the conjugates $u, wuw^{-1}, w^{-1}uw$ and their negatives constitute U. Hence $GL(\mathcal{T}) = \psi(H)$. $\square$

A nice corollary is that we can see another subgroup of $O(E_8)$ because of the ternary code viewpoint. This is analogous to the subgroup of shape $2^8{:}AGL(3,2)$ obtained from the binary Hamming code viewpoint.

6.2 The proofs 69

Corollary 6.2.10. *The semidirect product $Sym_3{}^4{:}GL(2,3) \cong Sym_3 \wr GL(2,3)$ is contained in $O(E_8)$.*

Proof. By (4.2.25), the direct product of the Weyl groups for the four summands Q_i preserves the overlattice S. The Weyl group $R(Q_i)$ of Q_i is isomorphic to Sym_3. The group $(\prod_{i=1}^4 (O(Q_i)/R(Q_i))){:}Sym_4 \cong \mathbb{Z}_2 \wr Sym_4 \cong Mon(4,3)$ acts on $\mathcal{D}(Q) \cong \mathbb{Z}_3^4$. The stabilizer of the overlattice S in this group therefore corresponds to the group of the tetracode, isomorphic gto $GL(2,3)$. $\square$

6.2.6 Another characterization of $[8,4,4]_2$

In the last section of binary codes, we showed that any two self-orthogonal $[8,4,4]_2$ codes are equivalent. We prove a stronger result.

Theorem 6.2.11. *If $\mathcal{C}$ and $\mathcal{C}'$ are linear codes in $\mathbb{F}_2^8$ such that $\dim(\mathcal{C}) = \dim(\mathcal{C}') = 4$, and the minimum weight of each code is more than 4. Then $\mathcal{C}$ and $\mathcal{C}'$ are equivalent.*

Proof. We show that such a code is self-orthogonal. Define X_i to be the set of vectors in $\mathbb{F}_2^8$ of weight i. Then its cardinality is $|X_i| = \binom{8}{i}$. We define $Y := X_0 \cup X_1 \cup X_2$, and ask when $y_1, y_2 \in Y$ satisfy $\mathcal{C} + y_1 = \mathcal{C} + y_2$. If so, $y_1 + y_2 \in \mathcal{C}$. Suppose $y_1 \neq y_2$, we have $0 < wt(y_1 + y_2) \leqslant wt(y_1) + wt(y_2) \leqslant 2 + 2 = 4$. Note that min $wt(\mathcal{C}) \geqslant 4$. So $supp(y_1), supp(y_2)$ are disjoint 2-element sets. Since $\dim(\mathcal{C}) = 4$, we have $\dim(\mathbb{F}_2^8/\mathcal{C}) = 8 - 4 = 4$. The set Y represents at least $\binom{8}{0} + \binom{8}{1} + \frac{1}{4}\binom{8}{2} = 16$ cosets of $\mathcal{C}$. But there are only 16 cosets. So, all inequalities are equal, and Y represents all cosets of $\mathcal{C}$. Take $y_1, y_2, y_3, y_4 \in Y$ of weight 2 so that $y_i + \mathcal{C} = y_j + \mathcal{C}$ for all i, j. We have a partition of $\Omega = \{1, 2, \ldots, 8\} = \cup_{i=1}^4 supp(y_i)$. Also, the all 1-vector $\mathbf{1} = (y_1 + y_2) + (y_3 + y_4) \in \mathcal{C}$. So if $u \in \mathcal{C}$, then its complement $u + \mathbf{1} \in \mathcal{C}$. If $u \neq 0, \mathbf{1}$, then both u and $u + \mathbf{1}$ have weight more than 4. Thus both have weight 4. Finally, it is trivial to deduce that $\mathcal{C}$ is self-annihilating from the formula $|A + B| = |A| + |B| - 2|A \cap B|$. $\square$

6.2.7 Uniqueness of the E_8-lattice implies uniqueness of the binary $[8,4,4]$ code

Let us consider the lattice $L = E_8$ and a sublattice $Q \leqslant L$, where $Q = \sum_{i=1}^8 \mathbb{Z}\alpha_i$ for a set of vectors α_i so that $(\alpha_i, \alpha_j) = 2\delta_{ij}$. For example, one may take such a sublattice Q in the standard model of the E_8 lattice to be generated by the roots $(1,1,0,0,0,0,0,0)$, $(1,-1,0,0,0,0,0,0)$, $(0,0,1,1,0,0,0,0)$, $(0,0,1,-1,0,0,0,0)$, $\ldots$, $(0,0,0,0,0,0,1,1)$, $(0,0,0,0,0,0,1,-1)$.

We claim that $Q \geqslant 2L$. Let $R = Q + 2L \geqslant 2L$. We have $(R, R) \subseteq 2\mathbb{Z}$. So in the quadratic space $L/2L$ over $\mathbb{F}_2$, $R/2L$ is totally singular. Hence $\dim_{\mathbb{F}_2}(R/2L) \leqslant \frac{1}{2}\dim_{\mathbb{F}_2}(L/2L)$ (since $\det(L) = 1$, the quadratic space $L/2L$ is nondegenerate). It follows that $\dim(R/2L) \leqslant 4$. So $|R : 2L|$ divides 2^4, and $|L : R|$ divides 2^4 since $|L : 2L| = 2^8$. We have $Q \leqslant R \leqslant L$. But $\det(Q) = 2^8$ implies $|L : Q| = 2^4$. So $Q = R \geqslant 2L$.

70   6 Constructions of Lattices by Use of Codes

Theorem 6.2.12. *Let L be an E_8-lattice, and let Q, Q' be sublattices of L isometric to A_1^8. Then there is an isometry $g \in O(L)$ so that $g(Q) = Q'$.*

Proof. Due to the previous analysis, we have $Q \geqslant 2L$ and $Q' \geqslant 2L$. So $Q/2L$ and $Q'/2L$ are 4-dimensional spaces in the nondegenerate quadratic space $L/2L$. They are isometric. By Witt's theorem (2.1.1), there is an orthogonal transformation $\overline{g}$ on $L/2L$ so that $\overline{g}(Q/2L) = Q'/2L$. But the isometry group $O(L)$ maps onto the isometry group of $L/2L$. So take any $g \in O(L)$, which will satisfy the property $g(Q) = Q'$. □

Exercise 6.2.13. *For $n = 1, 2, \ldots, 8$. Let $\mathcal{A}_n$ be the set of ordered sequences $\alpha_1, \alpha_2, \ldots, \alpha_8$ so that $(\alpha_i, \alpha_j) = 2\delta_{ij}$. Let $\mathcal{B}_n$ be the set of unordered sequences with such a property. Let $a(n) = |\mathcal{A}_n|$ and $b(n) = |\mathcal{B}_n|$. Determine the numbers $a(i)$ and $b(i)$. (Hint: $a(i) = b(i) = 1$ for $i = 1, 2, 3$, and $a(4) = b(4) = 2$.)*

Theorem 6.2.14. *The uniqueness of the E_8 lattice implies the uniqueness of the $[8, 4, 4]_2$ codes.*

Proof. We are given a pair of $[8, 4, 4]_2$ code $\mathcal{C}, \mathcal{C}'$.

Using $\mathcal{C}$, we construct a rank 8 lattice which is even and unimodular. Take $\alpha_1, \ldots, \alpha_8 \in \mathbb{R}^8$ such that $(\alpha_i, \alpha_j) = 2\delta_{ij}$. Define $Q = \sum \mathbb{Z}\alpha_i$. For a codeword $v \in \mathcal{C}$, let $\alpha_v = \sum_{i \in supp(v)} \alpha_i$, and $Q[\mathcal{C}] := Q + \sum_{v \in \mathcal{C}} \mathbb{Z}\frac{1}{2}\alpha_v$. Then $Q[\mathcal{C}]$ is even and unimodular. We can do the same for another such code $\mathcal{C}'$ and obtain $Q[\mathcal{C}']$. By the uniqueness of the E_8-lattice, there is an isometry $\alpha \in O(\mathbb{R}^8)$ so that $\alpha(Q[\mathcal{C}]) = Q[\mathcal{C}']$. So Q and $\alpha(Q)$ are sublattices of $Q[\mathcal{C}'] \cong E_8$. We have $\alpha(Q) \cong Q \cong A_1^8$. There exists an isometry $\beta \in O(Q[\mathcal{C}'])$, so that $\beta\alpha(Q) = Q$. Then $\gamma = \beta\alpha$ induces a permutation of the sets $\{\pm\alpha_1\}, \{\pm\alpha_2\}, \ldots, \{\pm\alpha_8\}$. So γ preserves Q and $\frac{1}{2}Q$, and so gives a permutation matrix on $\frac{1}{2}Q/Q \cong \oplus_{i=1}^8 (\mathbb{Z}\frac{1}{2}\alpha_i/\mathbb{Z}\alpha_i)$. It follows that γ takes the code $\mathcal{C}$ to $\mathcal{C}'$. Therefore, the two codes $\mathcal{C}$ and $\mathcal{C}'$ are equivalent. □

Uniqueness of the $[8, 4, 4]_2$ code implies uniqueness of the E_8-lattice, but that is harder to prove.

6.3 Codes over $\mathbb{F}_7$ and a (mod 7)-construction of E_8

We give another code-style construction of E_8. This time, we use a code over $\mathbb{F}_7$ and the sublattice we start from is not a root lattice.

The field $\mathbb{F}_7 = \{0, 1, \ldots, 6\}$ has squares $\{0, 1, 2, 4\}$. In $\mathbb{F}_7^n$, a singular vector has norm at least 3.

Definition 6.3.1. *We define the septacode $\mathcal{S}$ to be the span of the two vectors $(0, 1, 2, 4)$ and $(2, 1, 1, 1)$ in $\mathbb{F}_7^4$.*

Such a code $\mathcal{S}$ is a totally singular 2-space in which weights are just 0,3,4.

Define U to be the rank 2 lattice with basis α, β and Gram matrix $\begin{pmatrix} 2 & 1 \\ 1 & 4 \end{pmatrix}$, having determinant 7. The dual U^* has Gram matrix $\begin{pmatrix} 2 & 1 \\ 1 & 4 \end{pmatrix}^{-1} = \frac{1}{7}\begin{pmatrix} 4 & -1 \\ -1 & 2 \end{pmatrix}$.

So the dual basis is $\alpha^* = \frac{1}{7}(4,-1)$, $\beta^* = \frac{1}{7}(-1,2)$. Then $U^* = U + \gamma$, where $\gamma = \frac{1}{7}(-\alpha + 2\beta)$.

We study the lattices L so that $U_1 \perp \cdots \perp U_4 \leqslant L \leqslant U_1^* \perp \cdots \perp U_4^*$. So $L/U_1 \perp \cdots \perp U_4$ is a code in $U_1^* \perp \cdots \perp U_4^*/U_1 \perp \cdots \perp U_4 \cong (U_1^*/U_1) \perp \cdots \perp (U_4^*/U_4) \cong \mathbb{F}_7^4$. The 4-tuple $(a,b,c,d) \in \mathbb{F}_7^4$ corresponds to $a\gamma_1 + b\gamma_2 + c\gamma_3 + d\gamma_4$. We take the code $\mathcal{S}$ and let L correspond to it in the above sense. Since $\det(U_1 \perp \cdots \perp U_4) = 7^4$ and $\dim(\mathcal{S}) = 2$, we have $|L : U_1 \perp \cdots \perp U_4| = 7^2$. It follows that $\det(L) = 1$.

We claim that L is integral. Let $u = \sum \alpha_i \gamma_i$, $v = \sum \beta_i \gamma_i$ be two glue vectors. Note that $(\gamma, \gamma) = \frac{2}{7}$. We have $(u,v) = \sum a_i b_i (\gamma_i, \gamma_i) = \frac{2}{7} \sum a_i b_i \in 2\mathbb{Z}$ since $\sum a_i b_i$ is divisible by 7. Therefore, L is integral. The lattice L is even since the odd index sublattice $U_1 \perp \cdots \perp U_4$ is even. So $L \cong E_8$ by the uniqueness theorem (5.2.1).

6.3.1 The A_6-lattice

We have $L \supseteq U_1 \perp \cdots \perp U_4$ and the annihilator $\operatorname{ann}_L(U_i)$ is a lattice of rank 6 and determinant 7, for all i. It turns out to be the A_6-lattice. We can see this as follows.

We take $i = 1$ and note that $\operatorname{ann}_L(U_1) = U_1 + U_2 + U_3 + \mathbb{Z}u$, where u is the glue vector corresponding to the codeword $(0, 1, 2, 4)$.

We form the lattice $J := \operatorname{ann}_L(U_1) \perp \mathbb{Z}\delta$, where $(\delta, \delta) = 7$. Let γ_i be a vector in U_i^* of norm $\frac{2}{7}$ (see the previous definition of γ).

The vector $v := \gamma_2 + 2\gamma_3 + 4\gamma_4 + \frac{1}{7}\delta$ is in J^* and has norm 1. The lattice $K := J + \mathbb{Z}v$ is integral and unimodular, hence is isometric to $\mathbb{Z}^7$ by the classification (5.2.1). We may therefore realize an isometric copy of J in $\mathbb{Z}^7$ as the annihilator of a norm 7 vector, say w. Since J is even, every coordinate of w is nonzero. Therefore, up to coordinate signs, $w = (1, 1, 1, 1, 1, 1, 1)$. Its annihilator is the standard model of A_6, whence $J \cong A_6$. This discussion proves the following result.

Proposition 6.3.2. *An even integral lattice of rank 6 and determinant 7 is isometric to A_6.*

7

Group Theory and Representations

In this section, we will discuss finite groups and their representations. This will help with studies isometry groups of lattices.

First we begin with some general background, then specialize to extraspecial groups and certain upward extensions.

A *representation* of a group G is a homomorphism $\psi : G \to Aut(M)$, for some abelian group M. In case M is a module over the ring K and the image of ψ is contained in $Aut_K(M)$, the automorphisms of K which commute with the action of K, we say ψ is a K-*representation*.

The term *integral representation* usually means a representation on a free module of finite rank over K, a ring of integers in some number field. Our use of the term in this book almost always means a representation on some finite rank free abelian group (i.e., $K = \mathbb{Z}$).

7.1 Finite groups

There are many excellent books for an introduction to modern finite group theory and representation theory [23, 24, 29, 30, 31, 51, 52] We shall assume, often without comment, that certain topics can be looked up routinely. In this text, we shall treat some topics in detail, e.g., extraspecial groups and their upward extensions.

We now review a few definitions.

In a group G, we denote *conjugation of x by y* with $x^y = y^{-1}xy$. Conjugation by y is a group automorphism of G. The *commutator of x and y* is denoted by $[x,y] = x^{-1}y^{-1}xy$. The *commutator subgroup* of G is $G' := [G,G] := \langle [x,y] \mid x, y \in G \rangle$. The *center of G* is $Z(G) := \{x \in G \mid xy = yx \text{ for all } y \in G\}$. A group G is called *perfect* if $G = G' = [G,G]$. A homomorphism from a perfect group to an abelian group must be trivial.

The *derived series* is the descending chain of subgroups $G = G^{(0)} \geq G^{(1)} \geq \cdots$, where $G^{(1)} := G'$, and $G^{(k+1)} := [G^{(k)}, G^{(k)}]$. A group G is *solvable* if there is a k such that $G^{(k)} = 1$. The *lower central series* is the descending series defined by $\gamma_i(G) \leqslant \gamma_{i+1}(G)$, where $G = \gamma_1(G)$ and $\gamma_{i+1}(G) := [\gamma_i(G), G]$. The *upper central series* is the ascending chain defined by $Z_1(G) = Z(G)$ and

$Z_{i+1}(G)/Z_i(G) = Z(G/Z_i(G))$. A group G is *nilpotent* if and only if there is an i so that $Z_i(G) = G$, if and only if there is a j so that $\gamma_j(G) = 1$, if and only if G has a central series, i.e., $1 = G_0 \leqslant G_1 \leqslant \cdots \leqslant G_r = G$ with $G_{i+1}/G_i \leqslant Z(G/G_i)$ for all i.

The *exponent* of a group G is the smallest integer $n > 0$ so that $x^n = 1$ for all $x \in G$. We write $\exp(G) = n$. If no such n exists, we write $\exp(G) = \infty$.

We say a group is a *p-group* if it has order of a power of p, where p is a prime number. Such a group is nilpotent.

If G is a finite group. We define the *Frattini subgroup* $\Phi(G)$ to be the intersection of all maximal subgroups of G. Here maximal means $M \neq G$ and $M \leqslant H \leqslant G$ implies $H = M$ or $H = G$. The Frattini subgroup is the set of nongenerators of G, i.e., the set of all $x \in G$ such that, whenever a set $S \subseteq G$ and $G = \langle S, x \rangle$ we have $G = \langle S \rangle$.

Proposition 7.1.1. *If G is a p-group, then its Frattini subgroup has the decomposition $\Phi(G) = G'\langle x^p | x \in G \rangle$.*

Proposition 7.1.2. *We have the following commutator identities.*
(1) $[xy, z] = [x, z]^y [y, z]$; (2) $[x, yz] = [x, z][x, y]^z$.

Proof. Let a, b be two group elements. Then $a^b = b^{-1}ab = aa^{-1}b^{-1}ab = a[a, b]$. So $xy[xy, z] = (xy)^z = x^z y^z = x[x, z]y[y, z] = xyy^{-1}[x, z]y[y, z] = xy[x, z]^y[y, z]$. Similarly, we have $x[x, yz] = x^{yz} = (x^y)^z = (x[x, y])^z = x^z[x, y]^z = x[x, z][x, y]^z$. $\square$

Theorem 7.1.3. *If G has order dividing p^2, then G is abelian. A nonabelian p-group of order p^3 is isomorphic to*
(a) Dih_8 or $Quat_8$ if $p = 2$; otherwise
(b.1) $\langle x, y, z \mid x^p = y^p = z^p = 1, [z, x] = 1 = [y, z], x^y = xz \rangle$, or
(b.2) $\langle x, y, z \mid x^p = z, y^p = z^p = 1, [z, x] = 1 = [y, z], x^y = xz = x^{1+p} \rangle$.
They have exponents p and p^2, respectively.

Proof. See for instance [31], [51], etc. $\square$

Exercise 7.1.4. Let K be a field. Denote

$$U = \left\{ \begin{pmatrix} 1 & a & c \\ & 1 & b \\ & & 1 \end{pmatrix} \middle| a, b, c \in K \right\}.$$

Show that if $K = \mathbb{F}_p$, then we get the (b.1) case of the previous theorem when p is odd, and Dih_8 when $p = 2$.

Lemma 7.1.5. *Suppose x, y be group elements. Define $z := [x, y]$. Suppose $z \in Z(\langle x, y \rangle)$, the center of the subgroup generated by x and y. Then for all $i \geqslant 1$, we have $(xy)^i = x^i y^i z^{\binom{i}{2}}$.*

Proof. For the cases $i = 0$ and $i = 1$, the identity holds since $\binom{i}{2} = 0$. We assume this formula is true for $i \geqslant 1$. Let us calculate $(xy)^{i+1} = (xy)^i xy = x^i y^i z^{\binom{i}{2}} \cdot xy$. Note that $yx = xy[x, y] = xyz$, so $y^i x = xy^i[x, y^i] = xy^i z^i$ because $z \in Z(\langle x, y \rangle)$. We deduce $(xy)^{i+1} = x^i xy^i z^i z^{\binom{i}{2}} y = x^{i+1} y^{i+1} z^{\binom{i+1}{2}}$, since $i + \binom{i}{2} = \binom{i+1}{2}$. $\square$

Lemma 7.1.6. *Let x, y be group elements and $z := [x,y]$. Suppose $z \in Z(\langle x,y\rangle)$, and $z^n = 1$ for some odd integer $n \geq 1$. Then $(xy)^n = x^n y^n$, i.e. the n-th power map is a group endomorphism of $\langle x, y\rangle$.*

Proof. Due to the previous lemma, we have $(xy)^n = x^n y^n z^{\binom{n}{2}}$. Note that $\binom{n}{2} = \frac{n(n-1)}{2}$. So $z^{\binom{n}{2}} = (z^n)^{\frac{n-1}{2}} = 1$. $\square$

Remark 7.1.7. Observe that $(xy)^2 = xyxy = xx(x^{-1}yxy^{-1}) = xx(x^{-1}yxy^{-1})yy$. So $(xy)^2 = x^2 y^2$ if and only if $x^{-1}yxy^{-1} = 1$, or equivalently, x and y commute.

Corollary 7.1.8. *If P is a p-group of class ≤ 2 for an odd prime p and the center $Z(P)$ has exponent p, then the p-th power map is a homomorphism. In particular, the set $\{x \in P \mid x^p = 1\}$ is a subgroup of P, namely the kernel of the latter homomorphism.*

Exercise 7.1.9. Show that in a dihedral group of order $2m$, $m \geq 2$, there are m elements of order 2 if m is odd, and $m+1$ elements of order 2 if m is even. So the set of elements of order 1 or 2 is not a subgroup if $m \geq 3$.

Definition 7.1.10. *A group G is a central product of its subgroups H and K if $G = HK$ and $[H,K] = 1$. A central product of H and K is often written $H \circ K$.*

The definition implies that the elements in $H \cap K$ commute with $G = HK$, i.e. $H \cap K \leq Z(G)$. We have an epimorphism of groups $H \times K \to G$, sending $(x,y) \in H \times K$ to the product $xy \in G$. The kernel of this epimorphism is $\{(x,y) \mid xy = 1\} = \{(x, x^{-1}) \mid x \in H \cap K\}$. We may think of $G = H \circ K$ as $H \times K / \{(x, x^{-1}) \mid x \in H \cap K\}$.

7.2 Extraspecial p-groups

Definition 7.2.1. *The p-group P is called extraspecial if its center $Z(P) \cong \mathbb{Z}_p$ and the quotient $P/Z(P)$ is a nonidentity elementary abelian p-group.*

For such a group, the commutator subgroup has order p. If $z_i \in Z(P)$ for $i = 1, 2$, then we have $[xz_1, yz_2] = [x, yz_2]^{z_1}[z_1, yz_2] = [x, yz_2][z_1, yz_2] = [x, yz_2] = [x,y][x,z_2]^{z_2} = [x,y]$. Thus, we have a function $P/Z(P) \times P/Z(P) \to Z(P)$ so that $(xZ(P), yZ(P)) \mapsto [x, y] \in Z(P)$. This is bi-multiplicative (7.1.2) and so may be viewed as a bilinear map on the vector space $P/Z(P)$ to the field $\mathbb{Z}_p$. Also, the radical of the above form on the vector space $P/Z(P)$ is trivial. The form is alternating since $[y,x] = [x,y]^{-1}$. Hence the dimension of the vector space P/Z is even. We deduce that the order of the group P is $|P| = p^{1+2n}$ for some integer $n \geq 1$.

7.2.1 Extraspecial groups and central products

We give a construction of extraspecial groups. Suppose E and F are extraspecial p-groups for a prime number p, and their centers are $Z(E) = \langle u \rangle$, $Z(F) = \langle v \rangle$. Let $G := (E \times F)/D$, where $D = \langle (u, v^{-1}) \rangle$. We claim that the group G is extra special. Indeed, the commutator group $(E \times F)' = E' \times F' = \langle u \rangle \times \langle v \rangle$ has order p^2. So we have $G' = (E' \times F')D/D \cong \mathbb{Z}_p$, hence $|G'| = p$. Next we show that $Z(G) = G'$. Take $(x,y)D \in G$, where $x \in E$ and $y \in F$. If $x \notin Z(E)$, there is an $x' \in E$ such that $[x,x'] \neq 1 \in \langle u \rangle$. So in G, the commutator of $(x,y)D$ and $(x',1)D$ is not trivial. Similarly for $y \notin Z(F)$. Then $(x,y)D \notin Z(G)$. We conclude that $Z(G) = (Z(E) \times Z(F))/D \cong \mathbb{Z}_p$. So G is extraspecial of order $\frac{1}{p}|E||F|$.

Next we show how to decompose an extraspecial group G of order p^{1+2n} as a central product. There is a nontrivial such decomposition when $n \geq 2$.

Denote by $Z := Z(G)$ the center of G. Suppose that H is a subgroup, $Z \lneq H \lneq G$, such that H/Z is a nonsingular subspace of the vector space G/Z. Then we define K by $Z \lneq K \lneq G$ so that K/Z is the annihilator of H with respect to the commutation maps, *i.e.* $K = C_G(H)$. Due to the bilinear form on the space G/Z, we obtain a decomposition $G/Z = H/Z \times K/Z$ of vector spaces (where we are writing vector space addition as multiplication in the quotient group). So $G = H \circ K$ is a central product of H and K. We claim that the two groups H and K are also extraspecial. This follows since the subspaces H/Z and K/Z are nonsingular. The nonsingularity implies $Z(H)/Z = 1$, *i.e.* $Z(H) = Z$. This in turn implies that the commutator subgroup H' is not trivial. So $H' = Z(H)$. A similar argument applies to the group K.

Note that any orthogonal decomposition of G/Z gives a central product.

Theorem 7.2.2. *Let p be a prime number and $n \geq 1$. An extraspecial group of order p^{1+2n} is a central product of n extraspecial groups of order p^3.*

Remark 7.2.3. *If $P = P_1 \circ \cdots \circ P_n = Q_1 \circ \cdots \circ Q_n$, it is not generally true that the two sets of central factors $\{P_i \mid i = 1, \ldots, n\}$ and $\{Q_i \mid i = 1, \ldots, n\}$ are equal. It may note be true that these two set of isomorphism types are the same (even if we require all the P_i and Q_j to have order p^3). There are in general many central product decompositions for P.*

Example 7.2.4. *Let $n \geq 1$ and let p be a prime. Define $E_{2n+1}(p)$ to be the set of upper triangular matrices*

$$\begin{pmatrix} 1 & a_1 & a_2 & a_3 & \ldots & a_n & c \\ 0 & 1 & 0 & 0 & \ldots & 0 & b_1 \\ 0 & 0 & 1 & 0 & & 0 & b_2 \\ \vdots & \vdots & \vdots & \vdots & & \vdots & \vdots \\ 0 & 0 & 0 & 0 & \ldots & 0 & b_{n-1} \\ 0 & 0 & 0 & 0 & \ldots & 1 & b_n \\ 0 & 0 & 0 & 0 & \ldots & 0 & 1 \end{pmatrix}, \quad \text{all } a_i, b_j, c \in \mathbb{F}_p.$$

It is straightforward to check that the center of $E_{2n+1}(p)$ is the set of such matrices with all a_i and b_j equal to 0 and that the group is extraspecial. Find nontrivial central product decompositions of this group.

7.2.2 A normal form in an extraspecial group

The following procedure gives us a normal form for the elements of an extraspecial group G. Suppose the group G is extraspecial. Let $Z = Z(G) = \langle z \rangle$, and the elements $x_1, \ldots, x_{2n} \in G$ map to a basis of the space $G/Z \cong \mathbb{F}_p^{2n}$. Then every element of G has a unique expression $z^a x_1^{b_1} \cdots x_{2n}^{b_{2n}}$ for $a, b_1, \ldots, b_{2n} \in \{0, 1, 2, \ldots, p-1\}$. Suppose we have a product $(z^a x_1^{b_1} \cdots x_{2n}^{b_{2n}}) \cdot (z^{a'} x_1^{b'_1} \cdots x_{2n}^{b'_{2n}})$. We can put this in normal form $z^{a''} x_1^{b''_1} \cdots x_{2n}^{b''_{2n}}$ if we know all the powers x_i^p and all the commutators $[x_i, x_j]$.

It is clear that the above gives a *presentation* of G. That is, the free group F on symbols $z, x_1, \ldots, x_{2n}$, modulo the normal subgroup R generated by the relators $z^p, x_1^p, \ldots, [x_1, x_2] z^{-a(1,2)}, \ldots$ (for suitable integers $a(i,j)$) is isomorphic to G. There is an obvious homomorphism of F onto G with R contained in the kernel. The form of the relations makes it clear that every element of F/R may be put in a normal form, so that the induced homomorphism $F/R \to G$ is an isomorphism onto.

This preceding paragraph is useful for studying maps from an extraspecial group to another group. In particular, it will help us analyze $Aut(G)$. A function $f: G \to G$ which respects the p-th power map and commutation extends to a unique homomorphism. Given an isomorphism $g: P/Z \to P/Z$ which respects the p-th power map and commutation, there are p^{2n} automorphisms f of P, which satisfy $f(x_i) \in g(x_i Z)$. So we obtain a normal subgroup $K \triangleleft Aut(P)$ so that $|K| = p^{2n}$ and K acts trivially on P/Z. So $K = Inn(P) \cong P/Z \cong \mathbb{Z}_p^{2n}$.

Definition 7.2.5. *Define $Aut^0(P) := \{\alpha \in Aut(P) \mid \alpha \text{ is the identity on the center } Z\}$.*

We deduce that $Aut^0(P)/K$ is isomorphic to a subgroup of $Aut(P/Z) \cong GL(2n, p)$ which preserves the p-th power map and the alternating bilinear form on P/Z which comes from commutation.

7.2.3 A classification of extraspecial groups

We recall a definition made earlier (2.0.8). Let V be an n-dimensional quadratic space over a field. The *Witt index* is the maximum dimension of a totally singular subspace. When V is nonsingular, the Witt index is one of $\frac{n}{2}, \frac{n}{2} - 1, \frac{n-1}{2}$. Furthermore, given a finite field and dimension, the Witt index determines the similitude type of the quadratic space. For proofs, see [28].

A direct proof for the case of quadratic spaces over finite fields of characteristic 2 may be found in [37]. See also (2.1.6).

Let p be a prime and $n \geq 1$. We shall prove there are just two isomorphism classes of extraspecial p-groups of given order p^{1+2n}. When p is odd, they are distinguished by whether the group has exponent p or p^2. When $p = 2$, they are distinguished by whether there does or does not exist a subgroup of order 2^{n+1} which is elementary abelian.

Notation 7.2.6. *(7.2.6) For given p and n, we denote these respective isomorphism types of extraspecial groups by p_ε^{1+2n}, where $\varepsilon = +$ or $-$. That is, 2_+^{1+2n} is the order 2^{1+2n} extraspecial group which contains an elementary abelian group of rank $n+1$ and for odd primes p, p_+^{1+2n} is the extraspecial group of order p^{1+2n} of exponent p.*

For example, $2_+^{1+2} \cong Dih_8$ and $2_-^{1+2} \cong Quat_8$.

Let p be odd. Let $\psi : G \to Z$ be the p-th power map. Then ψ is a group homomorphism.

Suppose that $\psi(G) = 1$. Then the product of elements in the standard form depends only on the commutators. There is a basis of G/Z so that its Gram matrix is $\begin{pmatrix} 0 & I_n \\ -I_n & 0 \end{pmatrix}$ because this is the standard form for a Gram matrix of a nondegenerate alternating bilinear form. Correspondingly, we can choose such a sequence $x_1, \ldots, x_n, x_{n+1}, \ldots, x_{2n}$ in G whose image in the vector space G/Z gives this standard Gram matrix. Call $a(i,j)$ the i,j entry of this matrix. We summarize: in case $\psi(G) = 1$, there is a sequence $z, x_1, \ldots, x_{2n}$ so that $x_1, \ldots, x_{2n}$ maps to a basis of G/Z and satisfies the commutation rules $[x_i, x_j] = z^{a(i,j)}$ and the power rules $x_i^p = 1$ for $i \leq 2n$. In particular, the multiplication table is determined. We furthermore have an expression of G as a central product of the groups $\langle x_i, x_{n+i} \rangle$, for $i = 1, \ldots, n$ of order p^3. Each of these is isomorphic to the (unique) nonabelian group of order p^3 and exponent p.

Suppose that $\psi(G) \neq 1$. Let $H = Ker(\psi)$. Then H/Z is a $2n-1$ dimensional subspace of nullity 1. Take H_1, a subgroup $Z \leqslant H_1 \leqslant H$ of H, so that the vector space $H_1/Z \cong \mathbb{F}_p^{2(n-1)}$ is nonsingular. Take a sequence $x_1, \ldots, x_{2(n-1)}$ for H_1 as in the previous case. Let $x_{2n-1} \in H$, so that $x_{2n-1}Z$ spans the radical of H/Z. Let K satisfy $Z \leqslant K \leqslant G$ and $K/Z = ann_{G/Z}(H_1/Z)$, i.e., $K = C_G(H_1)$. Then $x_{2n-1} \in K$. Take $x_{2n} \in K$ so that $\langle x_{2n-1}, x_{2n}, Z \rangle = K$. Then there is $a \in \mathbb{F}_p, a \neq 0$, so that the Gram matrix of $x_1, \ldots, x_{2n}$ is

$$\begin{pmatrix} 0 & I_{n-1} & & \\ -I_{n-1} & 0 & & \\ & & 0 & a \\ & & -a & 0 \end{pmatrix}.$$

Call $a(i,j)$ the i,j entry of this matrix. Also, x_{2n} has order p^2. Define $k \in \mathbb{Z}$ by $x_{2n}^p = z^k$. Then $(k,p) = 1$. Replace x_{2n} by a power to make $k = 1$. Then replace x_{2n} by some x_{2n-1}^c, $(c,p) = 1$, to make $a = 1$. We summarize: in case $\psi(G) \neq 1$, there is a sequence $z, x_1, \ldots, x_{2n}$ so that $x_1, \ldots, x_{2n}$ maps to a basis of G/Z and satisfies the commutation rules $[x_i, x_j] = z^{a(i,j)}$ and the power rules $x_i^p = 1$ for $i \leq 2n-1$ and $x_{2n}^p = z$. In particular, the multiplication table is determined. We furthermore have an expression of G as a central product of the groups $\langle x_i, x_{n+i} \rangle$, for $i = 1, \ldots, n$ of order p^3. The first $n-1$ have exponent p and the last has exponent p^2.

Theorem 7.2.7. *An extraspecial p-group of order p^{1+2n} has a normal form for elements so that knowledge of commutators and p-powers on the set of genera-*

tors suffices to determine the multiplication table for the group. Commutators and powers are all powers of a fixed generator for the center.

7.2.4 An application to automorphism groups of extraspecial groups

Lemma 7.2.8. *Let p be a prime and let P be a nonabelian group of order p^3. Let z generate the center of P. Given an integer k so that $(k,p) = 1$, there exists an automorphism α of P so that $\alpha(z) = z^k$.*

Proof. Suppose first that P has exponent p. Then there are $x, y \in P$ so that xZ, yZ form a basis for P/Z. We use the normal form for P with respect to the sequence z, x, y. The function which takes z, x, y to z^k, x, y^k extends to a unique automorphism and this automorphism has the required property.

Suppose now that P has exponent p^2. We may take x, y as before but with x of order p and y of order p^2. The proof finishes as before, namely by taking the function which takes z, x, y to z^k, x, y^k and extending to an automorphism. □

Theorem 7.2.9. *If P is an extraspecial p-group with center Z, $Aut(P)/Aut^0(P) \cong \mathbb{Z}_{p-1}$. In other words, the action of $Aut(P)$ on Z induces $Aut(Z)$.*

Proof. We may assume that p is odd. Let k be an integer prime to p. Write $P = P_1 \circ \cdots \circ P_n$ as a central product of nonabelian groups of order p^3. For each i, we take an automorphism α_i of P_i as in (7.2.8). There is an automorphism of the direct product $P_1 \times \cdots \times P_n$ which extends the action of α_i on the factor P_i. This automorphism preserves the subgroup which is the kernel of the natural map of this direct product to $P = P_1 \circ \cdots \circ P_n$. The automorphism so induced on P takes z to z^k. □

7.3 Group representations

Definition 7.3.1. *Suppose that V is a vector space over a field F and that $\rho : G \to GL(V)$ is an F-representation of the group G. We say that ρ is reducible if there is a subspace $0 \neq W \lneq V$, so that $\rho(G)$ leaves W invariant. Otherwise ρ is called irreducible (provided that $V \neq 0$). The character of a representation ρ is the function χ_ρ. It satisfies $g \mapsto \mathrm{Tr}(\rho(g)) \in F$, for $g \in G$. (The notation $\mathrm{Tr}(\cdot)$ means the trace of a linear transformation.)*

Definition 7.3.2. *Let K be a commutative ring and G a group. The group algebra, written KG or $K[G]$, is the set of all formal sums $\sum_{g \in G} a_g g$, where $a_g \in K$ is zero for almost all $g \in G$. It is a K-algebra under the obvious vector space operations and product based on multiplication of basis elements.*

A K-representation of G on K-module M extends in an obvious way to a homomorphism of K-algebras $KG \to End_K(M)$, and conversely.

We give a summary of results in group representation theory.

(1) Suppose G is a finite group and $char(F) \nmid |G|$. Then any finite dimensional representation of G is completely reducible, i.e., if $\rho : G \to GL(V)$ is a representation, and W is an invariant subspace of V, then there exists an invariant subspace W' such that $V = W \oplus W'$.

(2) Suppose $\rho : G \to GL(V)$ and $\rho' : G \to GL(V')$ are two representations of a group G. We call them *equivalent* if there is a linear isomorphism $\sigma : V \to V'$ so that for any $x \in V$ and $g \in G$, we have $\sigma(\rho(g))(x) = \rho'(g)(\sigma(x))$.

(3) Suppose G is a finite group, and F is an algebraically closed field and $char(F) \nmid |G|$. Then the number of equivalent classes of irreducible representations is finite and is equal to the number of conjugacy classes in G. Two representations are equivalent if and only if they have the same character.

(4) Let F be a field. Suppose $H \leqslant G$ is a subgroup of the group G, and M is an FH-module of character ψ. Then the FG-module $FG \otimes_{FH} M$ has character $\psi^G(g) = \sum_{t \in T} \psi^\circ(tgt^{-1})$, where $\psi^\circ : G \to F$ is 0 if $g \in G \setminus H$ and is $\psi(g)$ if $g \in H$; and T is a set of left coset representatives for H in G. This representation is called the *induced representation*, or more fully, *the representation of G induced from the representation M on the subgroup H*. It is denoted $M_H \uparrow^G$. The corresponding induced character is written $\psi_H \uparrow^G$.

(5) Orthogonality relations of characters. We mention a few that we use. Suppose that $char(F)$ does not divide $|G|$ and that F is algebraically closed. Suppose that $\chi_1, \ldots, \chi_k$ are the characters of irreducible representations of the group G, and $g_1, \ldots, g_k$ represents the k conjugacy classes in G. Then we have $\delta_{rs}|G| = \sum_{i=1}^{k} \chi_i(g_r)\chi_i(g_s^{-1})$. In particular, $|G| = \sum_{i=1}^{k} \chi_i(1)^2$.

(6) The representation theories for G over any two algebraically closed fields of characteristics not dividing $|G|$ are "essentially the same". If we take a prime number p not dividing $|G|$, then a finite dimensional complex representation may be written over a ring of integers R and "reduced modulo $\mathfrak{P}$" (where $\mathfrak{P}$ is a prime ideal in R containing p), giving a representation of G in characteristic p. This procedure gives a bijection between isomorphism classes of modules for G over $\mathbb{C}$ and over the algebraic closure $\overline{\mathbb{F}_p}$. Their respective traces correspond under this reduction. Modules over both fields are completely reducible and the irreducibles are distinguished by their characters.

7.3.1 Representations of extraspecial p-groups

Let us consider the irreducible complex representation of an extraspecial group P of order p^{1+2n}, for some prime number p and integer n. The 1-dimensional irreducible representations correspond to $Hom(P, \mathbb{C}^\times)$, which is equivalent to $Hom(P/P', \mathbb{C}^\times) \cong Hom(\mathbb{Z}_p^{2n}, \mathbb{C}^\times) \cong \oplus_1^{2n} Hom(\mathbb{Z}_p, \mathbb{C}^\times) \cong \mathbb{Z}_p^{2n}$. If $\lambda \in Hom(P, \mathbb{C}^\times)$ is a character, then $\lambda(z^i) = 1$ for all $z \in P'$ because $P' \leqslant Ker(\lambda)$.

We now describe nonlinear irreducibles of P. Let $A \leqslant P$ be an abelian subgroup of order p^{n+1}. So A/Z corresponds to a totally singular subspace of P/Z. Take $\lambda \in Hom(A, \mathbb{C}^\times)$ so that $\lambda(a) = \zeta$ is a primitive p-th root of unity, i.e., $\zeta = e^{2\pi i j/p}$ where $(j, p) = 1$. We form the induced representation with character $\lambda_A \uparrow^P$. Note that $\mathbb{C}P \otimes_{\mathbb{C}A} M = \oplus_{t \in T}(\mathbb{C}t \otimes M)$, where M denotes a 1-dimensional module for A which affords λ and where T is a left transversal to A in P (i.e., a set of left coset representatives). For $g \in P, t \in T, m \in M$, we have $g(t \otimes m) = t' \otimes m'$ for some $t' \in T$ and $m' \in M$. We may take the basis $t \otimes m_0$, $t \in T$, to calculate the trace of g. The trace of g can get a nonzero contribution from only those basis elements with $t' = t$.

Lemma 7.3.3. *Suppose A is an $n \times n$ matrix, and $c \neq 1$ is a scalar. If A is conjugate to cA, then $\mathrm{Tr}(A) = 0$.*

Proof. We use the fact that if X, Y are two $n \times n$ matrices, then $\mathrm{Tr}(XY) = \mathrm{Tr}(YX)$. So $\mathrm{Tr}((B^{-1}A)B)) = \mathrm{Tr}(B(B^{-1}A)) = \mathrm{Tr}(A)$. Suppose A is conjugate to cA. Then $\mathrm{Tr}(A) = \mathrm{Tr}(cA) = c\mathrm{Tr}(A)$, or $(c-1)\mathrm{Tr}(A) = 0$. It follows that $\mathrm{Tr}(A) = 0$ since $c \neq 1$. □

Proposition 7.3.4. *The induced character $\lambda_A \uparrow^P$ is irreducible of degree p^n, and takes values 0 on $P \setminus Z$.*

Proof. If $x \in P \setminus Z$, there exists $y \in P$ so that $[x, y] = z$, or $x^y = xz$. Taking traces, and using the fact that z acts like the scalar ζ, we deduce from (7.3.3) that the trace of the matrix representing x is zero. To prove irreducibility, just check that $\sum_{g \in P} \mathrm{Tr}(g)\mathrm{Tr}(g^{-1}) = |P|$. Since the traces vanish off Z, the sum equals. $\sum_{g \in Z} \mathrm{Tr}(g)\mathrm{Tr}(g^{-1}) = p^{2n} \cdot |Z| = |P|$. □

Theorem 7.3.5. *The group P has p^{2n} linear characters and $p-1$ irreducible characters of degree p^n. The latter set forms a complete set of algebraically conjugate characters.*

Proof. Given a nontrivial homomorphism $\lambda : Z \to \langle \zeta \rangle$, we get an irreducible of degree p^n such that z acts by the scalar $\lambda(z)$. There are $p-1$ such characters, all distinct. They are clearly algebraically conjugate. By one of the orthogonality relations, $p^{2n+1} = |P| = \sum_{\chi \in Irr(P)} \chi(1)^2 \geq 1^2 p^{2n} + (p-1)(p^n)^2 = |P|$, whence we have accounted for all the irreducible characters. □

In the following, we give some character tables for the case the prime number p is 2 or 3. For the group $|P| = 8$, we obtain that P is isomorphic to the dihedral group Dih_8 or the quaternion group $Quat_8$. In this case, there are 5 irreducible representations, whose character table is below

Table 7.1. Character Table of P

	1	z	x, xz	y, yz	xy, xyz
χ_1	1	1	1	1	1
χ_2	1	1	1	-1	-1
χ_3	1	1	-1	-1	1
χ_4	1	1	-1	1	-1
χ_5	2	-2	0	0	0

Remark 7.3.6. *For the dihedral group Dih_8, the degree 2 irreducible may be afforded by a representation in $GL(2, \mathbb{Q})$. The degree 2 irreducible for the quaternion group $Quat_8$ is not afforded over the reals (this follows from the theory of*

the Schur index [52]). We have $Quat_8 \cong \left\langle \begin{pmatrix} & 1 \\ -1 & \end{pmatrix}, \begin{pmatrix} i & \\ & -i \end{pmatrix} \right\rangle$, while $Dih_8 \cong \left\langle \begin{pmatrix} & 1 \\ 1 & \end{pmatrix}, \begin{pmatrix} -1 & \\ & 1 \end{pmatrix} \right\rangle$.

7.3.2 Construction of the BRW groups

In this section, we shall prove that there exists a group G containing the extraspecial group P as a normal subgroup, so that the centralizer of P in G is $C_G(P) = Z(P)$, and G acts on P by conjugation inducing the group $Aut^0(P)$. These groups are called BRW-groups, after the authors Bolt, Room and Wall, who were probably the first to analyze them [7, 8, 9]. See (7.3.10) for historical remarks.

For a given group X, one may ask whether there exists a group Y which contains X as a normal subgroup such that $C_Y(X) = Z(X)$ and $Y/Z(X) \cong Aut(X)$. In general, the answer is no: one counterexample is the double cover $2 \cdot Alt_6$. Therefore, an affirmative answer for X extraspecial is not a trivial point.

Notation 7.3.7. *In what follows, $\mathbb{K}$ is a field of characteristic not p such that there exists a maximal abelian subgroup A of P and a linear character $\lambda : A \to \mathbb{K}^\times$ which is not trivial on Z. This means $\mathbb{K}$ contains a p-th or p^2-th root of unity, according to whether the exponent of A is p or p^2. The induced representation of P is absolutely irreducible of dimension p^n and, obviously, written over the field $\mathbb{K}$.*

Suppose we have a representation $\rho : P \to GL(p^n, \mathbb{C})$ of P which is nontrivial on the center. We claim that $Ker(\rho) = 1$. Indeed, $Ker(\rho) \triangleleft P$. In a p-group, a nonidentity normal subgroup has a nontrivial intersection with the center. But for the extraspecial group P, we have $Z(P) = \langle z \rangle$, which implies $\rho(z) \neq 1$. So the representation ρ is a monomorphism. By our classification, it is also an irreducible representation.

From now on, we identify P with its image $\rho(P)$. Let N denote the normalizer of P in $GL(p^n, \mathbb{C})$. Trivially, one has that $N \geq P \cdot \mathbb{C}^\times I$. Let H be a group, and $\mathbb{C}[H]$ its group algebra. If H is finite, then $\mathbb{C}[H] \cong A_1 \oplus \cdots \oplus A_k$, for a set of 2-sided ideals $A_i \cong Mat_{d_i}(\mathbb{C})$, where $d_1, d_2, \ldots$ are the degrees of the irreducible representations. The group $Aut(H)$ acts on $\mathbb{C}[H]$ as ring automorphisms. So $Aut(H)$ permutes the set $\{A_1, \ldots, A_k\}$. Take $H = P$. We have $\mathbb{C}[P] \cong \mathbb{C} \oplus \cdots \oplus \mathbb{C} \oplus Mat_{p^n}(\mathbb{C}) \oplus \cdots \oplus Mat_{p^n}(\mathbb{C})$, with p^{2n} copies of $\mathbb{C}$ and $p-1$ copies of $Mat_{p^n}(\mathbb{C})$. Say the 1-dimensional ideals are $A_1, \ldots, A_{p^{2n}}$ and the other indecomposable 2-sided ideals are $B_1, \ldots, B_{p-1}$ respectively, so that z acts on B_i as ζ^i, where ζ is a fixed primitive p-th root of unity. The action of $Aut(P)$ on $\{B_1, \ldots, B_{p-1}\}$ corresponds to the action on $\{z, z^2, \ldots, z^{p-1}\}$. By (7.2.8), this action is transitive. Also, $Aut^0(P)$ fixes each of the ideals isomorphic to $Mat_{p^n}(\mathbb{C})$. We now have a homomorphism $Aut^0(P) \to Aut(B)$, where $B = B_1$.

Theorem 7.3.8. *(Skolem-Nöther) Let F be a field, and let $n \geq 1$ be an integer. Every automorphism of $Mat_n(F)$ which fixes the set of scalar matrices elementwise* is inner, i.e., has the form $A \mapsto S^{-1}AS$, where $A \in Mat_n(F)$ for some fixed invertible matrix $S \in Mat_n(F)$.

Warning: Such an invertible matrix S is not unique if $|F| > 2$.

There is a function $\alpha : Aut^0(P) \to B^\times = GL(n, \mathbb{C})$ so that if $a \in Aut^0(P)$, then $a\rho(g)a^{-1} = \alpha(a)\rho(g)\alpha(a)^{-1}$ for all $g \in P$. For $a_1, a_2 \in Aut^0(P)$, there are scalars $c_{a_1,a_2} \in \mathbb{C}^\times$ so that $\alpha(a_1)\alpha(a_2) = c_{a_1,a_2}\alpha(a_1 a_2)$. Such a function c_{a_1,a_2} is called a 2-*cocycle* (the language comes from the theory of cohomology of groups).

Define $G_1 := \langle \alpha(a) | a \in Aut^0(P) \rangle$, which is a subgroup of N. We may, if we wish, take $\alpha(a)$ to be $\rho(g)$ when a is an inner automorphism of the form $x \mapsto gxg^{-1}$, where $x, g \in P$. So these $\alpha(a)$ lie in $\rho(P)$ and form a complete set of coset representatives for $\rho(Z)$ in $\rho(P)$.

Theorem 7.3.9. (Schur) *If H is a group and the index $|H : Z(H)|$ is finite, then the commutator subgroup H' is finite.*

Proof. This uses the transfer homomorphism (see [29, 48]). Let $n := |H : Z(H)|$.

Let $h \in H$. Suppose that $hZ(H)$ has order m in $H/Z(H)$. Then the action of h on the cosets of $Z(H)$ has the property that every orbit has length m. If $g_1, \ldots, g_r \in H$ so that $g_1 Z(H), \ldots, g_r Z(H)$ represent the distinct orbits, then the transfer $V : H \to Z(H)$ satisfies $V(h) = \prod_{i=1}^r g_i^m h g_i^{-m}$ which equals h^n since the terms $g_i^m h g_i^{-m}$ all lie in the center.

We have $H'Z/Z \cong H'/H' \cap Z$ finite (since $H'Z/Z \leq H/Z$) and the abelian group $H' \cap Z$ has exponent dividing n.

Now assume that H is finitely generated. Then the finite index abelian subgroup Z is finitely generated, whence so is any subgroup of Z. Therefore, $H' \cap Z$ is a finite abelian group. It follows that H' is finite.

Now we drop the assumption that H is finitely generated. Let S be any finite subset of H so that the image of S in H/Z generates the finite group H/Z. Define $K := \langle S \rangle$, a finitely generated group. Since $|K : K \cap Z|$ has finite index (dividing n in fact), then K' is finite by the previous paragraph. Since Z is central and $H = KZ$, $H' = K'$ is finite (by the commutator expansion rules (7.1.2)). □

The Schur result applies to G_1 since $Z(G_1)$ is the set of scalar matrices in G_1 and $G_1/Z(G_1) \cong Aut^0(P)$, which is finite. So the commutator group $G := G_1'$ is finite.

Definition 7.3.10. *We let p, n, ε, P be as before.*

(1) Suppose $p = 2$, $n \geq 3$ if $\varepsilon = +$ and $n \geq 2$ if $\varepsilon = -$.

We define the BRW group to be G_1', where G_1 is as in the previous paragraph. We use the notation $BRW(2^n, \varepsilon)$. In an extension field $\mathbb{K}[\sqrt{-2}]$, there exists a group $BRW^0(2^n, \varepsilon)$ which contains $BRW(2^n, \varepsilon)$ with index 2 and satisfies $BRW^0(2^n, \varepsilon)/P \cong Out(P)$. (The new coset contains elements which act on $P/Z(P)$ like transvections; see [40]).

(2) Now take p to be odd and let R be an extraspecial group of exponent p and order p^{1+2n}. For $n \geq 2$ or for $n = 1$ and $p \geq 5$, $Sp(2n, p)$ is a perfect group. Therefore, we may define the BRW group to be G_1', where G_1 is as in the previous paragraph. We use notation $BRW(p^n, \varepsilon)$ or $BRW^0(p^n, \varepsilon)$.

(3) We wish to define the BRW group for small n and arbitrary p and ε. Take $m \geq n$, $m \geq 3$. Then $G := BRW^\varepsilon(p^m)$ is a well-defined linear group. Let $R := O_p(G)$. There exists an extraspecial subgroup S of R so that $C_R(S)$ is extraspecial of order p^{1+2n} and type ε. The BRW group of dimension p^n and type ε is defined to be $C_G(S)$. It is a subgroup of $GL(p^n, \mathbb{K})$ since the S-module $\mathbb{K}^{p^m}$ is a direct sum of p^n isomorphic faithful, absolutely irreducible modules. When $p = 2$, we define BRW^0 as in (1), applied to the subgroup $C_G(S)$.

(4) Finally, take p to be odd and let R be an extraspecial group of exponent p^2 and order p^{1+2n}. We embed R in $GL(p^n, \mathbb{K})$, where $\mathbb{K}$ is the field generated by the rationals and the group of p^2-th roots of unity. Let Z be the group of scalar matrices of order p^2 in $\mathbb{K}^\times$.

Consider the central product RZ. Since p is odd and this group has class 2, the p-th power map is a homomorphism into the subgroup Z_1 of order p in Z. Let R_1 be its kernel. Then, $RZ = R_1 Z$.

We now apply the above procedures to the exponent p extraspecial group $R_1 \leq GL(p^n, \mathbb{K})$ to create a BRW group G containing it in $GL(p^n, \mathbb{J})$, where $\mathbb{J}$ is the field generated over $\mathbb{Q}$ by the p-th roots of unity. We define the BRW group for R to be $R_1 N_G(R_1)'$, a group of shape $p^{1+2n} \cdot p^{2(n-1)} Sp(2(n-1), p)$. The commutation was performed in order to make the BRW-group intersect the group of scalars Z in the subgroup Z_1. We call R_1 the exponent p deformation of R. We have a group homomorphism $N_{GL(p^n, \mathbb{K})}(R) \to Aut(R_1)$ since R_1 is characteristic in RZ. We use R_1 to study $N_{GL(p^n, \mathbb{K})}(R)$ in (7.4.10).

The above procedure requires knowing when $\Omega^\varepsilon(2n, 2)$ and $Sp(2n, p)$ are perfect. Proofs may be found in [26, 48].

Lemma 7.3.11. $C_G(\rho(P)) = \rho(Z(P))$.

Proof. Without loss, we may assume that $\mathbb{K} = \mathbb{Q}(\eta)$, and η is a primitive p-th or p^2-th root of unity, as needed for the representation of P. So the centralizer $C_G(\rho(P))$ is a finite group of scalars in $\mathbb{Q}(\eta)$, and it has order 2 or 4 if $p = 2$ and order $p, 2p, p^2$ or $2p^2$ if p is odd (this is a result from basic number theory; see [80]).

Suppose p is odd. Since $G = G'$, $\det(g) = 1$ for all $g \in G$. We have $-1 \notin G$. So $C_G(\rho(P))$ has odd order. If the exponent of the group P is p, then $\eta^p = 1$. We deduce that $C_G(\rho(P)) = \rho(Z(P)) \cong \mathbb{Z}_p$. □

An alternate proof uses the determination of Schur multipliers for the groups $Out(P)$ [33].

Remark 7.3.12. The BRW groups seem to have been studied first in [7] around the time of [5]. Without knowledge of these articles, this author [33] analyzed the normalizers in $GL(p^d, \mathbb{C})$ of subgroups isomorphic to the extraspecial group p_ε^{1+2d}. Both [7] and [33] contain nonsplitting results but the styles of proof are different. The nonsplitting criterion (7.4.11) and application (7.4.14) still appear to be new results. See also [8, 9].

Coincidentally, the lattices of [5] were (re)discovered in [15]. See our section on Barnes-Wall lattices.

7.3.3 Tensor products

We define the tensor product of representations of two groups. Suppose the group G_i has an irreducible finite dimensional module M_i over a field K, $i = 1, 2$. Then the tensor product $M_1 \otimes_K M_2$ of modules is an irreducible module for the product $G_1 \times G_2$ of groups. When $G_1 \cong G_2 \cong G$, this tensor product becomes a module for G upon identifying G with a diagonal subgroup of $G_1 \times G_2$, and is called the tensor product of modules M_1, M_2 for G.

We give a proof in a special case that the field K is the complex numbers $\mathbb{C}$ and each G_i is a finite group.

Let $\rho_i : G_i \to GL(M_i)$ be the representation, and let χ_i be the character of G_i on M_i for $i = 1, 2$. Let $p := \dim(M_1)$, $q := \dim(M_2)$. Suppose $A = (a_{ij})_{p \times p} = \rho_1(g_1)$, $B = \rho_2(g_2)_{q \times q}$. By choosing bases $x_1, \ldots, x_p$ of M_1 and $y_1, \ldots, y_q$ of M_2, we see that the representation $(\rho_1 \otimes \rho_2)(g)$ has the matrix

$$\begin{pmatrix} a_{11}B & \cdots & a_{1p}B \\ \vdots & & \vdots \\ a_{p1}B & \cdots & a_{pp}B \end{pmatrix},$$

with respect to the basis $\{x_i \otimes y_j\}$. The trace of such a matrix is $\text{Tr}(A)\text{Tr}(B)$. We deduce that the character of the tensor product is $\chi(g) = (\chi_1 \otimes \chi_2)(g_1 g_2) = \chi_1(g_1)\chi_2(g_2)$.

To prove irreducibility, we can calculate the inner product of $\chi_1 \otimes \chi_2$ with itself. The assertion follows the orthogonality relations.

$$\langle \chi_1 \otimes \chi_2, \chi_1 \otimes \chi_2 \rangle_{G_1 \times G_2}$$

$$= \frac{1}{|G_1 \times G_2|} \sum_{g \in G_1 \times G_2} \chi(g_1 g_2) \overline{\chi(g_1 g_2)}$$

$$= \frac{1}{|G_1|} \frac{1}{|G_2|} \sum_{g_1 \in G_1} \sum_{g_2 \in G_2} \chi_1(g_1) \overline{\chi_1(g_1)} \chi_2(g_2) \overline{\chi_2(g_2)}$$

$$= \left(\frac{1}{|G_1|} \sum_{g_1 \in G_1} \chi_1(g_1) \overline{\chi_1(g_1)} \right) \cdot \left(\frac{1}{|G_2|} \sum_{g_2 \in G_2} \chi_2(g_2) \overline{\chi_2(g_2)} \right)$$

$$= 1 \cdot 1 = 1.$$

Suppose M_1, M_2 are modules for a fixed group G. We can form a tensor product $M_1 \otimes M_2$ of modules and make G act diagonally $G \to G \times G$, $g \mapsto (g, g)$ on the tensor product. Such a module is often reducible but can be irreducible.

Lemma 7.3.13. *Let F be a field. Suppose S is a subset of the group $Mat_{n \times n}(F)$, so that S spans the whole algebra $Mat_{n \times n}(F)$. Let a be any element of $Mat_{n \times n}(F)$. Then the set aS contains an element of nonzero trace if $a \neq 0$.*

Proof. The function $f : Mat_{n \times n}(F) \times Mat_{n \times n}(F) \to F$, defined by $f(a, b) = \text{Tr}(ab)$, is a nondegenerate, symmetric bilinear form. The result follows. $\square$

7.4 Representation of the BRW group G

Fix a prime p, integer $n \geq 1$ and sign $\varepsilon \in \{\pm\}$. We use the notation G for the subgroup of $GL(p^n, \mathbb{K})$ we constructed in (7.3.10). The group G contains an extraspecial group P of order p^{1+2n} and type ε as a normal subgroup, so that $C_G(P) = Z(P)$ and $G/P \cong Aut^0(P)$.

Let ρ be the given faithful representation of G. Let χ be the character ($\chi(g) = \text{Tr}(\rho(g))$).

We wish to analyze character values for elements of G. By (7.3.4), we may replace $\mathbb{K}$ by $\mathbb{Q}(\eta)$ for $\eta = e^{2\pi i/p}$ or $\eta = e^{2\pi i/p^2}$.

Notation 7.4.1. *For any $g \in G$, g acts on $P \triangleleft G$ by conjugation and on the vector space $P/Z \cong \mathbb{F}_p^{2n}$. We let $d := d(g)$ be the dimension of the space of fixed points for g on P/Z, $0 \leqslant d \leqslant 2n$.*

Suppose H is a group acting on a set Ω. Let K be a commutative ring. The *permutation module* is the free K-module with basis Ω, i.e., $\oplus_{\alpha \in \Omega} K\alpha$ with the action of $h \in H$ on this free module defined by sending $\alpha \in \Omega$ to $h(\alpha)$. So, $h \in H$ is represented by a permutation matrix. The *permutation character* of H on Ω is the trace on this representation. Its value at $h \in H$ is the number of points fixed by h.

Lemma 7.4.2. *The product $\chi\overline{\chi}$ is a character of G with Z in its kernel.*

Proof. If $w \in Z$, then $\chi(w) = p^n \varepsilon$, $|\varepsilon| = 1$. Thus $(\chi\overline{\chi})(w) = \chi(w)\overline{\chi(w)} = p^n \varepsilon \cdot p^n \overline{\varepsilon} = p^{2n} = \chi(1)\overline{\chi(1)} = (\chi\overline{\chi})(1)$. So, $w \in Ker(\chi\overline{\chi})$. Therefore, the product $\chi\overline{\chi}$ indeed gives a representation. □

Lemma 7.4.3. (1) $\chi\overline{\chi}|_P = \sum_{\lambda \in Irr(\rho), \lambda(1)=1} \lambda$.
(2) *If $g \in G$, $0 \leq \chi\overline{\chi}(g) \leq p^d$.*

Proof. We let λ be any linear character of the quotient group P/Z. We shall prove that λ has multiplicity 1 in the character $\chi\overline{\chi}$ of G. Observe that we have an equality $\langle \chi\overline{\chi}, \lambda \rangle_P = \langle \chi, \chi\lambda \rangle_P$, by using the definition of the inner product $\langle \alpha, \beta \rangle_H = \frac{1}{|H|} \sum_{x \in H} \alpha(x)\overline{\beta(x)}$. We claim $\chi\lambda = \chi$ as characters of P. Recall that $\chi(g) = 0$ for all $g \in P$ unless $g \in Z = P'$, in which case $\lambda(g) = 1$. The claim follows. So $\chi\overline{\chi}|_P = \sum_{\lambda \in Irr(\rho), \lambda(1)=1} \lambda$. The action of G on the set of 1-spaces which afford the distinct linear characters on λ on $P \triangleleft G$ is equivalent to its action on the set $Hom(P, \mathbb{C}^\times) \cong P/Z$, which is in turn equivalent to the action of G on the set P/Z because of the G-invariant bilinear form on P/Z.

When we compute $\chi\overline{\chi}(g)$, we use a basis taken from the 1-spaces affording the characters λ. We get a nonzero contribution to the trace only if the 1-space is fixed by g. So $\chi\overline{\chi}$ is at most the number of elements in P/Z fixed by g, i.e., p^d. □

Let us go back to the study of $\chi\overline{\chi}$. This is the character for a module which is a direct sum of 1-dimensional spaces Ke_λ, where $\lambda \in Hom(P/Z, \mathbb{C}^\times)$.

So, $g \in G$ is represented by a monomial matrix with coefficients in K. As proved in (7.4.3), $|\chi\overline{\chi}(g)| \leqslant p^d$. For any element $a \in G$, we form the group $\langle P, a \rangle \leqslant G$. We consider how often $\chi(g)$ is 0 for $g \in aP$.

7.4 Representation of the BRW group G

Notation 7.4.4. *Define $s(a)$ to be the number of $g \in aP = Pa$, so that $\chi(g) \neq 0$. Observe that $s(a) > 0$ since (7.3.13) implies that aP contains elements with nonzero trace.*

If $g_1 \in aP$ and $\chi(g_1) \neq 0$, then χ takes nonzero value $\chi(g_1)$ on the set $\{xg_1x^{-1} | x \in P\}$. This set corresponds to the cosets in P of $C_P(g_1)$, which contains the center Z. The quotient $C_P(g_1)/Z$ is contained in the fixed space of g on P/Z, which we call Q/Z. It follows $Q = C_P(g_1)$ since g_1 is not conjugate to g_1z. Consequently, $|C_P(g_1)| = p^{d+1}$ (recall that d is the dimension of the fixed space for g on $P/Z \cong \mathbb{F}_p^{2n}$). We deduce that $|\{xg_1x^{-1} \mid x \in P\}| = (p^{2n+1}/(p^{d+1})) = p^{2n-d}$. Notice that we get similar disjoint sets for g_1 replaced by $g_1 z^j$, for $j = 1, 2, \ldots, p-1$, where z is a fixed generator for Z. Hence $s(a) \geqslant p^{2n-d+1}$.

Let ζ be the primitive p-th root of unity so that z acts by ζ.
We calculate

$$\frac{1}{|G|} \sum_{g \in G} \chi\overline{\chi}(g) = \frac{1}{|G|} \left(\sum_{\overline{g} \in G/P} \sum_{g \in \overline{g}} \chi(g)\overline{\chi(g)} \right)$$

$$= \frac{1}{|G|} \sum_{\overline{g} \in G/P} \left(\sum_{g \in \overline{g}, \chi(g) \neq 0} \chi(g)\overline{\chi(g)} \right)$$

$$\geq \frac{1}{|G|} \sum_{\overline{g} \in G/P} p^{2n-d+1} \chi(g_1)\overline{\chi(g_1)},$$

where $g_1 \in \overline{g}$ and $\chi(g_1) \neq 0$. Note that $\chi(g_1) = \chi(g_1')$ if g_1 and g_1' are conjugate and $\chi(g_1 z^j) = \zeta^j \chi(g_1)$. So we have

$$1 = \frac{1}{|G|} \sum_{g \in G} \chi\overline{\chi}(g) \geqslant \frac{1}{|G|} \sum_{\overline{g}} p^{2n-d+1} p^d = \frac{1}{|G|} \sum_{\overline{g}} p^{2n+1} = 1.$$

Thus, all the inequalities are equalities.

Notation 7.4.5. *Define $p_1 := \begin{cases} \sqrt{p} & \text{if } p \equiv 1 \bmod 4; \\ \sqrt{-p} & \text{if } p \equiv 3 \bmod 4; \\ \sqrt{2} & \text{if } p = 2. \end{cases}$*

If p is odd and ζ is a primitive p-th root of unity, then $\sum_{i=0}^{p-1} \zeta^{i^2} = p_1$ (this is a "Gauss sum"). See [80].

Observe that, if σ is in the Galois group $\text{Gal}(\mathbb{K}/\mathbb{Q})$, then the above discussion applies to the algebraically conjugate character χ^σ. We deduce that $|\varepsilon^\sigma| = 1$. An algebraic integer, all of whose conjugates have absolute value 1, is a root of unity. Therefore, ε is a root of unity.

In the ring of integers $\mathbb{Z}[\sqrt[n]{1}]$ of $\mathbb{Q}[\sqrt[n]{1}]$, the group of roots of unity has order n if n is even and has order $2n$ if n is odd. See [80].

We now have proved the following theorem.

Theorem 7.4.6. *Let $g \in G$, a BRW group in $GL(p^d, \mathbb{K})$. Let $P := O_p(G) \cong p_\varepsilon^{1+2d}$. Let $g \in G$, $d := d(g)$. Then $\chi(g)$ is 0 or εp_1^d, where ε is a ± 1 if $p = 2$ and is a $2p$-th root of 1 if p is odd.*

In the coset gP, the number of elements with nonzero trace is p^{2n-d+1}. The number of distinct nonzero traces is p and multiplication by Z acts transitively on the p sets of elements in gP with such traces.

Remark 7.4.7. *Suppose that $p = 2$. Take any element $g \in BRW(2^d, \varepsilon)$. Then $d(g) \in 2\mathbb{Z}$.*

Here is a sketch of the proof. The group $\Omega^\varepsilon(2n, 2)$ is the kernel of the Dickson homomorphism $D : O^\varepsilon(2n, 2) \to \mathbb{F}_2$. This map has the property that $D(r) = 1$ if r is a reflection (see the definition of the map D in [28]). An element of $O^\varepsilon(2n, 2)$ is in $\Omega^\varepsilon(2n, 2)$ if and only if its fixed point subspace is even dimensional. In particular, the image of the rational matrix groups $BRW(2^d, +)$ in the outer automorphism group of the normal extraspecial group corresponds to the index two subgroup $\Omega^\varepsilon(2n, 2)$ of $\Omega^\varepsilon(2n, 2)$.

7.4.1 BRW groups as group extensions

Abstract extension theory of finite groups has useful applications to the BRW groups.

Notation 7.4.8. *Let p be a prime number, $n \geq 1$, $\varepsilon = \pm$ and $P \cong p_\varepsilon^{1+2n}$ the corresponding extraspecial group (7.2.6). We let $\mathbb{K}$ be the field $\mathbb{Q}[\eta]$ where η is a p-th or p^2-th root of unity. We have a corresponding BRW group G in $GL(p^n, \mathbb{K})$ and two exact sequences of groups, namely*

$$1 \to P \to G \to G/P \to 1,$$

$$1 \to P/Z \to G/Z \to G/P \to 1.$$

In this section, we consider which of these sequences split.

Lemma 7.4.9. *Suppose that the finite group X has normal subgroups $Y > Z$ so that $(|Z|, |Y : Z|) = 1$ and if $1 \neq x \in Y$ and $(|x|, |Z|) = 1$, then $C_Z(x) = 1$.*

Then there exists $W \leq X$ so that $X = ZW, Z \cap W = 1$, i.e., X splits over Z.

Proof. By the Schur-Zassenhaus Theorem [31, 51], Y splits over Z and all complements are conjugate by Z. Let K be a complement in Y to Z. Then, by the Frattini argument, $X = Z \cdot N_X(K)$. Observe that $Z \cap N_X(K) \leq C_Z(K) = 1$ because $K \neq 1$ and $C_Z(x) = 1$ for any $1 \neq x \in K$. It follows that we may take $W = N_X(K)$. □

Proposition 7.4.10. *These sequences split for p odd.*

Proof. It suffices to do the case where $\exp(P) = p$ by (7.3.10)(4). We have $G/P \cong Sp(2n, p)$. Therefore, there exists $t \in G$ so that conjugation by t on P/Z is the inversion map. We apply the previous lemma to the normal subgroups

$P/Z \leq P\langle t \rangle/Z$ of G/Z to get existence of a subgroup K of so that $K \cap P = Z$ and $K/Z \cong Sp(2n, p)$.

According to (7.4.6), $\chi(t) = \pm 1$. Therefore, since $Z \leq C_G(t)$, Z preserves the eigenspaces of t, which have dimensions $\frac{1}{2}(p^n \pm 1)$. These numbers are relatively prime to p.

Let M be one of these eigenspaces. The representation of K is written over $\mathbb{K}$, whose group of roots of unity has order $2p$. The determinant of an element of K on M is a root of unity. In $\mathbb{K}$, the group of roots of unity has order $2p$ [80].

A generator for z has determinant not 1 on each of these spaces, whence Z is complemented in K by the subgroup $W := \{g \in K \mid \det(g) = \pm 1\}$ where here determinant means on one of the eigenspaces. □

For the case $p = 2$, we shall prove that they do not split if $n \geq 4$ or $n = 3$ and $\varepsilon = -1$. This will follow from a general nonsplltting result.

Proposition 7.4.11. *Suppose that a group K satisfies $P \leqslant K \leqslant G$, and that there is a subgroup $Q \leqslant P$, which is isomorphic to the dihedral group Dih_8 or the quaternion group $Quat_8$. Assume further that*

(1) there is an element $f \in Q$ of order 4 so that K fixes the cyclic group $\langle f \rangle$;

(2) there is an involution $t \in K/P$ so that it fixes Q and acts nontrivially on Q/Z;

(3) the quotient group K/P has no subgroup of index 2.

Then the group K does not split over P/Z, i.e., there does not exist a subgroup J so that $K = PJ$ and $P \cap J = Z$. In particular, K does not split over Z.

Proof. It suffices to prove the first statement. Deny the conclusion. Then there is a subgroup $J \leqslant K$ so that $K = PJ$ and $P \cap J = Z$. Let $u \in J$ so that u maps to t by the quotient $J \leqslant K \to K/P$.

Now, the element t fixes $\langle f \rangle$ under conjugation and conjugation by t has order 2. If we let $x \in Q - \langle f \rangle$, then $x^t \langle f^2 \rangle \neq x \langle f^2 \rangle$ by (2) since $Z = Z(P) = Z(Q) = \langle f^2 \rangle$. Also t fixes xx^t mod Z. So $xx^t = f$ or $-f$. Now we calculate $(xx^t)^t = x^t x^{t^2} = x^t x$ since t is an involution. Then we study $(xx^t)(x^t x) = x(x^2)^t x = x^2 (x^2)^t$, since $(x^2)^t \in Z$. Note that $x^2 = 1$ or f^2, so $(x^2)^t = 1$ or f^2. We deduce that $(xx^t)(x^t x) = 1$. Therefore, conjugation by t inverts f.

Observe that $xx^t \notin Z$ since $x^t Z \neq xZ$ and $xx^t = f^i$ for $i = 1$ or -1. So we have a homomorphism $\phi : J \to Aut(\langle f \rangle) \cong \mathbb{Z}_2$, and $t \notin Ker(\phi)$. That is, J has a normal subgroup of index 2, so does K. □

Corollary 7.4.12. *Under the same hypothesis as above. Then G does not split over P or even split over P modulo Z.*

Proof. Suppose there is a subgroup $H \leqslant G$, so that $G = PH$ and $P \cap H = Z$. We have $P \leqslant K \leqslant PH$. So by the Dedekind low, $K = P(H \cap K)$. Thus $H \cap K$ is a complement to P mod Z in K. This contradicts the above proposition. □

Theorem 7.4.13. *For $p = 2$, the BRW group for P does not split over P mod Z if $n \geq 4$ or if $n = 3$ and $\varepsilon = -$.*

Proof. Let $n \geq 4$ and let G be as above. Denote by $K = Stab_G(f)$ the stabilizer of $f \in P$ in G, an element of order 4. Then we have $G/P \cong \Omega^+(2n, 2)$ and $K/P \cong Sp(2(n-1), 2)$. The symplectic group $Sp(2(n-1), 2)$ has no subgroup of index 2 for $n - 1 \geq 3$.

Given a nonsingular quadratic space V over $\mathbb{F}_2$, a 2-dimensional subspace W which is nonsingular and a nonsingular vector $w_0 \in W$, there exists an involution t in the orthogonal group on V which fixes w_0 and interchanges the other two nonzero vectors. It follows that there exists an involution $t \in K$ so that conditions (1) and (2) of the proposition are satisfied. The previous paragraph implies that condition (3) is satisfied. The proposition now gives the conclusion for $n = 4$.

For $n = 3$ and $\varepsilon = -$, a special argument is required. For $n = 3$ and $\varepsilon = +$, the extension splits. This follows since a splitting may be seen by examining subgroups of the Weyl group of E_8. For details, see [30, 31]. $\square$

Remark 7.4.14. *The proposition proves that some other interesting extensions are nonsplit. For example, this is proved in [32, 33] for an extension K of P, where $P \cong 2^{1+24}$ and K/P is isomorphic to the isometry group of the Leech lattice modulo $\{\pm 1\}$ (the Leech lattice and its isometry group is discussed later in this book).*

Definition 7.4.15. *Suppose that P is an extraspecial group and G is a group such that*
  (1) $P \triangleleft G$;
  (2) $C_G(P) = Z(P)$;
  (3) $G/Z(P) \cong Aut^0(P)$.
We call G a holomorph of P. In case (3) is replaced by
  (3′) $G/Z(P)$ embeds in $Aut^0(P)$.
Then G is a partial holomorph.

Remark 7.4.16. *There is the possibility of nonequivalent group extensions, given the terms P and G/P as in (7.4.8). A partial holomorph which is contained in the BRW group is called a standard holomorph. Otherwise, it is called a twisted holomorph. See [35] for a discussion of the extension-theoretic issues and degree 2 cohomology. We note that partial holomorphs occur as centralizers of involutions in many finite groups, including sporadic simple groups where twisted ones occur. For a discussion of nonvanishing cohomology in finite group theory, see [36].*

8
Overview of the Barnes-Wall Lattices

This chapter previews aspects of the Barnes-Wall series and presents commutator density. The latter concept will be used in the next chapter which gives a formal construction and analysis of the series.

8.1 Some properties of the series

The Barnes-Wall lattice BW_{2^d}, for $d = 0, 1, 2, \ldots$, is an integral lattice in rank 2^d. Its discriminant group $\mathcal{D}BW_{2^d}$ is trivial if d is odd and isometric to the elementary abelian group $\mathbb{Z}_2^{2^{d-1}}$, otherwise. They have a remarkable group of automorphisms, rather high minimum norm and are connected to other important lattices.

Their minimum norms are $\mu(BW_{2^d}) = 2^{\lfloor \frac{d}{2} \rfloor}$. This goes to infinity roughly as the square root of the dimension. These were perhaps the first family of lattices with such a high growth property for minimum norms. The isometry group $O(BW_{2^d})$ contains our group $G_{2^d} = 2^{1+2d} \cdot \Omega^+(2d, 2)$. Moreover, the group G_{2^d} is transitive on the minimal vectors of the Barnes-Wall lattice *etc.* It is related to the Reed-Muller code $RM(e, d)$ on $\mathbb{F}_2^d$.

The original article [5] gave a multiparameter family of lattices in each dimension 2^d. The one of greatest interest is the one we treat here and call "the" Barnes-Wall lattice. It is essentially the only one with the large isometry group G_{2^d}.

The Barnes-Wall series was discovered independently by Broué and Enguehard [15], who used the Reed-Muller codes for construction and analysis of the lattice and its isometry group. These authors had been studying codes and quadratic forms in a series of articles [12, 13, 14].

Below is a look at the first few members of the Barnes-Wall series, BW_{2^d}, in dimension 2^d. The isometry group is $BRW^+(2^d)$ except at dimensions 1 and 8. See (4.2.1) for the definition of Weyl group.

8 Overview of the Barnes-Wall Lattices

Lattice (and notable isometry)	Rank	Min. norm	Det.	Isometry group (this is just $BRW^+(2^d)$ if $d \neq 3$)
$BW_{2^0} \cong \mathbb{Z}$	1	1	1	$\mathbb{Z}_2$
$BW_{2^1} \cong \mathbb{Z}^2$	2	1	1	Dih_8
$BW_{2^2} \cong D_4$	4	2	2^2	$Weyl(F_4)$
$BW_{2^3} \cong E_8$	8	2	1	$Weyl(E_8)$
BW_{2^4}	16	4	2^8	$2^{1+8}_+\Omega^+(8,2)$
BW_{2^5}	32	4	1	$2^{1+10}_+\Omega^+(10,2)$
BW_{2^6}	64	8	2^{32}	$2^{1+12}_+\Omega^+(12,2)$
BW_{2^7}	128	8	1	$2^{1+14}_+\Omega^+(14,2)$

Sketch of the construction procedure

Here, we discuss the construction for ranks up to 8. The general case will be treated, starting at (9.1.1).

In $\mathbb{Q}^{2^d}$, take an orthogonal direct sum $M_1 \perp M_2$, where

$$M_1 \cong M_2 \cong \begin{cases} BW_{2^d} & \text{if } d \text{ is odd} \\ \sqrt{2}BW_{2^d} & \text{if } d \text{ is even.} \end{cases}$$

Let t be an involution in the orthogonal group $O(\mathbb{Q}^{2^d})$ so that $t(M_1) = M_2$ and $t(M_2) = M_1$. Let $u \in O(\mathbb{Q}^{2^d})$ be -1 on M_1 and $+1$ on M_2. Then $\langle t, u \rangle \cong Dih_8$. We consider lattices L which satisfy $M_1 \perp M_2 \leqslant L \leqslant M_1^* \perp M_2^*$ so that $L \cap M_i^* = M_i$ for $i = 1, 2$. They are invariant under the actions of the group $\langle t, u \rangle$, and if p_i is the orthogonal projection to M_i, then $p_i(L)$ is invariant under $G_{2^{d-1}}(M_i)$ (this is a subgroup of $O(M_i)$ isomorphic to $BRW^+(2^d)$ and is equal to $O(M_i)$ if $d \neq 3$) and $p_i(L)/M_i \cong \mathbb{Z}_2^{2^{d-2}}$, for $i = 1, 2$. We shall obtain a proof of existence and uniqueness for this series of lattices.

Let us illustrate how above procedures are applied to the first few cases in the list. The general case will be covered in a later section.

For $d = 0$, $BW_1 = \mathbb{Z}$ and we have the isometry group $\{\pm 1\}$.

Now take $d = 1$, so that $BW_2 = \mathbb{Z}^2$. Its minimal vectors are its unit vectors, which are $(\pm 1, 0), (0, \pm 1)$. The isometry group is $O(\mathbb{Z}^2) = Mon(2, \{\pm 1\}) = \mathbb{Z}_2 \wr Sym_2 = \mathbb{Z}_2 \wr \mathbb{Z}_2 = Dih_8$, which is extraspecial. For the orthogonal direct sum $\mathbb{Z} \perp \mathbb{Z}$, we have the involutions defined by $t : (a, b) \mapsto (b, a)$ and $u : (a, b) \mapsto (-a, b)$.

For $d = 2$, we have $BW_4 \cong D_4$. The standard model of D_4 is all vectors with even coordinate sums in $\mathbb{Z}^4$. We show explicitly how D_4 may be constructed by the above procedure. Define $M_1 = \{(a_1, a_2, 0, 0) \mid a_1, a_2 \in \mathbb{Z}, a_1 + a_2 \in 2\mathbb{Z}\}$ and $M_2 = \{(0, 0, b_1, b_2) \mid b_1, b_2 \in \mathbb{Z}, b_1 + b_2 \in 2\mathbb{Z}\}$. Consider the norm-doubling map $\mathbb{Z}^2 \to M_1$, $(a, b) \mapsto (a + b, a - b, 0, 0)$. It is clear that these two lattices have the respective Gram matrices $\begin{pmatrix} 1 & \\ & 1 \end{pmatrix}$ and $\begin{pmatrix} 2 & \\ & 2 \end{pmatrix}$. So we have $M_1 \cong \sqrt{2}\mathbb{Z}^2$. Moreover, $M_i^* = \frac{1}{2}M_i$ for $i = 1, 2$. Consider the standard model $L \cong D_4$. It satisfies $M_1 \perp M_2 \leqslant L \leqslant M_1^* \perp M_2^* = \frac{1}{2}M_1 + \frac{1}{2}M_2$. We have the projection

$p_1(L) = \{(a_1, a_2, 0, 0) | a_1, a_2 \in \mathbb{Z}\}$, which is isometric to $\mathbb{Z}^2$ as a lattice. Note that $O(p_1(L)) = O(M_1)$, $p_1(L)/M_1 \cong \mathbb{Z}_2$ and that L is invariant by the group $\langle t, u \rangle$.

For $d = 3$, we have $BW_8 = E_8$ lattice. Recall that for standard models, $D_8 \leq \mathbb{Z}^8$ and $E_8 = D_8 + \mathbb{Z} \cdot \frac{1}{2}\nu$, where $\nu = (1, \ldots, 1) \in \mathbb{Z}^8$. Define M_1 to be all elements of E_8 with support in $\{1, 2, 3, 4\}$, and M_2 to be all elements of E_8 with support in $\{5, 6, 7, 8\}$. So $M_1 \cong M_2 \cong D_4$, each is a direct summand of E_8 as abelian groups, and $M_i^*/M_i \cong \mathbb{Z}_2 \times \mathbb{Z}_2$. We have $M_1 \perp M_2 \leqslant E_8 \leqslant M_1^* \perp M_2^*$ and $p_1(E_8) = M_1^*$ since $\det(E_8) = 1$.

8.2 Commutator density

Commutator density was introduced in [40]. It is quite useful in our theory of Barnes-Wall lattices.

We first review fixed points and commutation in modules for group algebras. Suppose G is a group and M is a $\mathbb{Z}G$-module.

Notation 8.2.1. *Define $M^G = \{x \in M \mid gx = x \text{ for all } g \in G\}$ the fixed point submodule of M, and $M_G = span\{(g-1)x \mid x \in M, g \in G\}$ the commutator submodule of M.*

Such modules have universal properties. If T is a trivial module and $\phi : T \to M$ is a G-homomorphism, then $\phi = \alpha\beta$, where $\alpha : M^G \to G$ and $\beta : T \to M^G$. If $\phi : M \to T$ is a G-homomorphism, then $\phi = \gamma\delta$, where $\delta : M \to M/M_G$ is a quotient and $\gamma : M/M_G \to T$ is a homomorphism of abelian groups.

Lemma 8.2.2. *Suppose the set $S \subseteq G$ generates G. Then $M_G = span\{(g-1)x \mid x \in M, g \in S\}$.*

Proof. The inclusion $M_G \supseteq span\{(g-1)x \mid x \in S, g \in G\}$ is obvious. To prove the opposite containment, define $K := span\{(g-1)x \mid x \in S, g \in G\}$. We have $(g-1)x = (1-g^{-1})(gx) = -(g^{-1}-1)(gx)$, and $(g^{-1}-1)x = (g^{-1}-1)(gy) = (1-g)y = -(g-1)y \in K$. So we may assume $S = S^{-1}$. Take $g, h \in G$, calculate $(gh-1)x = (gh-h+h-1)x = (g-1)(hx) + (h-1)x$. So it follows from $G = \langle S \rangle$ that $M_G \leqslant K$ so $M_G = K$. $\square$

Definition 8.2.3. *Suppose $S \subseteq G$, and M is a $\mathbb{Z}G$-module. We say S is a commutator dense set on M if $M_G = span\{(g-1)x \mid g \in S, x \in M\}$.*

We shall prove an interesting and useful case of commutator density for G an extraspecial 2-group and $S = \{f\}$, $|f| = 4$, for M in certain category of modules (8.2.19),(8.2.22). First, we treat the dihedral case.

8.2.1 Equivalence of 2/4-, 3/4-generation and commutator density for Dih_8

Notation 8.2.4. *Suppose that $D \cong Dih_8$, $D = \langle t, u \rangle$ where $|t| = |u| = 2$ and $f = tu$. Suppose $\mathcal{C}$ is the category of $\mathbb{Z}D$-module which are finite rank free abelian groups such that the involution in the center $Z(D)$ acts as -1.*

94   8 Overview of the Barnes-Wall Lattices

Definition 8.2.5. *Let M be a $\mathbb{Z}D$-module. We say the module M has 2/4-generation if whenever r, s are involutions which generate D, then $M = M^{\langle r \rangle} + M^{\langle s \rangle}$. We say that the module M has 3/4-generation means whenever r, s, t are distinct involutions in $D - Z(D)$, then $M = M^{\langle r \rangle} + M^{\langle s \rangle} + M^{\langle t \rangle}$ (the superscript means fixed points (8.2.1)).*

In a series of lemmas, we shall prove (8.2.19), the basic equivalence of 2/4-, 3/4-generation and commutator density for modules in the category $\mathcal{C}$.

Notation 8.2.6. We extend the action of D to the ambient vector space $\mathbb{Q} \otimes L$. For integers $\ell \le m$, let $Q(\ell, m) := 2^\ell Tel(t)/2^m Tel(t)$. Set $L_1 := L^-(t), L_2 := L^+(t)$.

Define integers d, e to be the number of Jordan blocks of size 2, 1 respectively, for the action of u on the elementary abelian group $L/Tel(t)$.

Lemma 8.2.7. $|L : Tel(t)| = 2^{2d+e}$ and $|L : 2Tel(t)| = 2^{2n+2d+e}$.

Notation 8.2.8. We have a chain of D-invariant abelian groups $2Tel(t) \le L \le \frac{1}{2}Tel(t)$. For $g \in D$, we denote by $A(g)$ and $B(g)$ the commutator modules $(g-1)Q(-1,1) = \frac{1}{2}(g-1)Tel(t) + 2Tel(t)/2Tel(t)$ and $(g-1)L + 2Tel(t)/2Tel(t)$, respectively.

Lemma 8.2.9. $A(u) \cap Q(0,1) = B(u) \cap Q(0,1) = Q(0,1)(u-1)$.

Proof. Clearly, $A(u) \cap Q(0,1) \ge B(u) \cap Q(0,1) \ge Q(0,1)(u-1)$. Now to prove the opposite containment. Since $A(u) \cap Q(0,1)$ consists of elements inverted by u, hence fixed by u, it is contained in the subgroup $(u-1)Q(0,1)$ of the free $\mathbb{F}_2\langle u \rangle$ module $Q(0,1)$. □

Lemma 8.2.10. $A(t) \cap A(u) = 0$.

Proof. Since $Tel(t)(t-1) = 2L_1$, the image of $A(t)$ is just $L_1 + 2Tel(t)/2Tel(t) \le Q(0,1)$. Also, $A(u) \cap Q(0,1)$ is exactly the image of the diagonal sublattice $\{(x, ux) \mid x \in L_1\}$ of $L_1 \oplus L_2$ in $Q(0,1)$. The result follows. □

Lemma 8.2.11. *A coset of $Tel(t)$ fixed by u contains an element fixed by u.*

Proof. Let $x + Tel(t)$ be such a coset. Since $Tel(t)$ is a free $\mathbb{Z}\langle u \rangle$-module, every element of $Tel(t)$ negated by u is a commutator. Therefore, there exists $v \in Tel(t)$ so that $(u-1)x = (u-1)v$. Then $x - v$ is fixed by u and is in $x + Tel(t)$. □

Lemma 8.2.12. $|B(u)| = 2^{n+d}$.

Proof. The right side is the product of $|Q(0,1)|^{\frac{1}{2}} = 2^n$ with $2^d = |(L/Tel(t))(u-1)|$. To evaluate the left side, use (8.2.9), (8.2.11). □

Lemma 8.2.13. *The kernel of the endomorphism induced by $t-1$ on $L/2Tel(t)$ is just $Q(0,1)$.*

Proof. If the kernel were larger, there would be $x \in L \setminus Tel(t)$ so that $(t-1)x \in 2Tel(t)$. Then there would be a unique $y \in L^-(t)$ so that $2y = (t-1)x$, whence $ty = -y$ and $t(x+y) = x+y$ and so $x + y \in L^+(t)$, a contradiction to $x \notin Tel(t)$. □

Corollary 8.2.14. $|B(t)| = 2^{2d+e}$.

Proof. Since $Q(0,1)$ has index 2^{2d+e} in $L/2\,Tel(t)$, the result follows from (8.2.13). □

Lemma 8.2.15. $B(t) \cap B(u) = 0$ and $|B(t) + B(u)| = 2^{n+3d+e}$.

Proof. For $B(t) \cap B(u) = 0$, use (8.2.10). The second statement follows from the formula $|B(t) + B(u)| = |B(t)||B(u)|/|B(t) \cap B(u)| = |B(t)||B(u)|$ and (8.2.12), (8.2.14). □

Lemma 8.2.16. $L(t-1) + L(u-1) = [L, D] \geq L(f-1) \geq 2L$.

Proof. First, $L(t-1) + L(u-1) = [L, D]$ holds because t and u generate D. The containment $[L, D] \geq (f-1)L$ is obvious. Since $[L, D]$ contains $(f-1)L$ and $(f-1)^2 L = 2L$, the final containment holds. □

Lemma 8.2.17. We have $|B(t)+B(u) : ((f-1)L/2\,Tel(t))| = 2^{n+3d+e-(2d+e+n)} = 2^d$. Therefore, $(f-1)L = (t-1)L+(u-1)L$ if and only if $d = 0$. In other language, commutator density is equivalent to 2/4-generation.

Proof. Observe that if M is any f-invariant subgroup of L, then $(f-1)^2 M = 2M$ and that for any integer j, $(*)$ $|(f-1)^j M : (f-1)^{j+1} M| = 2^n$. Since $(f-1)Tel(t) \leq Tel(t)$, we have $2\,Tel(t) = (f-1)^2 Tel(t) \leq (f-1)L \leq [L, D]$. Both $(f-1)L$ and $(t-1)L+(u-1)L = [L, D]$ contain $2\,Tel(t)$, whence a basic isomorphism theorem implies that $|(t-1)L + (u-1)L : (f-1)L| = |B(t) + B(u) : ((f-1)L/2\,Tel(t))|$. The statements follow from (8.2.7), (8.2.15) and $(*)$. □

Lemma 8.2.18. The properties 2/4-generation and 3/4-generation are equivalent.

Proof. Obviously, 2/4-generation implies 3/4-generation. Conversely, assume that $L = L^+(t) + L^-(t) + L^+(u)$. Using 8.2.16, we have $L^+(t) + L^+(u) \geq L(t+1) + L(u+1) = [L, D] \geq 2L$. Since $L^+(t) + 2L = L^-(t) + 2L = L^+(t) + L^-(t)$, $L = L^+(t) + L^-(t) + L^+(u) = L^+(t) + L^+(u) + 2L = L^+(t) + L^+(u)$, whence 2/4-generation. □

Theorem 8.2.19. If $L \in \mathcal{C}$, the following statements are equivalent:
(1) 2/4-generation;
(2) 3/4-generation;
(3) f has commutator density on L.

Proof. Lemmas (8.2.17) and (8.2.18) imply the required equivalence of commutator density, 2/4-generation and 3/4-generation.

Example 8.2.20. We verify commutator density and 2/4-generation in an example.

Suppose $D = O(\mathbb{Z}^2)$. The involutions $t : (a, b) \mapsto (b, a)$, $u : (a, b) \mapsto (-a, b)$ form a generating set for D. The module is $M = \mathbb{Z}^2$. We can easily check that $M^{\langle t \rangle} = \{(x, x) | x \in \mathbb{Z}\}$, $M^{\langle -t \rangle} = \{(x, -x) \mid x \in \mathbb{Z}\}$, $M^{\langle u \rangle} = \{(0, x) \mid x \in \mathbb{Z}\}$, $M^{\langle -u \rangle} = \{(x, 0) \mid x \in \mathbb{Z}\}$. We also have an inclusion $M_D \supseteq M_{\langle t \rangle} + M_{\langle u \rangle}$, where $M_{\langle t \rangle} = \{(x, -x) \mid x \in \mathbb{Z}\}$ and $M_{\langle u \rangle} = \{(2x, 0) \mid x \in \mathbb{Z}\}$. The sum $M_{\langle t \rangle} + M_{\langle u \rangle} = \{(a, b) | a + b \in 2\mathbb{Z}\}$ has index 2. This is M_D. So $M_D = M_{\langle t \rangle} + M_{\langle u \rangle}$ by the earlier lemma. For the commutator density, we have $f = tu : (a, b) \mapsto (-a, b) \mapsto (b, -a)$, where $(f-1)(a, b) = (b-a, -a-b)$. This is M_D.

Example 8.2.21. *In this example, we describe the failure of 2/4-generation.*

Suppose $D = Dih_8$ is the dihedral group. Denote the center by $\langle z \rangle = Z(D)$. We study the modules M where z acts as -1_M. We defined the 2/4-generation (equivalent to the commutator density). We can take the module $M = \mathbb{Z}[D]/(1+z)$ to be a quotient of the group algebra $\mathbb{Z}[D]$. Then on M, z acts as -1. If T is a transversal to $\langle z \rangle$ in D, then M has a $\mathbb{Z}$-basis $\bar{t} = t + (1 + z)$, where $t \in T$. Note that $\bar{z} = z + (1+z) = -1 + (1+z)$.

We can take involutions x, y which generate the group D so that $T = \{1, x, y, xy\}$. So $M = \oplus_{t \in T} \mathbb{Z}\bar{t}$. Observe that M is a free module for the subring $\mathbb{Z}[\bar{t}] \cong \mathbb{Z}[u]/(u^2 - 1)$. The module M has a basis $\{\bar{1}, \bar{y}\}$ as a module for $\mathbb{Z}[\bar{x}]$, which follows by the calculations $x \cdot \bar{1} = \bar{x}, x \cdot \bar{x} = \overline{x^2} = \bar{1}, x \cdot \bar{y} = \overline{xy}, x \cdot (\overline{xy}) = \bar{y}$. Similar argument shows that M has a basis $\{\bar{1}, \bar{x}\}$ as a free module for $\mathbb{Z}[\bar{y}]$.

So the submodule $Fix_M(x) := \{\bar{t} \in M | x\bar{t} = \bar{t}\}$ of M is the $\mathbb{Z}$-span of $\bar{1} + \bar{x}$ and $\bar{y} + \overline{xy}$, while $Fix_M(y)$ is the $\mathbb{Z}$-span of $\bar{1} + \bar{y}, \bar{x} + \overline{yx} = \bar{x} - \overline{xy}$. So the sum $Fix_M(x) + Fix_M(y)$ corresponds to the row space of

$$\begin{pmatrix} 1 & 1 & 0 & 0 \\ 0 & 0 & 1 & 1 \\ 1 & 0 & 1 & 0 \\ 0 & 1 & 0 & -1 \end{pmatrix}$$

with respect to $\{\bar{1}, \bar{x}, \bar{y}, \overline{xy}\}$, which is similar to the diagonal matrix

$$\begin{pmatrix} 1 & & & \\ & 1 & & \\ & & 1 & \\ & & & 2 \end{pmatrix}.$$

So the submodule $Fix_M(x) + Fix_M(y)$ has index 2 in M. Therefore, this module does not have the 2/4-generation property.

We form $M_1 \perp M_2$, where $M_i \cong BW_{2^{d-1}}$ or $\sqrt{2}BW_{2^{d-1}}$. We construct the BW series by induction. We assume that $d \geq 2$.

We need to know about the commutator sublattices. The group G_{2^d} we constructed contains a normal subgroup $R_{2^d} \triangleleft G_{2^d}$, where $R_{2^d} \cong 2_+^{1+2d}$ and $G_{2^d}/R_{2^d} \cong \Omega^+(2d, 2)$. Suppose L is a lattice of rank 2^d which is a faithful module for $G := G_{2^d}$ and $R := R_{2^d}$. Then if p is an odd prime, L/pL is an irreducible module for G since it is irreducible for R. In the following, we take the case $p = 2$ into consideration.

8.2.2 Extraspecial groups and commutator density

Commutator density for extraspecial groups, especially the dihedral group of order 8, is a very useful tool in our Barnes-Wall theory.

Theorem 8.2.22. *We now suppose that $R \cong 2_+^{1+2d}$ and that the free abelian group L is an R-module such that the central involution of R acts as -1 on L. We also* assume that L satisfies $2/4$ generation as a module for some subgroup D_1 of R, $D_1 \cong Dih_8$. Then

(1) $L_{D_1} = L_D$ and $L_R = L_D$ for every $D \leqslant R$, $D \cong Dih_8$ and $f \in D$; and

(2) for any fourvolution $f \in R$, $[L, R] = (f - 1)L$.

Proof. (1) Take $f \in D_1$, $|f| = 4$. Let $E(f) := \{D \leqslant R \mid f \in D \cong Dih_8\}$. We claim that the set $E(f)$ generates the group R. If this is true, then $L_R = (f-1)L$. Also, $(f - 1)L$ is a G-submodule.

Suppose that R has two subgroups R_1, R_2 which are isomorphic to $2^{1+2(n-1)}_+$ and $R_1 \cap R_2 \geq D_1$. By induction $R_i \cap E(f)$ generates R_i. If $R_1 R_2 = R$, then we are done. So, we are reduced to the case of $R \cong 2^{1+4}_+$, for which the verification is an exercise.

(2) follows from (1) and (8.2.19). □

9

Construction and Properties of the Barnes-Wall Lattices

In this chapter, we give a formal construction and analysis of the Barnes-Wall series. The initial terms of the series (familiar lattices of ranks 2, 4 and 8) were analyzed in the last chapter (and earlier).

9.1 The Barnes-Wall series and their minimal vectors

To build the series BW_{2^d} by induction, we take two copies $L_1 \cong L_2$ of $\sqrt{s}BW_{2^{d-1}}$ on L_i, where $s = \begin{cases} 1 \text{ if } d \text{ is even} \\ 2 \text{ if } d \text{ is odd} \end{cases}$. We have a group $F_i \cong G_{2^{d-1}} \cong 2^{1+2(d-1)}$. $\Omega^+(2(d-1),2)$, $F_i \leq O(L_i)$. If $R_i = O_2(F_i) \cong 2^{1+2(d-1)}$ is the normal subgroup, take $f_i \in R_i$ such that $f_i^2 = -1$. Then $F_1 \times F_2$ fixes $L_1 \perp L_2$. We have an involution t interchanges L_1 and L_2. We assume that F_i fixes $(f_i - 1)L_i$ and $tf_1t = f_2$. So the wreath product $F_1 \wr \langle t \rangle$ fixes $L_1 \perp L_2$.

Notation 9.1.1. *Define* $L := L_1 \perp L_2 + \frac{1}{2}(f_{12} - 1)L_{12}$ (∗). *This lattice is the standard* BW_{2^d} *built from* $BW_{2^{d-1}}$.

We have $L_1 \perp L_2 \leq L \leq \frac{1}{2}(f_1 - 1)L_1 + \frac{1}{2}(f_2 - 1)L_2$. Also t interchanges L_1 and L_2 so that $L/(L_1 \perp L_2)$ is just the set of fixed points for t on $(f_1-1)L_1 \oplus (f_2-1)L_2$, which is just the "diagonal" and covered by $\frac{1}{2}(f_{12} - 1)L_{12}$.

The stabilizer $Stab_{F_1F_2\langle t \rangle}(L) \geq R_1R_2\langle t \rangle$ and, modulo $R_{12}\langle t, u \rangle \cong 2^{1+2d}$, the quotient is $\Omega^+(2(d-1), 2)$. Note that $F_1 \times F_2 \not\leq Stab_{F_1F_2\langle t \rangle}(L)$.

We shall prove that in BW_{2^d} constructed above, the minimal norm is $\mu(BW_{2^d}) = 2^{\lfloor \frac{d}{2} \rfloor}$ and we shall display the set of minimal vectors, and give their number $mv(2^d) = (2+2)(2+4)(2+8)\cdots(2+2^d)$.

Notation 9.1.2. *Let* $mv(U)$ *denote the minimum norm for nonzero vectors of the lattice* U *and let* $MV(U)$ *be the set of minimal vectors.*

Then we obtain $MV(L_1 \perp L_2) = MV(L_1) \cup MV(L_2)$, a disjoint union. Let p_i be the orthogonal projection to V_i, the $\mathbb{Q}$-span of L_i for $i = 1, 2$. Take a minimal vector $x \in MV(L)$, $x \notin L_1 \perp L_2$. Then $x_i := p_i(x) \in \frac{1}{2}(f_i - 1)L_i$. Note that

$\mu(\frac{1}{2}(f_i - 1)L_i) = \frac{1}{2}\mu(L_i)$. Since $x_i \neq 0$, $(x_i, x_i) \geqslant \frac{1}{2}\mu$. So $\mu = (x, x) \geqslant (x_1, x_2) + (x_2, x_2) \geqslant \frac{1}{2}\mu + \frac{1}{2}\mu = \mu$. Thus all the inequalities are equalities. Therefore, $x_i \in MV(\frac{1}{2}(f_i - 1)L_i)$.

Now, we must study which pairs $y_1 + y_2 \in L$ for $y_i \in MV(\frac{1}{2}(f_i-1)L_i) = MV_i$.

Notation 9.1.3. For $i = 1, 2$, define $MV_i := MV(\frac{1}{2}(f_i - 1)L_i)$. For $y_1 \in MV_1$, we study $B(y_1) = \{y_2 \in MV_2 \mid y_1 + y_2 \in L\}$.

Lemma 9.1.4. *For any two vectors $y_2', y_2'' \in B(y_1)$, we have*
(1) $y_2' - y_2'' \in L \cap V_2 = L_2$; *and*
(2) y_2', y_2'' *are orthogonal or proportional.*

Proof. (1) This is obvious by taking the difference $(y_1 + y_2') - (y_1 + y_2'')$.

(2) We suppose that y_2' and y_2'' are not proportional. Since $L/(L_1 + L_2)$ is elementary abelian, all of $(y_1 \pm y_2') - (y_1 \pm y_2'')$ are in L_2. Thus, each of $y_2' \pm y_2''$ has norm at least $\mu := \mu(L_2)$. Since $y_2' + y_2''$ and $y_2' - y_2''$ are orthogonal, their sum $2y_2'$ has norm at least 2μ. In fact, y_2' has norm $\frac{1}{2}\mu$, so all inequalities in this paragraph are equalities. Therefore y_2' and y_2'' are orthogonal. $\square$

Proposition 9.1.5. *We have $|B(y_1)| = 2^d$ and in fact $B(y_1)$ is an orbit of R_2.*

Proof. Suppose $y_2 \in B(y_1)$, then $\{gy_2 | g \in R_2\} = B(y_1)$, where $R_2 = O_2(F_2)$ fixes y. Let $R_{2,y_2} = \mathrm{Stab}_{R_2}(y_2)$ be the stabilizer. This does not contain -1 so it is isomorphic to its image modulo $Z(R_2)$, i.e., is elementary abelian. Since $R_2 \cong 2^{1+2(d-1)}$, the stabilizer has order at most 2^{d-1}. So the orbit length is a multiple of 2^d.

By (9.1.4), if $y_2', y_2'' \in B(y_1)$, then y_2', y_2'' are orthogonal or proportional. It then follows that the index is not more than 2^{d-1} (in $\mathbb{R}^n$, $2n$ is the maximal size of a set of equal length vectors which are pairwise orthogonal or proportional). The result follows. $\square$

Corollary 9.1.6. *The orbit of a minimal vector of BW_{2^d} under the lower group R is a set of 2^{d+1} vectors, two of which are orthogonal or proportional. Two members of such an orbit are congruent modulo $(f - 1)BW_{2^d}$, for any $f \in R$.*

Proof. Imitate the argument of (9.1.4). $\square$

Definition 9.1.7. *In BW_{2^d}, a set of minimal vectors for which two members are equal or orthogonal and which spans the ambient rational space is called a frame. If such a set is an orbit of the lower group R, it is called a lower frame.*

Theorem 9.1.8. *Let $d \geq 2$. The number of minimal vectors in BW_{2^d} is $mv(2^d) = \prod_{i=1}^{d}(2 + 2^i)$. The lattices BW_{2^d} are indecomposable and are spanned by their minimal vectors.*

Proof. For $y \in MV(L) \setminus (MV(L_1) \cup MV(L_2))$, we have $y = y_1 + y_2$. The number of y_1 is $mv(2^{d-1})$. Given y_1, the number of y_2 so that $y_1 + y_2 \in MV(L)$ is 2^d. Therefore, we have $mv(2^d) = mv(2^{d-1}) \cdot (2 + 2^d)$. The first statement follows. For the second, we use (3.5.3) and the fact that the set $MV(L)$ is orthogonally indecomposable. The third statement follows from (9.1.1) and induction. $\square$

In (9.5.1), we shall describe minimal vectors in detail.

9.2 Uniqueness for the BW lattices

We prove uniqueness of the Barnes-Wall lattices by an induction argument. The low rank cases $D_4 = BW_4$ and $E_8 = BW_8$ are root lattices and are easily characterized by minimum norm, rank and discriminant group. For the higher rank cases, we introduce the $X(2^d)$ condition for $d \geqslant 2$.

The induction argument allows us to obtain properties of the lattices and their isometry groups with relatively little matrix work and special counting.

Definition 9.2.1. *Let $s \in \{0,1\}$ be the remainder of $d+1$ modulo 2. We say that (L, L_1, L_2, t) is an X-quadruple if it satisfies the condition $X(2^d)$:*

(a) *the lattice L is even integral of rank 2^d containing $L_1 \perp L_2$, where $L_1 \cong L_2$ and both have rank 2^{d-1};*

(b) *for $d = 2$, $L \cong D_4$ and $L_1 \cong L_2 \cong A_1^2$; for $d \geqslant 3$, $2^{-s/2}L_1$ and $2^{-s/2}L_2$ are first entries of an $X(2^{d-1})$-quadruple;*

(c) *the number of minimal vectors in L is $\mu(L) = 2^{\lfloor \frac{d}{2} \rfloor}$;*

(d) *the determinant group is $\mathcal{D}L = \begin{cases} 1 & \text{if } d \text{ is odd} \\ 2^{2^{d-1}} & \text{if } d \text{ is even} \end{cases}$;*

(e) *there is an involution $t \in O(L)$, so that $t : L_1 \to L_2 \to L_1$ and t acts trivially on $L/(L_1 \perp L_2)$;*

(f) *the projection of L to each V_i, the $\mathbb{Q}$-span of L_i, is a sublattice which is isometric to a scaling of a $BW_{2^{d-1}}$ sublattice and which is invariant under a subgroup of $O(L)$ which fixes each L_i, commutes with t and is isomorphic to $G_{2^{d-1}}$.*

We will prove the characterization (9.2.3) of the Barnes-Wall series by the above properties and give applications to coding theory.

For the condition $X(2^d)$, we denote by $\mathfrak{X} = \mathfrak{X}(L_1, L_2)$ the set of all X-quadruples of the form (L, L_1, L_2, t), provided $d \geqslant 3$ and L_1, L_2 as before. Let $V = \mathbb{Q} \otimes_\mathbb{Z} L$ be the $\mathbb{Q}$-span of the Barnes-Wall lattice L.

Proposition 9.2.2. *Let $d \geq 2$ and let $U := \mathbb{F}_2^{2d}$ be the natural module for the group $J \cong \Omega^\varepsilon(2d, 2)$. Let U_i, for $i = 1, 2$, be a pair of distinct singular 1-dimensional subspaces which are not orthogonal. Let $J_i := Stab_J(U_i)$. Then $J = \langle J_1, J_2 \rangle$.*

Proof. The first proof is to use a standard result in group theory. We quote the theory of parabolic subgroups in Chevalley groups to see that each J_i is a maximal subgroup of J [16]. The second proof is to use the relatively elementary classification of finite dimensional irreducible groups generated by transvections [65]. The groups J_i do not contain transvections. However, there exists a transvection r which interchanges U_1 and U_2, namely the transvection at the nonsingular point of $U_1 + U_2$. It is easy to prove that $\langle J_1, J_2 \rangle$ is irreducible and that the subgroup of $\langle r, J_1, J_2 \rangle$ generated by transvections is irreducible, so that [65] may be used to identify $\langle r, J_1, J_2 \rangle$ as $O^\varepsilon(2d, 2)$. $\square$

Theorem 9.2.3. *(1) The set $\mathfrak{X}$ is an orbit under the natural action of $F_1 \times F_2$, where $F_i = Stab_{O(L_i)}((f_i - 1)L_i)$. Define $Q_i := C_{F_i}(L_i/(f_i - 1)L_i) \triangleleft F_i$. The element of $\mathfrak{X}$ are in bijection with each of*

(a) *$F_1/Q_1 \cong \Omega^+(2(d-1), 2)$;*

(b) F_2/Q_2;

(c) the pairs of involutions $\{u, -u\}$ in $O(V)$ which interchange L_1, L_2 and $(f_1 - 1)L_1, (f_2 - 1)L_2$;

(d) $(Q_1 \times Q_2)$-orbits on the set of dihedral groups of order 8, which are generated by the SSD involutions associated to L_1, L_2 and pairs of involutions as in (c).

(2) The stabilizer $Stab_{O(L_1 \perp L_2)}(L)$ has the form $(2^{1+2(d-1)} \times 2^{1+2(d-1)}) \cdot (\Omega^+(2(d-1), 2) \times 2)$. The subgroup of this which acts trivially on the quotient $L/(f_{12} - 1)L$ is $Q := \langle Q_{12}, s, t_i \rangle \cong 2^{1+2(d-1)}$, where s as in (c) and t_i is -1 on L_i or $+1$ on L_i.

Proof. We use induction. We may assume $d \geqslant 4$. Take an element $(L, L_1, L_2, t) \in \mathfrak{X}$. Then the determinants $\det(L_i)$, $\det(L)$ and the index $|L : L_1 \perp L_2|$ are determined.

Let p_i be the orthogonal projection $V \to V_i = \mathbb{Q} \otimes_{\mathbb{Z}} L_i$ for $i = 1, 2$. Then $p_i(L) \leqslant L_i^*$ since L is integral.

If d is odd, the indices imply $p_i(L) = L_i^*$.

Suppose d is even. We have a strict inclusion $p_i(L) < L_i^*$. There is a subgroup $H \leqslant C_G(t)$, $H \cong F_i$ so that H acts on both L_1, L_2 as F_1, F_2 respectively (e.g. take $H = C_{F_1 \times F_2}(t)$). By the hypothesis $X(2^d)(f)$, we have $p_i(L) = \frac{1}{2}(f_i - 1)L_i$. The group $\langle Q_1, Q_2, t \rangle$ stabilizes the lattice L. Set $Q := \langle Q_{12}, Z(Q_1), Z(Q_2), t \rangle = Q_{12} \circ \langle t, Z(Q_1) \rangle \cong 2^{1+2d}$. So L determines the dihedral group $D = \langle t, Z(Q_1) \rangle$ up to conjugacy, namely up to the choice of t. This is equivalent to making a choice of isometry from L_1 to L_2 which takes $\frac{1}{2}(f_i - 1)L_1$ to $\frac{1}{2}(f_i - 1)L_2$. Now use (**??**).

Now, suppose we are given a such group D, we get L as follows: $L/L_1 \perp L_2$ is the set of fixed points for D on $\frac{1}{2}(f_1 - 1)L_1/L_1 \oplus \frac{1}{2}(f_2 - 1)L_2/L_2$. We ask to what extent L_1, L_2, L determine t (or the group D). The subgroup of $F_1 \times F_2$ which acts trivially on $L/L_1 \perp L_2$ is $(Q_1 \times Q_2)D = (Q_1 \times Q_2)\langle t \rangle$, whence L determines t and D up to conjugacy by $(Q_1 \times Q_2)$. □

We study the automorphism group of the Barnes-Wall lattice $L = L_1 + L_2 + \frac{1}{2}(f_{12} - 1)L_{12}$. As in the proof of (9.2.3) have a group H_1 of the form $(2^{1+2(d-1)} \times 2^{1+2(d-1)}) \cdot (\Omega^+(2(d-1), 2) \times 2)$ in $O(L_1 \perp L_2)$. Then $L = \frac{1}{2}(f_{12} - 1)L_{12} + \frac{1}{2}(f_{12} - 1)L'_{12} + L_1$, where $L_{12} = \{x + tx \mid x \in L_1\}$ and $L'_{12} = \{x - tx \mid x \in L_1\}$. By the theorem of the commutator density, we get a group $H_2 \cong H_1$. Let $H_2 \leqslant O(L_{12} \perp L'_{12})$. We can check that $R \triangleleft \langle H_1, H_2 \rangle \cong G_{2^d}$ (9.2.2)

9.3 Properties of the BRW groups

We list some applications of our BRW theory. For proofs, see [40] and [41].

Theorem 9.3.1. *We are given the Barnes-Wall lattice $L = BW_{2^d}$ and associated group G_{2^d}.*

If M is a G_{2^d}-invariant lattice M, then

(a) there is a rational number $r \neq 0$, so that $rM = L$ or $(f - 1)L$;

(b) if p is an odd prime, then M/pM is an irreducible module;

(c) the quotient $M/2M$ has just one submodule $(f - 1)M/2M$ besides $0 = 2M/2M$ and $M/2M$.

Theorem 9.3.2. *The action of $BRW^+(2^d)$ on the minimal vectors of BW_{2^d} is transitive.*

Theorem 9.3.3. *For $d \neq 3$, $O(BW_{2^d}) = BRW^+(2^d)$. Also, $O(BW_{2^3}) \cong Weyl(E_8)$.*

See also (5.2.2),(4.2.39).

Remark 9.3.4. *There are several results about transitivity of the action of $BRW^+(2^d)$ on certain isometry types of RSSD sublattices in BW_{2^d}. See (13.3)[40] and (4.1)[41] for details.*

In [40], the Ypslianti lattices were constructed using variations on the methods of our constructions of the BW series. The Ypsilanti lattices are explicitly defined families of indecomposable even unimodular lattices which are pairwise nonisomorphic. These lattices represent roughly the square root of the number of isometry types given by the mass formula estimate. These are one of the the largest such families.

9.4 Applications to coding theory

In the original constructions of the BW series [5, 15], error-correcting codes were used to build up the lattices in layers. Our theory implies coding theory results.

Take a lower frame, i.e. a set Φ of 2^{d+1} minimal vectors in L which form an orbit under R (9.1.7). Define S to be the $\mathbb{Z}$-span of Φ. We consider $2^{-k}S \cap L$ for $k = 0, 1, 2, \ldots$ (note that the intersection has finite index in both L and $2^{-k}S$). We have a map $2^{-k}S \cap L \to 2^{-k}S/2^{-k+1}S$. Let Q_k be the image of the map. Then it is a subcode of the space $\mathbb{F}_2^\Omega$.

We want to label $\Omega := \Phi/\{\pm 1\}$ by $\mathbb{F}_2^d$, so that Q_k looks like a Reed-Muller code.

Example 9.4.1. *For background, see items (6.2.2), (6.2.3). Recall that the Reed-Muller code $RM(e, d)$ is the span of all codimension e affine subspace of $\mathbb{F}_2^d$. In the low dimensional cases, we have for example $d = 3$, the dimension $0, 1, 2, 3$ subspaces are points, any 2-set, certain 2-set and $\mathbb{F}_2^3$, respectively. We wish to label the 8 points the index set by $\mathbb{F}_2^3$, so that the 4-sets, which occur are affine subspaces. We have shown that the code $[8,4,4]_2$ is unique up to equivalence, which is spanned by the four vectors $(11110000), (11001100), (10101010), (11111111)$. For the elements $(000), (001), (010), (011), (100), (101), (110), (111)$ in $\mathbb{F}_2^3$, we obtain that the vectors $(11110000), (11001100), (10101010)$ are the kernels of the 1st, 2nd and 3rd coordinate functions, respectively, and the vector (11111111) is the kernel of the zero function. So it follows that one code $\mathcal{C}$ is contained in $RM(1, 3)$ which is of dimension 4.*

Theorem 9.4.2. *Let $d \geq 3$. There is a way to label $\Omega := \Phi/\{\pm 1\}$ by the set $\mathbb{F}_2^d$ so that for all k, Q_k is the Reed-Muller code $RM(d - 2k, d)$.*

Proof. We use induction for $d \geqslant 3$. We study the lattice $L = L_1 \perp L_2 + \frac{1}{2}(f_{12} - 1)L_{12}$ (9.1.1). Take a minimal vector $v \in L_1$, then $\Phi = \{gv \mid g \in R\}$. So $\Phi =$

$\Phi_1 \cup \Phi_2$, where $\Phi_i = L_i \cap \Phi$. Define $\Omega_1 := \Phi_1/\{\pm 1\}$ There is a good identification of Ω_1 with $\mathbb{F}_2^{d-1}$ so that the image of $L_1 \cap 2^{-k}S_1$ in Q_k is $RM(d-2k,d)$.

We use a set isomorphism $\Omega_1 \to \Omega_1' := \{(x_1,\ldots,x_d) \in \mathbb{F}_2^d \mid x_d = 0\}$ and we set $\Omega_2' := \{(x_1,\ldots,x_d) \in \mathbb{F}_2^d \mid x_d = 1\}$. We note that a lower involution t associated to the expression $L = L_1 \perp L_2 + \frac{1}{2}(f_{12}-1)L_{12}$ (9.1.1) which interchanges L_1 and L_2 takes a basis $\{v_i \mid i \in \Omega_1\}$ to a subset of Φ_2 which is linearly independent. For $i \in \Omega_1$, we label $\pm tv_i$ by $(x_1,\ldots,x_{d-1},1)$ if $\pm v_i$ has the label $(x_1,\ldots,x_{d-1},0)$.

For the intersection $2^{-k}S \cap L$, we have a sequence $S \cap L \subseteq \frac{1}{2}S \cap L \subseteq \frac{1}{4}S \cap L \subseteq \cdots \subseteq L$. Notice that the quotient L/S is a finite 2-group (by determinant considerations). There is a smallest integer m so that $2^{-m}S \geq L$, hence the above sequence ends with L.

We observe that $L = L_1 + \frac{1}{2}(f_{12}-1)L_{12} = L_1 \oplus \frac{1}{2}(f_{12}-1)L_{12}$. The first equality comes from the equivalence of the 2/4-generation with the 3/4-generation, and the second one is due to the fact that $\langle u, t \rangle \cong Dih_8$ and its center is -1.

Each vector v_i with $i \in \Omega_1'$ is in L_1. What about the vectors v_i with $i \in \Omega_2'$? We can see certain linear combinations of the vectors v_j with $j \in \Omega_2'$, which lie in $T := \frac{1}{2}(f_{12}-1)L_{12}$. If $B \subseteq \Omega$, define $v_B := \sum_{i \in B} v_i$. Write $\dot{v}_B$ for some linear combination $\sum_{i \in B} \dot{v}_i$, where $\dot{v}_i = \pm v_i$. By induction, we know that $2^{-l}S \cap L_1$ is spanned modulo $2^{-l+1}S \cap L_1$ by certain vectors of the form $2^{-l}\dot{v}_B$, where B ranges over the affine subspaces of Ω_1 of dimension $2l$. So $2^{-l}S \cap L_{12}$ is spanned modulo $2^{-l+1}S \cap L_{12}$ by all $2^{-l}(\dot{v}_B + t\dot{v}_B)$.

The result is that we get a set of generators for $2^{-l}S \cap L$ of the form $2^{-l}\dot{v}_C$ where C is an affine subspace of dimension $2 + \dim(B)$. This proves the theorem. □

Our procedure creates lattice vectors $2^{-m}\dot{v}_B$, where $m = \lfloor \frac{d}{2} \rfloor$ and $\dim(B) = 2m$. In this case, $(v_B, v_B) = (2^{-m})^2(v_i, v_i) \cdot 2^{2m} = (v_i, v_i) = 2^m$ is the minimal norm. Compare the procedure "$[u \mid u+v]$" on page 76 [64].

There is a formula for the dimension of the Reed-Muller code: $\dim(RM(j,d)) = \sum_{i=0}^{j} \binom{d}{i}$ [64].

9.5 More about minimum vectors

We describe the minimum vectors in BW_{2^d} and in $(f-1)BW_{2^d}$, where f is a lower fourvolution. These results should be compared to (9.1.8). We continue to let Φ denote a lower frame (9.1.7) and we a labeling of $\Omega := \Phi/\{\pm 1\}$ as in (9.4.2).

Theorem 9.5.1. *There exists a basis $\{v_i \mid i \in \Omega\}$ contained in Φ and labeling of Ω by $\mathbb{F}_2^d$ so that the following properties hold:*

(1) The set of minimum vectors of BW_{2^d} is all vectors of the form $\varepsilon_S 2^{-k}v_A$, where A is a $2k$-dimensional space and $S \in RM(2,d)$, for $k = 0, 1, \ldots, \lfloor \frac{d}{2} \rfloor$.

(2) Let f be a lower involution. The set of minimum vectors of $(f-1)BW_{2^d}$ is all vectors of the form $\varepsilon_S 2^{-k}v_A$, where A is a $2k+1$-dimensional space and $S \in RM(2,d)$, for $k = 0, 1, \ldots, \lfloor \frac{d-1}{2} \rfloor$.

Proof. We prove both (1) and (2) by induction. To an index $i \in \Omega$ corresponds a pair consisting of a minimal vector and its negative. In this proof, we intend to choose a member of each pair to label with a vector from $\mathbb{F}_2^d$, whereas in (9.4.2), we labeled the pair with a vector from $\mathbb{F}_2^d$.

This is trivial to prove for the first few values of d. Now suppose that such a labeling exists for $BW_{2^{d-1}}$, for $d \geq 3$. We use the labeling procedure of (9.4.2). In that notation of that proof, L_1 and L_2 have suitable labelings by copies of $\mathbb{F}_2^{d-1}$ which form complementary affine subspaces in a copy of $\mathbb{F}_2^d$ in which the action of t is interpreted as a translation which interchanged the two complements. It is then clear that the minimal vectors of $L_1 + L_2$ have the right form. For y a minimal vector in L but $y \notin L_1 + L_2$, we write $y = y_1 + y_2$, where y_i is the orthogonal projection of y to the rational span of L_i, for $i = 1, 2$. There exists an affine space B in Ω so that $supp(y) = B$. For each i, $B \cap \Omega_i$ is an affine subspace, of dimension $\dim(B) - 1$. By induction, there is a minimal vector y_1' of $\frac{1}{2}(f_1 - 1)L_1$ of the form $2^{-k}v_C$, for an affine subspace C of dimension $2k - 1$ so that $y_1 - y_1' \in L_1$. Also, $y_2 - ty_1 \in L_2$. Now, $y_1 + ty_1 = 2^{-k}v_D$, where $D = C \cup tC$ is an affine space of dimension $2k$. By induction, $y_1 - y_1' = v_E$, where E is in the Reed-Muller code of type $RM(2, d-1)$ based on the affine space Ω_1. Therefore, $E + tE$ is in the Reed-Muller code of type $RM(2, d)$ based on the affine space Ω. Therefore $y = \varepsilon_{E+tE} 2^{-k} v_D$, as required. This completes the induction step for the case of minimal vectors in BW_{2^d}.

Now consider a minimal vector y of $(f - 1)L$. If it lies in one of L_1 or L_2, it has the right form.

Suppose otherwise and let y_i be the orthogonal projection of y to the rational span of L_i, for $i = 1, 2$. Each y_i is nonzero and has norm $\frac{1}{2}(y, y)$. If y_1 is in L, it is a minimal vector of L and so is y_2. Thus y_1 and y_2 are members of a lower frame (9.1.7)

Now suppose that neither y_1 nor y_2 is in L.

Since $d \geq 2$, we may choose the lower fourvolution to commute with the orthogonal projections $p_i : L \to \mathbb{Q} \otimes L_i$. Therefore, each y_i is in $(f - 1)p_i(L) = (f-1)\frac{1}{2}(f-1)L_i = L_i$. This means that y_i is a minimal vector of L_i, for $i = 1, 2$. Since y_1 and y_2 are congruent modulo $(f-1)L$, y_1, y_2 are orthogonal members of a lower frame. The action of the lower group is transitive on the frame.

Now, y_1 has the form $2^{-k}\varepsilon_S v_A$ as in (1). There is a lower element $g = g_\tau \varepsilon_T$, which takes y_1 to y_2, where τ is a translation on Ω and $T \in RM(1, d)$.

Case 1: τ fixes A. Then $k \geq 1$. Since y_1 and y_2 are orthogonal, their nonzero coordinates differ exactly 2^{-k-1} times. This means that the hyperplane T meets A transversely, i.e., $T \cap A$ is an affine space of codimension 1 in A. Thus, $y_1 + y_2$ is a vector of the form $2^{-k+1}\varepsilon_{T+\tau(S)} v_A$, as required.

Case 2: τ does not fix A. Then the support of $y_1 + y_2$ has the form $A + \tau A$, an affine space of dimension $1 + \dim(A)$. Furthermore, $y_1 + y_2 = \varepsilon_S 2^{-k} v_A + 2^{-k}\varepsilon_{T+\tau S} v_{\tau A} = 2^{-k}\varepsilon_S w$, where $w = v_A + \varepsilon_{T+S+\tau S} v_{\tau A}$. To show that $y_1 + y_2$ has the right form, it suffices to show that $2^{-k}w$ has the right form.

There exists T', equal to the empty set or an affine hyperplane, so that $A \cap T' = \emptyset$ and $T' \cap \tau A = T \cap \tau A$. It follows that $w = \varepsilon_{T'}(v_A + \varepsilon_{S+\tau S} v_{\tau A})$. Since $S \in RM(2, d)$, $S + \tau(S) \in RM(1, d)$. There exists T'', equal to the empty set or an affine hyperplane, so that $A \cap T'' = \emptyset$ and $T'' \cap \tau A = (S + \tau(S)) \cap \tau A$.

Consequently, $w = \varepsilon_{T'}(v_A + \varepsilon_{S+\tau S} v_{\tau A}) = \varepsilon_{T'+T''}(v_A + v_{\tau A}) = \varepsilon_{T'+T''} v_{A+\tau A}$. Therefore, $2^{-k} w = 2^{-k} \varepsilon_{T'+T''} v_{A+\tau A}$ has the required form. □

10

Even unimodular lattices in small dimensions

We can make a very useful series from the number of lattice vectors which lie on each sphere centered at the origin.

Definition 10.0.1. *Let L be an even integral lattice. The theta series of L is defined by $\theta_L(\tau) = \sum_{x \in L} q^{\frac{1}{2}(x,x)} = \sum_{n=0}^{\infty} u_n q^n$, where $u_n = |\{x \in L \mid (x,x) = 2n\}|$, $\tau \in \{z \in \mathbb{C} \mid \text{Im}(z) > 0\} \cup \{i\infty\}$ and $q := e^{2\pi i \tau}$.*

If L is also unimodular, the function θ_L is a modular form. See e.g., [71, 72]. We shall quote a few results from this theory. In particular, $\text{rank}(L) \in 8\mathbb{Z}$.

Unimodular even lattices are of special interest for many reasons. Their theta functions (10.0.1) are invariant by a large group, $SL(2, \mathbb{Z})$, and so satisfy especially useful arithmetic conditions. Classification results are more developed for unimodular lattices than arbitrary integral lattices. Certain unimodular lattices are especially interesting for finite group theory. The E_8-lattice contains root lattices for all the finite-dimensional exceptional complex Lie algebras and so can give a viewpoint for work with any exceptional Lie algebra. The Leech lattice represents a common completion of many interesting arithmetic, group theoretic and combinatorial themes. Recent results of Cohn and Kumar [60] show that the densest packings in 8-space are essentially E_8-lattice packings and that the densest lattice packings in 24-space are essentially Leech lattice packings. In short, a lot of interesting and varied mathematics occurs around unimodular lattices.

An equivariant embedding result (10.8) shows that any lattice can be embedded in a unimodular lattices so that the automorphism group is respected. This result has a useful application to VOA theory [42].

10.1 Classifications of even unimodular lattices

The Niemeier lattices with roots can be constructed in a fairly direct way by the method of starting with the root sublattice, then extending upward by glue vectors corresponding to a suitable code over some finite ring (usually a quotient of the integers). We shall sketch constructions of several Niemeier lattices and give detailed treatment to the Leech lattice, the case of an empty root system.

Recall that for a given rank and discriminant, there are only finitely many positive definite integral lattices (5.0.5). Classifications are complete only for dimensions 8, 16 and 24. We summarize the results.

For $n = 8$, there is only one such lattice up to equivalence, namely E_8 (5.2.1). See [39] for historical remarks.

For $n = 16$, there are two equivalent classes, $E_8 \perp E_8$ and HS_{16} (2.3.6). This was proved by Witt [81].

For $n = 24$, we have the famous 1968 classification of Niemeier [66]. He finds exactly 24 equivalent classes of such lattices.

Definition 10.1.1. *We use the term Niemeier lattice for a rank* 24 *even unimodular lattice.*

A shorter proof of the Niemeier classification was given by Venkov in 1978 [78], using theta functions to study configurations of roots. It turns out that if L is even, unimodular of rank 24, its isometry type is determined by its root system $\Phi(L) := \{x \in L \mid (x,x) = 2\}$. Also, $\Phi(L)$ has rank 0 or 24, and each connected component of $\Phi(L)$ has the same Coxeter number (defined below). This beautiful fact was established and used quite effectively in [78].

Definition 10.1.2. *Let S be an indecomposable root system of rank l. Then l divides $|S|$ and $h := |S|/l$ is the Coxeter number. See* [11].

Table 10.1.

lattice	number of roots	Coxeter number h
A_n	$(n+1)n$	$n+1$
D_n	$2n(n-1)$	$2(n-1)$
E_6	72	12
E_7	126	18
E_8	240	30

Theorem 10.1.3. *The* 24 *root systems which give the* 24 *Niemeier lattices are the following:* $\emptyset$;

$24A_1, 12A_2, 8A_3, 6A_4, 4A_6, 3A_8, 2A_{12}, A_{24}$;

$6D_4, 4D_6, 3D_8, 2D_{12}, D_{24}$;

$4E_6, 3E_8$;

$4A_5 + D_4, 2A_7 + 2D_5, 2A_9 + D_6, A_{16} + D_9, E_8 + D_{16}, 2E_7 + D_{10}, E_7 + A_{17}, E_6 + D_7 + A_{11}$.

For $n = 32$, the number of such lattices is over 10^7 [21, 71]. A complete enumeration seems unlikely, but see [56].

10.2 Constructions of some Niemeier lattices

The Niemeier lattices with roots can be constructed in a fairly direct way by the method of starting with the root sublattice, then extending upward by glue

10.2 Constructions of some Niemeier lattices

vectors corresponding to a suitable code over some finite ring (usually a quotient of the integers). We shall sketch constructions of several Niemeier lattices and give detailed treatment to the Leech lattice, the case of an empty root system.

Notation 10.2.1. Let Φ be a rank 24 root system. Denote by $\mathcal{N}(\Phi)$ the Niemeier lattice with root system Φ.

Definition 10.2.2. A Leech lattice is a lattice $L \leqslant \mathbb{R}^{24}$ which is integral, even, unimodular and has no roots. In particular, a Leech lattice is a Niemeier lattice without roots.

By Niemeier's theorem, $\mathcal{N}(\Phi)$ exists only for the root systems listed in (10.1.3).
For the root system $\Phi = E_8^3$, we have the Niemeier lattice $\mathcal{N}(E_8^3) = E_8 \perp E_8 \perp E_8$.

For $\Phi = D_{24}$ we have $\mathcal{N}(D_{24}) = HS_{24}$.

The lattice $\mathcal{N}(E_6^4)$ is between E_6^4 and its dual. The discriminant group of D is $\mathcal{D}E_6^4 = \mathbb{F}_3^4$. It corresponds to a code which is self-orthogonal of dimension 2 (since $\mathcal{N}$ has determinant 1). The self-orthogonality is required to make $\mathcal{N}$ integral. The code is the ternary code $\{(s, a, a+s, a+2s) \mid a, s \in \mathbb{F}_3\}$, up to equivalence (6.2.5).

The lattice $\mathcal{N}(A_1^{24})$ is between A_1^{24} and its dual, so corresponds to a code in $\mathbb{F}_2^{24}$, which has dimension 12 and is self-orthogonal. Take $\{\alpha_1, \ldots, \alpha_{24}\}$ an orthogonal set of roots, with index set $\Omega = \{1, \ldots, 24\}$. For a subset $A \subseteq \Omega$, define $\alpha_A := \sum_{i \in A} \alpha_i$. Then the lattice $\mathcal{N}$ contains certain vector $\frac{1}{2}\alpha_A$, $A \subseteq \Omega$, with $(\frac{1}{2}\alpha_A, \frac{1}{2}\alpha_A) = \frac{1}{4} \cdot 2 \cdot |A|$. Since $\mathcal{N}$ is an even lattice, $|A| \in 4\mathbb{Z}$. Since $\pm\alpha_i$ are the only roots, we require $|A| \geqslant 8$ whenever $\frac{1}{2}\alpha_A \in \mathcal{N}$. The corresponding code $\mathcal{C}$ in $\mathbb{F}_2^{24}$ has parameters $[24, 12, w]_2$ for some integer $w \geqslant 8$.

10.2.1 Construction of a Leech lattice

We construct a Leech lattice (10.2.2) such lattice here. Later, we prove that such a lattice is unique up to isometry.

Definition 10.2.3. A Golay code is a $[24, 12, w]_2$-code for some integer $w \geqslant 8$.

We shall prove existence and uniqueness for Golay codes.

Lemma 10.2.4. The space $\mathbb{F}_2^8$ contains two $[8, 4, 4]_2$-codes, H and H', so that $H \cap H' = \mathbb{F}_2 \cdot (1, 1, 1, 1, 1, 1, 1, 1)$ (equivalently, that $H + H'$ is the space of even sets).

Proof. One can simply write down two such subspaces and check the conditions directly. Instead, we give a proof inspired by the theory of cyclic codes.

Note that $x^7 - 1 \in \mathbb{F}_2[x]$ has the factorization $x^7 - 1 = (x+1)(x^3 + x + 1)(x^3 + x^2 + 1)$ into irreducibles. Think of a 7-cycle σ on the first 7 of the 8-coordinates The polynomial $a + bx + cx^2 + \cdots$ shall correspond to the 8-tuple $(a, b, c, \ldots)$. Then the product $p(x) := (x+1)(x^3 + x + 1) = x^4 + x^3 + x^2 + 1$ corresponds to the sum $(1, 1, 0, 1, 0, 0, 0, 0) + (0, 1, 1, 0, 1, 0, 0, 0) = (1, 0, 1, 1, 1, 0, 0, 0)$. Define $q(x) := (x+1)(x^3 + x^2 + 1) = x^4 + x^2 + x^1 + 1$.

Define H_1 to be the image of the linear transformation $p(\sigma)$ and H'_1 to be the image of the polynomial $q(\sigma)$. Since the greatest common divisor of p

110   10 Even unimodular lattices in small dimensions

and q is $(x+1)$, $H_1 + H_1'$ is the image of $\sigma + 1$, the space of even weight polynomials which have eighth coordinate zero. This is 6-dimensional. Define $H := H_1 + \mathbb{F}_2 \cdot (1,1,1,1,1,1,1,1)$, and $H' := H_1' + \mathbb{F}_2 \cdot (1,1,1,1,1,1,1,1)$. Then $H + H'$ is the 7-dimensional space of even norm vectors. It follows that $\dim(H \cap H') = 1$ and so $H \cap H' = \mathbb{F}_2 \cdot (1,1,1,1,1,1,1,1)$. □

The idea of the following proof is due to Turyn [4]; [64], Ch. 18, Sec. 6, Th. 12, p 588.

Theorem 10.2.5. *There exists a Golay code with minimum weight 8. It is a doubly even code.*

Proof. We proved that $[8,4,4]_2$-codes exist and are unique up to equivalence (6.2.2). The standard one is spanned by $(1,1,1,1,0,0,0,0)$, $(1,1,0,0,1,1,0,0)$, $(1,0,1,0,1,0,1,0)$, $(1,1,1,1,1,1,1,1)$. We take two $[8,4,4]_2$-codes H, H' in $\mathbb{F}_2^8$ such that $H \cap H' = \mathbb{F}_2 \cdot (1,1,1,1,1,1,1,1)$. Existence of such a pair was proved in the preceding lemma.

Define $K := \{(x \mid y \mid z) \in H \oplus H \oplus H \leqslant \mathbb{F}_2^{24} \mid x + y + z = 0\}$, which has dimension 8. Define $K' = \{(w,w,w) \in H \oplus H \oplus H \leqslant \mathbb{F}_2^{24}\}$, which has dimension 4. Note that $K \cap K' = 0$ ($3w = 0$ implies $w = 0$). Define $C := K + K'$, which has dimension 12. We want to show that C is doubly even with minimal weight 8.

First we prove that $C \subseteq C^\perp$. For $(x,y,z) \in K$ and $(w,w,w), (w',w',w') \in K'$, there are three cases to check: $(w,w,w) \cdot (w',w',w') = 3(w,w') = 0$ since H' is doubly even; $(x,y,z) \cdot (w,w,w) = (x+y+z,w) = (0,w) = 0$; $(x,y,z) \cdot (x',y',z') = (x,x') + (y,y') + (z,z') = 0 + 0 + 0 = 0$ since H is doubly even.

Since $C \subseteq C^\perp$, and C is spanned by a set of doubly even vectors, so C is doubly even. If C does not have minimal weight 8, C has a vector v of weight 4. Since $v = (x,y,z) + (w,w,w) = (x+w \mid y+w \mid z+w)$ with three even parts, $wt(v) = 4$ implies one component is 0. Say $x+w = 0$. Then $x = w \in H \cap H'$, whence $w = (0^8)$ or (1^8). If $w = 0$, $v = (x,y,z)$ has weight at least 8 since at least two of x,y,z are not 0. Now assume that $w = (1,1,1,1,1,1,1,1)$. Then at least one of y,z has weight 4. so the condition $x+y+z = 0$ implies that both x,y have weight 4. Then v plainly has weight 8. □

Notation 10.2.6. We have the Niemeier lattice $\mathcal{N}(A_1^{24}) = A_1^{24} + \sum_{A \in \mathcal{G}} \mathbb{Z} \cdot \frac{1}{2}\alpha_A$, where $\mathcal{G} \leqslant \mathbb{F}_2^{24}$ is a Golay code. Define a homomorphism $\theta : \mathbb{Q}^{24} \to \mathbb{Q}$ by $\theta(\alpha_i) = \frac{1}{2}$ for all i. We have $\theta(x) = (\frac{1}{4}\alpha_\Omega, x)$ for all $x \in \mathbb{Q}^{24}$. We have $\theta(\alpha_A) \in 2\mathbb{Z}$, or $\theta(\frac{1}{2}\alpha_A) \in \mathbb{Z}$, whence $\theta(\mathcal{N}(A_1^{24})) = \frac{1}{2}\mathbb{Z}$. Define $M := \mathcal{N}(A_1^{24}) \cap \theta^{-1}(\mathbb{Z})$. Then it has index 2 in $\mathcal{N}(A_1^{24})$ and contains no roots. Define $\nu_i := \frac{1}{4}\sum_{j \neq i}\alpha_j - \frac{3}{4}\alpha_i$, a vector of norm 4.

Definition 10.2.7. *Let $\nu := \nu_i$ for some i. Define $\Lambda := M + \mathbb{Z}\nu$. This is the standard model of the Leech lattice.*

Lemma 10.2.8. *Λ is integral even unimodular and contains no roots, so is a Leech lattice. Furthermore, it is independent of the choice of i.*

Proof. Clearly, $\Lambda \neq M$ since $\nu \notin M$; but $2\nu = (\frac{1}{2}\sum_{j \in \Omega}\alpha_j) - 2\alpha_i$. Also, $\nu_i - \nu_j = -\alpha_i + \alpha_j$, thus $\theta(\nu_i - \nu_j) = \theta(-\alpha_i + \alpha_j) = 0 \in \mathbb{Z}$. So $\nu_i - \nu_j \in M$. The independence of i follows. To prove that Λ is integral, it is sufficient to prove $(\nu, M) \subseteq \mathbb{Z}$. By definition of θ, $\theta(x) = (\frac{1}{4}\alpha_\Omega, x)$, which is integral for $x \in M$. So, we are reduced to showing that $(\alpha_i, M) \in \mathbb{Z}$. But this is clear since $\mathcal{N}(A_1^{24})$ is an integral lattice. Finally, since $|\Lambda : M| = 2 = |\mathcal{N}(A_1^{24}) : M|$, $\det(\Lambda) = \det(\mathcal{N}(A_1^{24}))$, which is 1. □

10.3 Basic theory of the Golay code

We shall analyze weight distribution and certain configurations in the Golay code, and prove a uniqueness result. The Reed-Muller codes $RM(r,3)$ and $RM(r,4)$ are closely related to the Golay code, so we begin with a few results about Reed-Muller codes.

10.3.1 Characterization of certain Reed-Muller codes

Proposition 10.3.1. *In $\mathbb{F}_2^{2^d}$, for $k = 1, 2, \ldots, d$, let v_k be the binary vector $(A, B, \ldots A, B)$ where A is the all-1 vector of length 2^{d-k} and B is the all-0 vector of length 2^{d-k} Let $\mathcal{C}$ be a linear code in $\mathbb{F}_2^{2^d}$ such that for any $0 \neq x \in \mathcal{C}$, we have $wt(x) = 2^{d-1}$. Then $r := \dim(\mathcal{C}) \leqslant d$ and $\mathcal{C}$ is equivalent to the code spanned by the r vectors $v_1, \ldots, v_r$.*

Proof. Let Ω be the index set, $|\Omega| = 2^d$. Take a basis $x_1, \ldots, x_r$ of $\mathcal{C}$. We may assume that $x_1 = v_1$. The result is therefore true for $r \leq 1$. Assume $r \geq 2$. If $y \in \mathcal{C}$, $y \notin span\{x_1\}$, then $wt(y \cap x_1) = 2^{d-2}$ (because $wt(y) = wt(y+x) = 2^{d-1}$). Therefore, if $\lambda : \mathcal{P}(\Omega) \to \mathcal{P}(x_1)$ is the linear map $y \mapsto y \cap x_1$, then $\lambda(\mathcal{C})$ satisfies the hypothesis for $d - 1, r - 1 = \dim(\lambda(\mathcal{C}_1))$, where $\mathcal{C}_1$ is an $(r-1)$-dimensional subspace of $\mathcal{C}$, so that $\mathcal{C} = \mathcal{C}_1 \oplus span\{x_1\}$. By induction, $r - 1 \leqslant d - 1$, and we may assume that $\lambda(x_2), \ldots, \lambda(x_r)$ have the correct form. Now we do the same for the map $\lambda' : \mathcal{P}(\Omega) \to \mathcal{P}(x_1 + \Omega)$. We again use induction to get $\lambda'(x_2), \ldots, \lambda'(x_r)$ in the right form. Finally, we use coordinate permutations to put together the results for λ and λ'. □

Corollary 10.3.2. *Suppose that the code $\mathcal{C} \leqslant \mathbb{F}_2^{2^d}$ has the property that $x \in \mathcal{C}$ implies $wt(x) \in \{0, 2^{d-1}, 2^d\}$. Then $\mathcal{C}$ is equivalent to the code $\mathcal{C}_1$ of the previous proposition (for $r = \dim(\mathcal{C})$) if $\Omega \notin \mathcal{C}$. If $\Omega \in \mathcal{C}$, then $\mathcal{C}$ is equivalent to $span\{\mathcal{C}_1, \Omega\}$ (for $r = \dim(\mathcal{C}) - 1$).*

We recall that $RM(1,d)$ has dimension $1+d$. In $\mathbb{F}_2^d$, $RM(i,d) = ann_{\mathcal{P}(\Omega)}RM(j,d)$, where $i + j = d + 1$ [64].

Proposition 10.3.3. *Suppose $\mathcal{C} \leqslant \mathbb{F}_2^{2^d}$, has minimal weight at least 4, $\mathcal{C}$ is even, and $\dim(\mathcal{C}) \geqslant 2^d - d - 1$. Then $\dim(\mathcal{C}) = 2^d - d - 1$ and $ann(\mathcal{C})$ is equivalent to $RM(1,d)$. Thus, $\mathcal{C}$ is equivalent to the Reed-Muller code $RM(d-2,d)$.*

Proof. The case $d = 2$ trivial and the case $d = 3$ was done before (6.2.2). We shall prove that $w \in \mathcal{C}^\perp, w \neq 0, \Omega$ implies $wt(w) = 2^{d-1}$. Then we use (10.3.2).

Since $\mathcal{C}$ is even, $\Omega \in \mathcal{C}^\perp$. So w or $w + \Omega$ is an element of $\mathcal{C}^\perp$ of weight $\leqslant 2^{d-1}$. Suppose $|w| \leqslant 2^{d-1}$. Define $\lambda : \mathcal{C} \to \mathcal{P}(w)$ by $A \mapsto A \cap w$. Since $w \in \mathcal{C}^\perp$, $\text{Im}(\lambda) \leqslant$

$\mathcal{PE}(w)$, so $\dim(\text{Im}(\lambda)) \leq |w| - 1$ and $\dim(Ker(\lambda)) = \dim(\mathcal{C}) - \dim(Ker(\lambda)) \geq 2^d - d - 1 - (|w| - 1) = 2^d - d - |w|$. Let $a_1, \ldots, a_k$ (for $k := 2^{d-1} - |w|$) be distinct points of $\Omega + w$ so that $w' := w \dot\cup \{a_1, \ldots, a_k\}$ is a 2^{d-1}-set. Define $\lambda' : \mathcal{C} \to \mathcal{P}(w')$. Then $\dim(Ker(\lambda')) \geq (2^d - d - |w| - k) = 2^{d-1} - (d-1) - 1$. Let $\mathcal{C}'' = Ker(\lambda')$. So $\mathcal{C}''$ projects isomorphically to a code in $\mathbb{F}_2^{2^{d-1}}$. Also $\mathcal{C}'' \leq \mathcal{C}$, so $\mathcal{C}''$ is even, with $\min wt(\mathcal{C}'') \geq 4$. We use induction to get $\dim(\mathcal{C}'') = 2^{d-1} - (d-1) - 1$. We have $2^d - d - 1 \leq \text{rank}(\mathcal{C}) \leq \text{rank}(\lambda) + nullity(\lambda) \leq |w| - 1 + (2^{d-1} - 1)$. It follows that $|w| \geq 2^{d-1}$. So $|w| = 2^{d-1}$, and $k = 0$. □

10.3.2 About the Golay code

Let $\mathcal{G} \leq \mathbb{F}_2^{24}$ be a Golay code, which has dimension 12 and minimum weight 8. As usual, we may identify a codeword with a subset of $\mathcal{P}(\Omega)$.

We shall study the weight distribution in $\mathcal{G}$ and the cocode $\mathbb{F}_2^{24}/\mathcal{G}$. We consider two distinct vectors therein, $u, v \in \mathbb{F}_2^{24}$ of weight less than 4 and ask when $u \equiv v \pmod{\mathcal{G}}$. Note that $wt(u - v) \leq wt(u) + wt(v)$. If $u \equiv v \pmod{\mathcal{G}}$ and $u \neq v$, then $wt(u + v) \geq 8$. Since u, v small weights, this means $wt(u) = wt(v) = 4$ and $supp(u) \cap supp(v) = \emptyset$.

Let W_n be the set of weight n vectors in $\mathbb{F}_2^{24}$. The above argument shows that $W := W_0 \cup W_2 \cup W_3 \cup W_4$ covers at least $\binom{24}{0} + \binom{24}{1} + \binom{24}{2} + \binom{24}{3} + \frac{1}{6}\binom{24}{4} = 4096 = 2^{12}$

cosets. So we have covered all cosets of $\mathcal{G}$ in $\mathbb{F}_2^{24}$. We summarize:

Theorem 10.3.4. *Let W_n be the set of weight n vectors in $\mathbb{F}_2^{24}$ and define $W := W_0 \cup W_2 \cup W_3 \cup W_4$. A coset in the Golay cocode contains a vector of W. If K is any coset and $K \cap W$ is not a 1-set, then it is a 6-set consisting of weight 4 vectors, the sum of any two of which is an octad. We have $\binom{24}{0} + \binom{24}{1} + \binom{24}{2} + \binom{24}{3} + \frac{1}{6}\binom{24}{4} = 4096 = 2^{12}$.*

Definition 10.3.5. *A Steiner system $\mathcal{S}(a, b, c)$ with parameters a, b, c is a c-set Ω, and a family $\mathcal{B}$ of b-element subsets, so that, when A is any a-set in Ω, there is a unique $B \in \mathcal{B}$ so that $A \subseteq B$.*

Proposition 10.3.6. *The set of weight 8 vectors in a Golay code gives a Steiner system $\mathcal{S}(5, 8, 24)$.*

Proof. Let $\Omega = \{1, 2, \ldots, 24\}$ be the index set. Let $F = \{a_1, \ldots, a_5\} \in \binom{\Omega}{5}$. We claim that there exists a unique $w \in \mathcal{G}$ with weight 8 such that $F \subseteq supp(w)$. Take $E \subseteq F$, $|E| = 4$. Let $e \in \mathbb{F}_2^{24}$, $supp(e) = E$. There are exactly $e_1 = e, e_2, \ldots, e_6$ weight 4 vectors so that $e_i \in e + \mathcal{G}$. If $i \neq j$, $e_i + e_j \in \mathcal{G}$ and $wt(e_i + e_j) = 8$. So it is clear that some $e_i + e_j$ has $F \subseteq supp(e_i + e_j)$. Also, $e_i + e_j \in \mathcal{G}$.

For the uniqueness, suppose $w, w' \in \mathcal{G}$ are subsets of weight 8 and $F \subseteq supp(w) \cap supp(w')$. Then, $wt(w + w') \leq 3 + 3 = 6$ and so $w + w' = 0$ since the minimal weight of $\mathcal{G}$ is 8. Hence $w = w'$. □

Definition 10.3.7. *A weight 8 codeword in a Golay code is called an octad.*

The name *octad* comes from the work of John Todd [76].

Example 10.3.8. *We mention a few more Steiner systems which come up naturally in geometry.*

(1) *Let V be a vector space of dimension at least 3 over $\mathbb{F}_q$, $\Omega =$ set of all 1-spaces, $\mathcal{B} =$ set of all 2-spaces give a Steiner system $\mathcal{S}(2, q+1, \frac{q^n-1}{q-1})$.*

(2) *For an index set $\Omega = \mathbb{F}_2^d$. Any 3 points determine a unique affine 2-space, hence gives a Steiner system $\mathcal{S}(3, 4, 2^d)$.*

10.3.3 The octad triangle and dodecads

We shall study configurations of octads in the 24-set.

Notation 10.3.9. *Suppose $\mathcal{O}$ is an octad in a Steiner system $\mathcal{S}(5, 8, 24)$, or an index set Ω. Let $A \subseteq B \subseteq \Omega$. Define $N_{B,A}$ to be the number of octads $\mathcal{O}'$ so that $\mathcal{O}' \cap B = A$. When $N_{B,A}$ depends only on $i = |B|$ and $j = |A|$, we let $N_{i,j} = N_{B,A}$.*

Proposition 10.3.10. (The Octad Triangle) *The values $N_{i,j}$ are found in the table below.*

$$
\begin{array}{ccccccccccccccccc}
 & & & & & & & & 759 & & & & & & & & \\
 & & & & & & & 506 & & 253 & & & & & & & \\
 & & & & & & 330 & & 176 & & 77 & & & & & & \\
 & & & & & 210 & & 120 & & 56 & & 21 & & & & & \\
 & & & & 130 & & 80 & & 40 & & 16 & & 5 & & & & \\
 & & & 78 & & 52 & & 78 & & 12 & & 4 & & 1 & & & \\
 & & 46 & & 32 & & 20 & & 8 & & 4 & & 0 & & 1 & & \\
 & 30 & & 16 & & 16 & & 4 & & 4 & & 0 & & 0 & & 1 & \\
30 & & 0 & & 16 & & 0 & & 4 & & 0 & & 0 & & 0 & & 1 \\
\end{array}
$$

The proof follows from (10.3.12) and (10.3.13).

Notation 10.3.11. *Suppose $S \subseteq \Omega$, we denote by $(-1)^S := (-1)^{|S|}$ and $N_S := N_{S,S}$ for abbreviations.*

Lemma 10.3.12. *If $A \subseteq B \subseteq \mathcal{O}$, an octad, and $|A| = j$, $|B| = i$, then $N_{B,A}$ depends only on i and j.*

Proof. Given an i-set, $A \subseteq \mathcal{O}$, $N_A = 1$ if $|A| \geq 5$. Suppose $i \leq 5$. Then $N_A = \binom{24-i}{5-i}/\binom{8-i}{5-i}$.

These numbers form the right edge of the octad triangle. □

Lemma 10.3.13. *Let $A \subseteq B \subseteq \Omega$, $y \in \Omega - B$. Let $B' := B \cup \{y\}$ and $A' := A \cup \{x\}$, where $x \in B' - A$. Then*

(1) *If $x = y$ or B' is contained in an octad, then $N_{B,A} = N_{B',A} + N_{B',A'}$;*

(2) $N_{B,A} = \sum_{S: A \subseteq S \subseteq B'} (-1)^{S-A} N_S$.

Proof. For (1). In case $x = y$, it is clear. For (2). First, let priming denote the union of A and B with a common point $z \in \Omega - B$. Then $|B - A| = |B' - A'|$. We have $N_{B',A} = \sum_{S: A \subseteq S \subseteq B'} (-1)^{S-A} N_S = \sum_{A \subseteq S \subseteq B} (-1)^{S-A} N_S + \sum_{A' \subseteq S \subseteq B'} (-1)^{S-A} N_S = N_{B,A} - N_{B',A'}$. If we assume the result for $|B - A| =$

$|B' - A'|$, the result follows for $|B' - A|$. Finally when $A \subseteq S \subseteq B \subseteq \mathcal{O}$, then $N_{B,A}$ depends only on $|B|$ and $|A|$, so formula (2) implies (1) for general x,y. □

Definition 10.3.14. *A dodecad is a 12-set in $\mathcal{G}$.*

Lemma 10.3.15. *The dodecads exist. In fact, there are 2576 dodecads which are sums $\mathcal{O}_1 + \mathcal{O}_2$, where $\mathcal{O}_1, \mathcal{O}_2$ are octads.*

Proof. Let $\mathcal{A}$ be the set of dodecads which may be written as $\mathcal{O}_1 + \mathcal{O}_2$, with $\mathcal{O}_1, \mathcal{O}_2$ octads. Let $\mathcal{B} = \{(\mathcal{O}_1, \mathcal{O}_2) \mid \mathcal{O}_1, \mathcal{O}_2 \text{ octads}, \mathcal{O}_1 \cap \mathcal{O}_2 \text{ 2-set }\}$. We have a map $f : \mathcal{B} \to \mathcal{A}$ defined by $(\mathcal{O}_1, \mathcal{O}_2) \mapsto \mathcal{O}_1 + \mathcal{O}_2$. Let $\mathcal{O}$ be a fixed octad, $\mathcal{B}_\mathcal{O} := \{(\mathcal{O}_1, \mathcal{O}_2) \in \mathcal{B} \mid \mathcal{O}_1 = \mathcal{O}\}$. Then $|\mathcal{B}| = 759 \cdot \binom{8}{2} \cdot 16 = 340032$. Take $D \in \mathcal{A}$, we may study its inverse image $f^{-1}(D)$. If $D = \mathcal{O}_1 + \mathcal{O}_2$. Then $D \cap \mathcal{O}_1$ is a 6-set which is contained in an octad. We let n be the number of octads which are contained in D. So $|f^{-1}(D)| = |\binom{12}{5} - \binom{8}{5}| \cdot n \cdot \frac{1}{6} \leq 132$. It follows that $|\mathcal{A}| \geq \frac{340032}{132} = 2576$. Note that $2^{12} - (1 + 759 + 759 + 1) = 2576$. So $n = 0$ and $|\mathcal{A}| = 2576$. □

Corollary 10.3.16. *If $\mathcal{G}$ is the span of the set of octads of $\mathcal{S}(5, 8, 24)$ in $\mathcal{P}(\Omega)$. Then $\mathcal{G} \subseteq \mathcal{G}^\perp$ and so $\dim(\mathcal{G}) \leq 12$.*

Proof. The octad triangle bottom row implies that the span is self-orthogonal, hence of dimension at most 12. The dimension is at least 12 by (10.3.15). □

Corollary 10.3.17. *The Golay code is doubly even with weight distribution*

$$0^1 8^{759} 12^{2576} 16^{759} 24^1.$$

Proof. We have shown that the distribution of a Golay code must be at least $0^1 8^{759} 12^{2576}$. The octad triangle indicates that the 5-dimensional space W of Golay sets disjoint from the octad $\mathcal{O}$ contains 15 pairs of disjoint octads, hence contains the 16-set $\Omega + \mathcal{O}$. Therefore Ω is in any Golay code. This forces the distribution to be $0^1 8^{759} 12^{2576} 16^{759} 24^1$ since a Golay code has cardinality $2^{12} = 4096$. □

Corollary 10.3.18. *We have a one-to-one correspondence between the set of Golay codes in $\mathbb{F}_2^{24}$ and the set of Steiner systems $\mathcal{S}(5, 8, 24)$ on the index set $\{1, 2, \ldots, 24\}$. The correspondence is as follows. Given a Golay code, the set of weight 8 vectors in it forms a Steiner system. Given a Steiner system, its linear span is a Golay code.*

Proof. (10.3.6),(10.3.16),(10.3.17). □

Remark 10.3.19. *In 1937, E. Witt constructed a Steiner system whose underlying 24-set is $\mathbb{P}^2(\mathbb{F}_4) \cup \{I, II, III\}$, where $\mathbb{P}^2(\mathbb{F}_4)$ is the projective plane over the field of 4 elements [82].*

In 1966, J. Todd constructed a Steiner system whose underlying set is $\mathbb{P}^1(\mathbb{F}_{23}) = \{\infty, 0, 1, \ldots, 21, 22\}$ and which has an isometry group $PSL(2, 23)$ [76].

Notation 10.3.20. *We shall need to study the linear transformations $\rho_S : \mathcal{P}(\Omega) \to \mathcal{P}(S)$ defined by $A \mapsto A \cap S$, where $S \subseteq \Omega$ is a subset. Define $\lambda_S = \rho_S|\mathcal{G}$ where $\mathcal{G}$ is the Golay code.*

10.3 Basic theory of the Golay code

Theorem 10.3.21. *Suppose $S \in \mathcal{G}$, then the map $\lambda_S : \mathcal{G} \to \mathcal{P}(S)$ satisfies*
  (1) *if $|S| = 8$, then $\mathrm{Im}(\lambda_S) = \mathcal{PE}(\lambda_S)$ and $Ker(\lambda_S)$ is an $RM(1,4)$ code in $\mathcal{P}(\Omega + S)$.*
  (2) *if $|S| = 12$, then $\mathrm{Im}(S) = \mathcal{PE}(S)$ and $Ker(\lambda_S) = \{0, S\}$.*
  (3) *if $|S| = 16$, then $Ker(\lambda_S) = \{0, S + \Omega\}$ and $\mathrm{Im}(\lambda_S)$ has dimension 11 and is an $RM(2,4)$ code in $\mathcal{P}(S)$.*

Proof. Let $A \in Ker(\lambda_S) \supseteq \{0, S+\Omega\}$. Suppose $A \neq 0, S+\Omega$. Then $0 < |A| < |S|$. Since $\mathcal{G}$ has minimal weight 8, we have a contradiction if $|S| = 16$. If $|S| = 12$, $|\Omega + S| = 12$ and then either A or $\Omega + S + A$ has cardinality $\leqslant 6$. Now assume $|S| = 8$, $A \subseteq \Omega + S$, $0 < |A| < 16$. Also, note $|A| \geqslant 8$ and $|\Omega + S + A| \geqslant 8$. Since A and $\Omega+S+A$ partition the 16-set $S+A$, we get $|A| = 8$. Note that $\mathrm{Im}(\lambda_S) \subseteq \mathcal{PE}(S)$ has dimension 7. So $Ker(\lambda_S)$ has dimension $\geqslant 5$. Given an octad, S, there are 30 octads are disjoint from it and no 12-sets (10.3.17). This proves that $Ker(\lambda_S)$ has dimension just 5.

To identify these spaces as codes, we note that intersections of Golay sets are even and use (10.3.2),(10.3.3). □

Corollary 10.3.22. *The radical of the 11-dimensional quadratic space $\mathrm{Im}(\lambda_S)$ is the 5-dimensional space W. The quotient $\mathrm{Im}(\lambda_S)/W$ has maximal Witt index.*

Proof. Since the quadratic space $\mathcal{B}(\Omega)$ is nonsingular and $\mathrm{Im}(\lambda_S)$ has codimension 5, the radical of $\mathrm{Im}(\lambda_S)$ has dimension at most 5. Since W annihilates $\mathrm{Im}(\lambda_S)$, the radical equals W. □

Theorem 10.3.23. *There is a bijection between the set of $\mathcal{A}$ of Steiner systems with parameters $(5, 8, 24)$, with underlying set Ω and $\mathcal{B}$ the set of Golay codes in $\mathcal{P}(\Omega)$.*

Proof. We showed that there is a function $\mathcal{B} \to \mathcal{A}$. To get the inverse one $\mathcal{A} \to \mathcal{B}$, we take $S \in \mathcal{A}$. As in preceding results, we show that the span of S has weight distribution $0^1 8^{759} 12^{2576} 16^{759} 24^1$ and so is a Golay code. □

Definition 10.3.24. *A trio of octads is a partition $\Omega = \mathcal{O}_1 \cup \mathcal{O}_2 \cup \mathcal{O}_3$, where $\mathcal{O}_1, \mathcal{O}_2, \mathcal{O}_3$ are octads.*

Theorem 10.3.25. *Trios exist. If $\mathcal{O}$ is an octad, it is contained in 15 trios.*

Proof. This follows from (10.3.21)(1). □

Remark 10.3.26. *The group $PGL(2, F)$ has a left action on $\mathbb{P}^1(F)$ by the formula $\begin{pmatrix} a & b \\ c & d \end{pmatrix} : z \mapsto \frac{az+b}{cz+d}$.*

Remark 10.3.27. (The Trio Labeling) *There is a labeling of the 24 points as below so that the natural linear fractional group preserves the members of the trio, and a commuting group isomorphic to Sym_3 permutes the three blocks faithfully such that the labeling is respected. For a proof, see [37], which contains details about many useful labelings of the 24-set which exhibit actions of various finite groups.*

Table 10.2. Trio Labeling

$\mathcal{O}_1$		$\mathcal{O}_2$		$\mathcal{O}_3$	
∞	0	∞	0	∞	0
3	2	3	2	3	2
5	1	5	1	5	1
6	4	6	4	6	4

10.3.4 A uniqueness theorem for the Golay code

Uniqueness proofs were given in [25, 34, 69]. Here, we give a uniqueness proof which may be new.

Notation 10.3.28. *We let $\mathcal{G}$ be a Golay code. In it, let $\mathcal{O}$ be an octad and let W be the subspace of Golay sets which are disjoint from $\mathcal{O}$. By (10.3.21)(1), $\dim(W) = 5$ and W has weight distribution $0^1 8^{30} 16^1$. Let S be the 16-set which complements $\mathcal{O}$.*

Lemma 10.3.29. (1) *A Golay code in $\mathcal{B}(\Omega)$ which contains $span\{\mathcal{O}\} + W$ is given by a gluing of the quadratic spaces $\mathcal{PE}(\mathcal{O})/span\{\mathcal{O}\}$ with $\mathrm{Im}(\lambda_S)/W$. Such a gluing is an isometry.*

(2) *We have an action of a natural subgroup $Sym_{\mathcal{O}} \times AGL(4,2)$ of Sym_{Ω} acting on $\mathbb{F}_2^{\Omega}$ preserving $span\{\mathcal{O}\} + W$ and $\mathcal{B}(\mathcal{O}) \oplus \mathrm{Im}(\lambda_S)$. This action is transitive on the labelings as in (1). A stabilizer S is a group isomorphic to $AGL(4,2)$ which meets the direct factor isomorphic to $Sym_{\mathcal{O}}$ trivially and which meets the second factor in $O_2(S) \cong 2^4$.*

Proof. (1) Let C be such a code. Then $C \cap span\mathcal{B}(\Omega) = span\mathcal{B}(\Omega)$ because the minimum weight in C is 8. Also, $C \cap \mathrm{Im}(\lambda_S) = W$ (10.3.21). It follows that the doubly even glue vectors for the containment of $span\{\mathcal{O}\} + W$ in C represent an isometry of quadratic spaces.

(2) This follows from the the action of the first factor of $Sym_{\mathcal{O}} \times AGL(4,2)$ on $\mathcal{PE}(\mathcal{O})/span\{\mathcal{O}\}$ as $Sym_8 \cong \Omega^+(6,2)$. □

Theorem 10.3.30. *Any two Golay codes are equivalent. They are doubly even and have weight distribution $0^1 8^{759} 12^{2576} 16^{759} 24^1$.*

Proof. Uniqueness follows from (10.3.29)(2). Since the Turyn-style construction of a Golay code (10.2.5) creates a doubly even code, all Golay codes are doubly even. The weight distribution result follows from (10.3.17). □

10.4 Minimal vectors in the Leech lattice

Recall our notation that $MV(J)$ denotes the set of minimal vectors of the positive definite lattice J (9.1.2).

Theorem 10.4.1. *(10.4.1) $MV(\Lambda)$ consists of the $2^2 \binom{24}{2}$ vectors $\pm \alpha_i \pm \alpha_j$ (for $i \neq j$ in Ω), the $759 \cdot 2^7$ vectors of the form $\frac{1}{2} \sum_{i \in \mathcal{O}} \varepsilon_S \alpha_i$ (for an octad $\mathcal{O}$ and a*

Golay set S) and the $2^{12} \cdot 24$ vectors of the form $\varepsilon_S \nu_i$ (for $i \in \Omega$ and a Golay set S). We have $\binom{24}{2} \cdot 2^2 + 759 \cdot 2^7 + 24 \cdot 2^{12} = 196560$ minimal vectors in all.

Proof. Let $v \in \Lambda$, $(v,v) = 4$. Suppose $v \in M = \mathcal{N}(A_1^{24}) \cap \theta^{-1}(\mathbb{Z})$. If $v \in A_1^{24}$, $v = \sum_{i \in \Omega} c_i \alpha_i$, with $\sum c_i$ even. Note that $(v,v) = 4$ implies just two c_i are nonzero and if such c_i are nonzero, they equal ± 1.

If A is a nonzero Golay set and $v \in \frac{1}{2}\alpha_A + A_1^{24}$, $v = \sum c_i \alpha_i$, $c_i \in \frac{1}{2} + \mathbb{Z}$ for $i \in A$. Then $(v,v) \geq \frac{1}{4} \cdot 2 \cdot |A|$. Hence $|A| = 8$. So A is octad and $v = \frac{1}{2} \sum_{i \in A} c_i \alpha_i$, $c_i = \pm 1$. We have 2^7 minimal vectors of the form $\frac{1}{2}\varepsilon_S(\alpha_A)$ (for $S \in \mathcal{G}$ because the intersection of two Golay sets is even. Since the number of octads is 759, the number of such v is $759 \cdot 2^7$. If $S \notin \mathcal{G}$, there exists $A \in \mathcal{G}$ so that $|S \cap A|$ is odd, which implies that $(\varepsilon_S - 1)\frac{1}{2}v_A = -v_{S \cap A}$, which is not in L. If follows that only the $\frac{1}{2}\varepsilon_S(\alpha_A)$ for $S \in \mathcal{G}$ are in L.

Now assume $v \notin M$. So $v \in \nu + M$ and $v = \sum_{i \in \Omega} c_i \alpha_i$ and $c_i \in \frac{1}{4} + \frac{1}{2}\mathbb{Z}$ for all i. Then $(v,v) \geq (\frac{1}{4})^2 \cdot 2 \cdot 24 = 3$, and so $\pm \frac{1}{4}$ occurs at 23 places and $\pm \frac{3}{4}$ occurs once. There are 24 places for $\pm \frac{3}{4}$. Since all coefficients of $\nu_i = \frac{1}{4}\alpha_\Omega - \alpha_i$ are nonzero, its orbit under $\{\varepsilon_S \mid S \in \mathcal{G}\}$ has length 2^{12}. So we have $24 \cdot 2^{12}$ vectors in L, .

We now show that this list is complete. So let $\varepsilon_S \nu_i$ be a norm 4 vector in Λ, we wish to show $S \in \mathcal{G}$. For $A \in \mathcal{G}$, $\varepsilon_A \nu_i \in \Lambda$. So $(\varepsilon_A \nu_i, \varepsilon_S \nu_i) \in \mathbb{Z}$ for all $A \in \mathcal{G}$. So $(\varepsilon_{A+S} \nu_i, \nu_i) \in \mathbb{Z}$. We calculate $(\varepsilon_{A+S} \nu_i, \nu_i) = (\varepsilon_{A+S}(\frac{1}{4}\alpha_\Omega - \alpha_i), \frac{1}{4}\alpha_\Omega - \alpha_i) = \frac{1}{16}(\varepsilon_{A+S}\alpha_\Omega, \alpha_\Omega) - \frac{1}{4}(\varepsilon_{A+S}\alpha_\Omega, \alpha_i) - \frac{1}{4}(\varepsilon_{A+S}\alpha_i, \alpha_\Omega) + (\varepsilon_{A+S}\alpha_i, \alpha_i)$. The latter three terms are integers $1, 1, \pm 2$, and for the first one we have $\frac{1}{16}(\varepsilon_{A+S}\alpha_\Omega, \alpha_\Omega) = \frac{1}{16}(\alpha_\Omega - 2\alpha_{A+S}, \alpha_\Omega) = \frac{1}{16}(\alpha_\Omega, \alpha_\Omega) - \frac{1}{8}(\alpha_{A+S}, \alpha_\Omega)$. Note that $\frac{1}{8}|A+S| \cdot 2 \in \mathbb{Z}$ and $(\alpha_\Omega, \alpha_\Omega) = 48$. This implies that $|A+S| \in 4\mathbb{Z}$ for all $A \in \mathcal{G}$ and so $S \in \mathcal{G}^\perp = \mathcal{G}$. □

Corollary 10.4.2. Let $S \subseteq \Omega$. Then ε_S is in $O(\Lambda)$ if and only if $S \in \mathcal{G}$.

10.5 First proof of uniqueness of the Leech lattice

When L is a rank 24 Niemeier lattice, the function θ_L lies in a 2-dimensional space of modular forms. The constant term must be 1, so θ_L is determined by the number of roots. Examination of the relevant space of modular forms proves the following:

Theorem 10.5.1. If L is a Leech lattice, θ_L begins $1 + 196560q^2 + 16773120q^3 + 398034000q^4 + \cdots$.

For the rest of this section, L is a Leech lattice, i.e., some rootless Niemeier lattice. The following result is an analogue of a result for the E_8-lattice (4.2.33).

Theorem 10.5.2. A coset of $2L$ in L contains an element of $A = L_0 \cup L_2 \cup L_3 \cup L_4$. Also, if $x, y \in A$ so that $x \equiv y \bmod 2L$, then $x = \pm y$ or $x, y \in L_4$ and $(x,y) = 0$. Also we have $u_0 + \frac{1}{2}u_2 + \frac{1}{2}u_3 + \frac{1}{48}u_4 = 2^{24}$.

Proof. Define $L_n := \{x \in L \mid (x,x) = 2n\}$; $A := L_0 \cup L_2 \cup L_3 \cup L_4$. Take $x, y \in A$, assume $x + 2L = y + 2L$. Assume above, but $x \neq \pm y$, then $x - y \neq 0$, $x + y \neq 0$. So $(x-y, x-y) \geq 16$, which implies that $(x,y) \leq 0$, since $(x,x) \leq 8$ and $(y,y) \leq 8$. Similar argument with $(x+y, x+y) \geq 16$ shows that $(x,y) \geq 0$, hence $(x,y) = 0$, and $x, y \in L_4$. Therefore, the number of cosets of $2L$ represented by A is at least than $u_0 + \frac{1}{2}u_2 + \frac{1}{2}u_3 + \frac{1}{48}u_4$, which coincidently is 2^{24}. □

Lemma 10.5.3. *Let K be an integral lattice containing roots r and s so that (r,s) is odd. Suppose J is a sublattice of index 2 in K. Then J contains a root of $\mathbb{Z}r + \mathbb{Z}s \cong A_2$.*

Proof. We may assume $(r,s) = -1$. Set $t = -r - s$. It is a root and $t + r + s = 0$. Let $\phi : K \to K/J$ be the quotient map. Then $0 = \phi(0) = \phi(r) + \phi(s) + \phi(t)$. One of these summands is zero. □

We give the first uniqueness proof. It is a slightly modified form of the original uniqueness proof of Conway [20].

Theorem 10.5.4. *Any two Leech lattices are isometric.*

Proof. Let L be a Leech lattice. The theta series is determined, so there is a frame of norm 8 vectors, say F ($|F| = 48$; $x, y \in F$ implies that $x - y \in 2L$). Take $v \in F$, let $U := \{x \in L \mid (v, x) \in 2\mathbb{Z}\}$. (Note that v is primitive, and $\det(L) = 1$, so the map $L \to \mathcal{D}\mathbb{Z}v$ is onto.) Define the neighbor $N := U + \sum_{u \in F} \mathbb{Z}\frac{1}{2}u = U + \mathbb{Z}\frac{1}{2}v$. Then the root system in N is nonempty since it contains $\frac{1}{2}F$.

Suppose there is a root $r \in N - \frac{1}{2}F$. Then there is $y \in \frac{1}{2}F$ so that $(r, y) \neq 0$. Since r, y are not proportional, $(r, y) = \pm 1$. So the lemma implies that U contains a root since $U \subseteq L$. This is a contradiction. Therefore, $\frac{1}{2}F$ is the entire set of roots in N.

Define $X := \{x \in N \mid (x, y) = \pm 1 \text{ for all } y \in \frac{1}{2}F\}$. Since N is unimodular, $|X| = 2^{24}$. If $x \in L$, $(x, x) = \frac{24}{2} = 12$. We let S be the $\mathbb{Z}$-span of $\frac{1}{2}F$. So $S \leqslant N$, $S^* = \frac{1}{2}S$, $S \leqslant N \leqslant S^*$. So N corresponds to a binary code of dimension 12, with minimum weight at least 8. (Since N is even and all roots of N are in S.) The uniqueness of the Golay code (10.3.30) means N is unique up to isometry. We now consider finding L, when given N.

Take $x \in X$. Define $Q_x := \{v \in N \mid (v, x) \in 2\mathbb{Z}\}$. We claim that Q_x is a sublattice of index 2 in L. This follows because $\mathbb{Z}v$ is a direct summand and $\det(L) = 1$. Observe that X is an orbit under the group $O(N)$, even under the diagonal group with respect to the frame $\frac{1}{2}F$. Take $y \in \frac{1}{2}F$, define $N_{x,y} := Q_x + \mathbb{Z}(\frac{1}{2}x + y)$. Note that $x = (\pm\frac{1}{2}, \ldots, \pm\frac{1}{2})$, $y = (0, \ldots, 0, \pm 1, 0, \ldots, 0)$, which implies that $x \pm 2y$ has form $(\pm\frac{1}{2}^{23}, \frac{3}{2})$. Trivially, if $y' \in \frac{1}{2}F$, then $N_{x,y} = N_{x,y'}$. This completes the proof. □

A second uniqueness proof will be given in (11.1.9).

10.6 Initial results about the Leech lattice

We list here some basic results about the Leech lattice Λ and its isometry group $O(\Lambda)$.

10.6.1 An automorphism which moves the standard frame

We now resume our study of $\mathbb{F}_2^{24}/\mathcal{G}$, the cocode for $\mathcal{G}$. A coset has a codeword of cardinality no more than 4 (10.3.4). If $x, y \in \mathbb{F}_2^{24}$, $wt(x), wt(y) \leqslant 4$ so that $x + y \in \mathcal{G}$, then $x = y$ or x and y have weight 4 and $supp(x) \cap supp(y) = \emptyset$.

Let $T_1, \ldots, T_6$ be a sextet (partition of the index set Ω into 4-sets, so that $T_i + T_j \in \mathcal{G}$ for all i, j.) Let

$$B := \frac{1}{2} \begin{pmatrix} -1 & 1 & 1 & 1 \\ 1 & -1 & 1 & 1 \\ 1 & 1 & -1 & 1 \\ 1 & 1 & 1 & -1 \end{pmatrix}.$$

Define $\xi \in O(\mathbb{R}^{24})$ by

$$\xi = \begin{pmatrix} -B & & & & & \\ & B & & & & \\ & & B & & & \\ & & & B & & \\ & & & & B & \\ & & & & & B \end{pmatrix},$$

indexed by Ω, blocked by a sextet.

Theorem 10.6.1. $\xi \in O(L)$, where L is the standard Leech lattice.

Proof. It is sufficient to check the map ξ act on each member of the standard generating set $X := \{\alpha_i, \frac{1}{2}\alpha_A \text{ for } A \in \mathcal{G}, \nu_i = -\alpha_i + \frac{1}{4}\alpha_\Omega\}$. We verify that $\xi(x) \in \Lambda$, or equivalently, that $(\xi(x), y) \in \mathbb{Z}$, for y in a generating set for Λ.

This is a straightforward task which we omit. See [37]. $\square$

10.7 Turyn-style construction of a Leech lattice

Recall the Turyn-style construction of a Golay code (10.2.5). If Γ is an 8-set, we took H, K $[8, 4, 4]_2$-codes in $\mathcal{P}(\Gamma)$, such that $H \cap K = span(\Gamma)$. The Golay code is $\mathcal{G} := \{(w+x, w+y, w+z) \in \mathcal{P}(\Gamma) \oplus \mathcal{P}(\Gamma) \oplus \mathcal{P}(\Gamma) \mid w \in K; x, y, z \in H, x+y+z = 0\}$. We shall make an analogous construction for a Leech lattice, following the ideas of Tits, Lepowsky and Meurman [63, 75].

Notation 10.7.1. We let $L \cong E_8$ lattice, $M \leq L$ and $M \cong \sqrt{2}E_8$.

Lemma 10.7.2. Let $L \cong E_8$. If $2L \leq M \leq L$ and $M/2L$ is a totally singular 4-dimensional subspace, then $M \cong \sqrt{2}E_8$. Furthermore, there exist 135 such sublattices.

Proof. The hypotheses imply that $(M, M) \leq 2\mathbb{Z}$ and that every vector in M has norm in $4\mathbb{Z}$, so the first statement is clear. For the second, this follows since $L/2L$ has maximal Witt index.

Another proof is to use $f \in O(L)$ such that $f^2 = -1$. Then the map $f - 1 : L \to L$ doubles the norms, i.e. $((f-1)x, (f-1)y) = 2(x, y)$. We can take $M = (f-1)L$. To prove the existence of such f, take $L \supset K \cong D_8 = \{(x_i) \in \mathbb{Z}^8 \mid \sum x_i \in 2\mathbb{Z}\}$. Let

$$f = \begin{pmatrix} B & & & \\ & B & & \\ & & B & \\ & & & B \end{pmatrix},$$

where $B = \begin{pmatrix} 0 & 1 \\ -1 & 0 \end{pmatrix}$. We claim that f is a product of reflections at roots. Take

$$p = \begin{pmatrix} 0 & 1 & & & & & & \\ 1 & 0 & & & & & & \\ & & 0 & 1 & & & & \\ & & 1 & 0 & & & & \\ & & & & 0 & 1 & & \\ & & & & 1 & 0 & & \\ & & & & & & 0 & 1 \\ & & & & & & 1 & 0 \end{pmatrix}$$

and a permutation

$$d = \begin{pmatrix} -1 & & & & & & & \\ & 1 & & & & & & \\ & & -1 & & & & & \\ & & & 1 & & & & \\ & & & & -1 & & & \\ & & & & & 1 & & \\ & & & & & & -1 & \\ & & & & & & & 1 \end{pmatrix}.$$

Then the product $f = dp$ is in the standard reflection group on K, so is in $O(L)$. The count of 135 such subspaces is left as an exercise. $\square$

Lemma 10.7.3. *Suppose $L = E_8$, and J is a sublattice. Then $J \cong \sqrt{2}E_8$ if and only if $J \geqslant 2L$, and $J/2L$ is a totally singular 4-dimensional subspace of $L/2L$.*

Proof. The if statement was proved in (10.7.2), we now prove the converse. So, suppose $J \cong \sqrt{2}E_8$. Then $J^* = \frac{1}{2}J$, so J^*/J is an elementary abelian 2-group. Since $J \leqslant L \leqslant J^*$, L/J is elementary abelian. So $J \geqslant 2L$. By determinant considerations, $|L : J| = 2^4$. Since J is a doubly even lattice, $J/2L$ is a totally singular subspace of $L/2L$. $\square$

Lemma 10.7.4. *Let V be a finite dimensional vector space over a field F, with a nonsingular quadratic form, Q. Suppose W is a totally singular subspace with dimension d. There exists a totally singular subspace W' of dimension d so that $W + W'$ is nonsingular and has dimension $2d$.*

Proof. It suffices to prove for case $d = 1$, because if $W_1 \leqslant W$, $\dim(W_1) = 1$ and there is a totally singular W_1' of dimension 1 so that $W_1 + W_1'$ is nonsingular, then we may work in $(W_1 + W_1')^\perp$ and use induction on the subspace $W \cap (W_1 + W_1')^\perp$ of dimension $d - 1$.

We have for any $0 \neq x \in W$, $Qx = 0$. Take any $y \in V$ so that $b(x, y) = 1$, where $b(x, y) = Q(x + y) - Qx - Qy$. Then $U = \text{span}\{x, y\}$ is nonsingular, (in fact, it has Gram matrix $\begin{pmatrix} 0 & 1 \\ 1 & a \end{pmatrix}$, where $a = Qy$). Consider a vector $z = cx + y$, calculate $Qz = b(cx, y) + Q(cx) + Qy = cb(x, y) + c^2 \cdot 0 + a = c + a$. Take $c = -a$. Then x, z is a basis of U, and x, z are singular. $\square$

Corollary 10.7.5. *If* $L \cong E_8, J \leqslant L, J \cong \sqrt{2}E_8$, *then there is* $K \leqslant L, K \cong \sqrt{2}E_8$ *and* $J \cap K = 2L$.

Notation 10.7.6. *We take three orthogonal copies of L and isometries* $\theta_i : L \to L_i, i = 1, 2, 3$. *Define* $\varepsilon : L_1 \perp L_2 \perp L_3 \to L \supseteq M$ *by* $\varepsilon(\theta_1 x_1, \theta_2 x_2, \theta_3 x_3) = x_1 + x_2 + x_3$. *Then* $\varepsilon^{-1}(M)$ *is a sublattice of L of index* 2^4, *and it has no roots. We have* $M \leqslant L \leqslant M^*$. *In the dual M^*, we take K so that $K \cap L = M$ and $K \cong L$ (use (10.7.4) and $M^* \cong \frac{1}{2}M$). Define* $K_{123} := \{\theta_1 x + \theta_2 x + \theta_3 x \mid x \in K\}$, *where* $\theta_i : \mathbb{Q} \otimes L \to \mathbb{Q} \otimes L_i$ *are extensions to the rational vector spaces* $\mathbb{Q} \otimes L_i$. *We set* $T := \varepsilon^{-1}(M) + K_{123}$.

We shall prove that T is a Leech lattice. From now on, we may identify L_i with L by θ_i.

Theorem 10.7.7. *The lattice T is a Leech lattice.*

Proof. Observe that $T \geqslant T_1 \oplus T_2$, where $T_1 = \{(x+w, y+w, z+w) \mid x, y, z \in L, x+y+z = 0\}$ and $T_2 = \{(w, w, w) \mid w \in K\}$. The projection $T_1 \to L \oplus L$, $(x, y, z) \mapsto (x, y)$ is onto, so $\mathrm{rank}(T_1) \geqslant 16$, and $T_2 \cong K$, so $\mathrm{rank}(T_2) = 8$. Also, $T_1 \cap T_2 = \{(w, w, w) \mid w \in K\} = 0$ since $3w = 0$ implies $w = 0$. Therefore $T_1 + T_2$ has rank 24, so $\mathrm{rank}(T) = 24$.

Note that $T_2 \cong \sqrt{3}K$ and $S_1 := \{(x, y, z) \in J \mid x+y+z \in L_1)\} \leqslant J \perp J \perp J$ is even integral. We calculate that $(x, y, z) \cdot (w, w, w) = (x, w) + (y, w) + (z, w)$. Write $x+y+z = v \in M$. So $(w, x+y+z) \in (K, M) \leqslant (K, K) = \mathbb{Z}$. It follows that T is even and integral.

Next, we show the lattice T is unimodular. We use ε as in (10.7.6). We have $\varepsilon(T) = 3K$. Also, $T \geqslant M \perp M \perp M$. Now we study $L/M \perp L/M \perp L/M$. The image $\overline{T}_1$ of T_1 here is of dimension 8+8=16. The image $\overline{T}_2$ of T_2 here is of dimension 8. It is easy to prove that $\overline{T}_1 \cap \overline{T}_2 = 0$. So $|L \perp L \perp L : T| = 2^{12}$, and $\det(L) = 2^{-8}$. It follows that $\det(L \perp L \perp L) = 2^{-24}$, which implies $\det(T) = 1$.

Finally we prove $\mu(T) > 2$. Since $\mu(L) = 2$, if r is a root in T, then at least one component of $r = (x+w, y+w, z+w)$ is 0. Say $x+w = 0$, so $x = -w \in L \cap K = M$, which implies $y+w \in M$ and $z+w \in M$. So $r \in 0 \perp J \perp J$. Since $(r, r) = 2$, one more component is 0, say $y+w = 0$. So $y = -w$ and $z+w$ is a root. We have $x+y+z = -2w+z \in M$. Since $w \in M, z \in M, z+w \in M$ has norm at least 4 or is 0, a contradiction. $\square$

10.8 Equivariant unimodularizations of even lattices

This section is essentially an appendix item in [42].

Sometimes it is convenient to have an integral lattice embedded in an integral lattice whose determinant is ± 1 or at least avoids certain primes. Furthermore, it can be useful to do this in a way which respects automorphisms and signature. Our main results are (10.8.5), (10.8.6). Applications to VOA theory are in [42].

First, we recall some basic facts concerning extensions of lattices (cf. [67]). An even lattice L defines a *quadratic space* (A, q), where $A = L^*/L$, $L^* = \{x \in L \otimes \mathbb{Q} \mid (x, y) \in \mathbb{Z}$ for all $y \in L\}$ the dual lattice, and $q : L^*/L \longrightarrow \mathbb{Q}/2\mathbb{Z}$ is the quadratic form $x \pmod{L} \mapsto (x,x) \pmod{2\mathbb{Z}}$. Even overlattices M of L define isotropic subspaces $C = M/L$ of (A, q) and this correspondence is one to one. An automorphism g of L extends to an automorphism of M if and only if the induced automorphism $\bar{g} \in O(A, q)$ fixes the subspace C. A subgroup C of A generated by a set of elements is isotropic if the generating elements are isotropic and orthogonal to each other with respect to the $\mathbb{Q}/\mathbb{Z}$-valued bilinear form obtained by taking the values of $b(x, y) = \frac{1}{2}(q(x + y) - q(x) - q(y))$. The determinant $\det(L)$ of L is the order of A. If A has exponent N, then q takes values in $\frac{1}{2N}(2\mathbb{Z})/2\mathbb{Z}$. There is an orthogonal decomposition $(A, q) = \bigoplus_{p \mid \det(L)} (A_p, q_p)$ of quadratic spaces, where A_p is the Sylow p-subgroup of A. A sublattice L of a lattice M is called *primitive* if M/L is free. Let $K = L_M^\perp$ the orthogonal complement of L in M. Then, L is primitive exactly if the projection of $M/(L \oplus K)$ to K^*/K is injective.

Definition 10.8.1. Let M be a lattice and L a sublattice. We say that an automorphism α of L *extends (weakly)* to M if there is $\beta \in O(M)$ so that $\beta|_L = \alpha$. We say that a subgroup $S \leq O(L)$ *extends (weakly)* to M if every element extends and we say that it *extends strongly* if there is a subgroup $R \leq O(M)$ which leaves L invariant and the restriction of R to L gives an isomorphism of R onto S. In this case, call such R a *strong extension of S to M*.

Definition 10.8.2. Let L be an even lattice. An *equivariant unimodularization* of L is an unimodular lattice M containing L as a primitive sublattice such that $O(L)$ extends strongly to M.

Theorem 10.8.3. [54] *An equivariant unimodularization M of an even lattice L exists of rank at most $2 \cdot \mathrm{rank}(L) + 2$.*

One can take for M the orthogonal sum of $\mathrm{rank}(L) + 1$ hyperbolic planes. A somewhat stronger result can be found in [67] (see Prop. 1.14.1 and Th. 1.14.2).

The above unimodularizations from [54, 67] are all indefinite. The next theorem shows that one can get equivariant definite unimodularizations of a definite lattice. For its proof we need:

Lemma 10.8.4. (1) *Let p be an odd prime and $r \geq 0$. Then, -1 is the sum of two squares in $\mathbb{Z}/p^r\mathbb{Z}$.* (2) *For all $r \geq 0$, one can write -1 is a sum of four squares in $\mathbb{Z}/2^r\mathbb{Z}$.*

Proof. (1) When $r = 1$, we quote the well known fact that every element in a finite field is a sum of two squares. Part (1) is now proved by induction on r: Assume that $r \geq 1$ and that a, b are integers such that $a^2 + b^2 = -1 + p^r m$, for some integer m. Let x, y be integers and consider $(a + p^r x)^2 + (b + p^r y)^2 = -1 + p^r m + 2p^r[ax + by] + p^{2r}e$, for some integer e. Since not both a and b can be divisible by p we can solve $2[ax + by] \equiv -m \pmod{p}$ for integers x, y. Thus, -1 is a sum of two squares modulo p^{r+1}.

Part (2) follows from a similar argument, or from Lagrange's theorem that every nonnegative integer is a sum of four integer squares. □

Theorem 10.8.5. *Let L be an even lattice of signature (n_1, n_2). Then there exists an equivariant unimodularization of the lattice L whose rank is $8 \cdot \mathrm{rank}(L)$ and signature is $(8n_1, 8n_2)$. If $\det(L)$ is odd, there is one whose rank is $4 \cdot \mathrm{rank}(L)$ and* signature is $(4n_1, 4n_2)$. In particular, if the lattice is definite, this unimodularization is also definite.

Proof. Assume first that $\det(L)$ is odd and let (A, q) be the finite quadratic space associated L. Let $K = L \perp L \perp L \perp L$ having the associated quadratic space $(B, q') = (A, q) \oplus (A, q) \oplus (A, q) \oplus (A, q)$. We decompose (A, q) as the orthogonal sum
$$(A, q) = \bigoplus_{p \mid \det(L)} (A_p, q_p),$$
where $A_p \cong \mathbb{Z}/p^{a_{p,1}}\mathbb{Z} + \mathbb{Z}/p^{a_{p,2}}\mathbb{Z} + \cdots + \mathbb{Z}/p^{a_{p,n_p}}\mathbb{Z}$ is an abelian p-group with $a_{p,1} \geq a_{p,2} \geq \cdots \geq a_{p,n_p}$ of order p^{a_p}, where $a_p := a_{p,1} + a_{p,2} + \cdots + a_{p,n_p}$.

Fix a prime $p \mid \det(L)$. Using Lemma 10.8.4, let $r, s \in \mathbb{Z}$ so that $r^2 + s^2 \equiv -1 \pmod{p^{a_{p,1}}}$. We let
$$D_p = \{(rx, sx, 0, x) \mid x \in A_p\} \quad \text{and} \quad E_p = \{(sx, -rx, x, 0) \mid x \in A_p\}.$$
Since $q_p(\pm rx) + q_p(\pm sx) + q_p(\pm x) = p^{a_{p,1}} q_p(x) \in 2\mathbb{Z}/2\mathbb{Z}$, the groups D_p and E_p are isotropic subspaces of $(A_p, q_p) \oplus (A_p, q_p) \oplus (A_p, q_p) \oplus (A_p, q_p)$. They are orthogonal to each other, so that $C_p = D_p + E_p$ is also isotropic and has order p^{2a_p}.

Finally, let $C = \bigoplus_{p \mid \det(L)} C_p$. It is an isotropic subspace of (B, q') with $|A|^2$ elements and it is invariant under the diagonal action of $O(L)$ induced on (B, q'). Since $|C|^2 = |B|$, the overlattice M of K belonging to $C = M/K \leq B$ is a definite even unimodular lattice having an automorphism group which contains a strong extension of $O(L)$ and L is also primitive.

Now, we do the case of even $\det(L)$. This time we take $K = L \perp \cdots \perp L$ (8 times) with associated quadratic space $(B, q') = (A, q) \oplus \cdots \oplus (A, q)$ (8 times). We proceed in a similar spirit:

For $p = 2$, let r, s, t, u be integers such that $r^2 + s^2 + t^2 + u^2 \equiv -1 \pmod{2^{a_{2,1}+1}}$ and define
$$D_2 = \{(rx,sx,tx,ux,x,0,0,0) \mid x \in A_2\},$$
$$E_2 = \{(sx,-rx,ux,-tx,0,x,0,0) \mid x \in A_2\},$$
$$F_2 = \{(-x,0,0,0,rx,sx,tx,ux) \mid x \in A_2\} \text{ and}$$
$$G_2 = \{(0,-x,0,0,sx,-rx,ux,-tx) \mid x \in A_2\}.$$

Since $q_2(\pm rx) + q_2(\pm sx) + q_2(\pm tx) + q_2(\pm ux) + q_2(\pm x) = 2^{a_{2,1}+1} q_2(x) = 2\mathbb{Z}/2\mathbb{Z} \in \mathbb{Q}/2\mathbb{Z}$, the groups D_2, E_2, F_2 and G_2 are totally isotropic subspaces of $(A_2, q_2) \oplus \cdots \oplus (A_2, q_2)$. They are pairwise orthogonal, so that $C_2 := D_2 + E_2 + F_2 + G_2$ is also isotropic and has order 2^{4a_2}.

For the odd primes, we let $C_p = (D_p + E_p) \oplus (D_p + E_p)$. As in the preceding cases, we see that the overlattice M of K belonging to $C = \bigoplus_{p \mid \det(L)} C_p$ has all the required properties. $\square$

If we try only to double the rank of L, we may not find an unimodularization in general, but we can achieve the following:

Theorem 10.8.6. *Let L be an even lattice with signature (n_1, n_2). Then there exists an even lattice M of signature $(2n_2, 2n_2)$ containing L as a primitive sublattice such that $O(L)$ can be strongly extended to a subgroup of $O(M)$ and $\det(M)$ is a power of an arbitrarily large prime.*

Proof. The Dirichlet Theorem implies that there are infinitely many primes s satisfying $s \equiv -1 \pmod{2 \det(L)}$. Let s be such a prime.

Let $L[s]$ be a lattice which as a group is isomorphic to L by $\psi : L \longrightarrow L[s]$ with bilinear form defined by $(\psi(x), \psi(y)) = s \cdot (x, y)$. Then, $\det(L[s]) = s^n \det(L)$, where $n = \mathrm{rank}(L)$. Extend ψ to maps between the rational vector spaces spanned by L and $L[s]$.

We will define M as an overlattice of $K = L \perp L[s]$. Proceeding as in the proof of the last theorem, let $C = \{(x, \psi(x)) \mid x \in A\} \leq (B, q')$. We have $q'((x, \psi(x))) = (1+s)q(x) \in 2\mathbb{Z}/2\mathbb{Z}$, i.e., C is isotropic. The determinant of the overlattice M belonging to C is $\det(K)/|C|^2 = s^n \det(L)^2 / \det(L)^2 = s^n$. We get an extension of $O(L)$ to M by taking the diagonal subgroup of $O(L) \times O(L[s])$, with respect to the isomorphism ψ. This diagonal subgroup preserves both $L \perp L[s]$ and C hence also M. □

11

Pieces of Eight

We now describe the Pieces of Eight[1] theory, which gives a new uniqueness proof for the Leech lattice and analysis of its isometry group.

This approach has the advantages of avoiding many of the special arguments with matrix work, counting, Sylow theory, permutations, etc. which are part of the earlier theories.

In particular, the Pieces of Eight theory implies theories of the Golay code and Mathieu groups, instead of depending on these theories, as was the case with the earlier treatments.

Pieces of Eight theory starts with a study of Leech trios (11.1.1) in a Leech lattice and analysis of overlattices. Such analysis should seem natural from our experience with Barnes-Wall lattices. The uniqueness theories used along the way control certain outcomes and get many basic results with relative ease. We give new proofs of results on the Leech lattice already given in this text. See [38] for more details.

11.1 Leech trios and overlattices

We give the Pieces of Eight uniqueness proof for the Leech lattice using a study of Leech trios.

Definition 11.1.1. *Define a Leech triple (or a Leech trio) in $\mathbb{R}^{24}$ to be an ordered triple M_1, M_2, M_3, where $M_i \cong \sqrt{2}E_8$ for all i and $(M_i, M_j) = 0$ for $i \neq j$.*

We shall demonstrate that a Leech lattice (10.2.7) has Leech triples.

Notation 11.1.2. *Let L be a Leech lattice Let F be a frame of norm 8 vectors (it exists because of (10.5.2)). Define $L(F, k) := 2^k \, span(F) \cap L$, for $k \in \mathbb{Z}$. (Here, span means span over the integers, $\mathbb{Z}$).*

---

[1] "Pieces of eight" refers to both the role of E_8-lattices in Leech lattice theory and to an old Spanish silver coin worth eight reals.

Lemma 11.1.3. *We have* $L(F,k) = 2^k L$ *for* $k \geq 1$, $L(F,0) = span(F)$, $L(F,-1) \cong \sqrt{2}\, D_{24}$, $(L(F,-2) + \frac{1}{2}span(F))/\frac{1}{2}span(F)$ *corresponds to a Golay code in* $\frac{1}{4}span(F)/\frac{1}{2}span(F) \cong \mathbb{F}_2^{24}$, *and* $L(F,k) = L$ *for* $k \leq -3$.

Proof. The statements for $k \geq 0$ are trivial. For $k = -1$, use the property that any two elements of F are congruent modulo $2L$ and L has minimum norm 4.

The statements for $k \leq -3$ are trivial since L is an integral lattice and $span(F)^* = \frac{1}{8}span(F)$.

The facts that $det(L) = 1$ and that any two elements of F are congruent modulo $2L$ imply that $|L{:}L(F,-2)| = 2$. Therefore, the code $(L(F,-2) + \frac{1}{2}span(F))/\frac{1}{2}span(F)$ in $\frac{1}{4}span(F)/\frac{1}{2}span(F)$ has dimension 12. Since L has minimum norm 4 and is even, this code is doubly even and has minimum weight at least 8. Therefore, this code is a Golay code (10.2.3). □

Lemma 11.1.4. *If* M *is a sublattice of a Leech lattice* L *and* $M \cong \sqrt{2}E_8$, *then* M *is a direct summand.*

Proof. The reason is that any nontrivial coset of M in M^* contains a vector of norm 1 or 2, whereas the minimum norm in L is 4. □

Theorem 11.1.5. *A Leech lattice contains a Leech trio.*

Proof. An elementary argument shows that a Golay code has trios (10.3.21)(1). This does not require a uniqueness result.

Let $\{\mathcal{O}_1, \mathcal{O}_2, \mathcal{O}_3\}$ be such a trio. If $\mathcal{O}$ is any octad, define $M(\mathcal{O}) := \{x \in L \mid supp(x) \subseteq \mathcal{O}\}$. Here. we are using a frame (see (11.1.2)) as a double basis of $\mathbb{Q} \otimes L$. Take a 24-set Ω and write $\pm \alpha_i, i \in \Omega$ for the elements of F. We shall prove that $M(\mathcal{O}) \cong \sqrt{2}E_8$.

We The sublattice $M(\mathcal{O})$ contains $2\alpha_i$ and $\pm \alpha_i \pm \alpha_j$ for all $i,j \in \mathcal{O}$, because $\frac{1}{2}(\pm x \pm y) \in L$ for any $x,y \in F$. Finally, $\frac{1}{2}\alpha_\mathcal{O} \in M(\mathcal{O})$. Let R be the $\mathbb{Z}$-span of these vectors $2\alpha_i, \pm \alpha_i \pm \alpha_j, \frac{1}{2}\alpha_\mathcal{O}$. Then $R \cong \sqrt{2}E_8$. So $M(\mathcal{O}) \geqslant R \cong \sqrt{2}E_8$. From (11.1.4), we deduce $M(\mathcal{O}) = R$.

To complete the proof, we note that the sublattices $M(\mathcal{O}_1), M(\mathcal{O}_2), M(\mathcal{O}_3)$ are pairwise orthogonal. □

Notation 11.1.6. *Now take a Leech trio in* L: M_1, M_2, M_3. *Let* $\{i,j,k\} = \{1,2,3\}$. *Define* $T_i := ann_L(M_i)$. *It contains* $M_j \perp M_k$.

The natural map $L \to \mathcal{D}(M_i) \cong 2^8$ is onto since $det(L) = 1$ and M_i is a summand. So $\mathcal{D}(T_i) \cong 2^8$, which has rank 16. We shall prove that $T_i \cong BW_{16}$ by using our uniqueness theory of Barnes-Wall lattices (9.2.3).

Let p_l be the projection of $\mathbb{Q} \otimes L$ to $\mathbb{Q} \otimes M_l$ for $l = i1,2,3$. For all l, $p_l(L) = M_l^*$ by (11.1.4) since $det(L) = 1$.

Notice that $M_j \perp M_k \leqslant L \leqslant M_j^* \perp M_k^*$, and $p_j(L) \leqslant M_j^*, p_k(L) \leqslant M_k^*$. We must show that there are groups G_l for $l = j,k$, $G_l \cong BRW(2^3) \cong 2^{1+6}\Omega^+(6,2)$ in $O(\mathbb{Q} \otimes M_l)$ and fourvolutions $f_l \in O_2(G_l)$ so that $p_l(L) \leqslant (f_l - 1)^{-1} M_l$ and that there exists an involution $t \in O(\mathbb{Q} \otimes (M_j \perp M_k))$ so that $t: M_j \to M_k$ interchanges the two sublattices, taking L to L and acting trivially on $L/(M_j \perp M_k)$.

Lemma 11.1.7. *Let* $J \leq K$ *be lattices,* $K \cong E_8$ *and* $J \cong \sqrt{2}E_8$. *Then* $O(J) \cap O(K) \cong 2^{1+6}GL(4,2)$.

Proof. This follows from the shape of the stabilizer of a maximal totally singular subspace in the orthogonal group. □

For $l = 1, 2, 3$, we take G_l to be the intersection $O(M_l) \cap O(p_l(L))$ (11.1.7).

Consider $L/(M_j \perp M_k) \leqslant (M_j^* \perp M_k^*)/(M_j \perp M_k) \cong M_j^*/M_j \perp M_k^*/M_k$. Then $T_i/(M_j \perp M_k)$ projects isomorphically onto M_l^*/M_l for each $l = j, k$. This unique isometry from M_j^*/M_j to M_k^*/M_k. Such an isometry lifts to a pair of isometries from M_j to M_k (members of the pair are negatives of each other)

each being the negative of the other (4.2.40). Any member of this pair gives an involution t interchanging M_j and M_k.

We take any element f_j of order 4 in $O_2(G_j)$. Define f_k in $O_2(G_k)$ by $f_k = -f_j^t$. This data, with (8.2.22) and (9.2.3), proves the following.

Theorem 11.1.8. *For* $i = 1, 2, 3$, $T_i \cong BW_{2^4}$.

Observe that, on the set of all $\sqrt{2}E_8$-sublattices of E_8, the isometry group $O(E_8)$ acts transitively (4.2.42). Also, $T_i/M_j \perp M_k$ is diagonally embedded in the direct sum $(M_j^* + M_k)/(M_j + M_k) \oplus (M_j + M_k^*)/(M_j + M_k)$ of vector spaces, as a totally singular subspace with respect to the natural $\mathbb{Q}/\mathbb{Z}$-valued form. Therefore, there exists an isometry of the four-dimensional spaces $p_j(T_i)/M_j$ to $p_k(T_i)/M_k$. The group $O(M_j) \cap O(p_j(T_i)) \cong 2^{1+6} \cdot GL(4, 2)$ acts transitively on this set of maps. Therefore, the Leech trio determines each T_i, up to the action of $O(M_1) \times O(M_2) \times O(M_3)$ (*).

Next, we study the overlattice L of $T_i \perp M_i$.

Note that $\mathcal{D}(T_i) \cong 2^8 \cong \mathcal{D}(M_i)$. We have $L \cap \mathbb{Q}M_i = M_i$ (since L has minimum norm 4). Since the natural map $L \to \mathcal{D}(J)$ and $L \to \mathcal{D}(ann_L(J))$ are onto, for any direct summand J of L, It follows from determinant considerations that $L \cap \mathbb{Q}T_i = T_i$.

So we get a linear isometry $\theta : T_i^*/T_i \to M_i^*/M_i$ so that $L/(T_i \perp M_i) = \{x + \theta(x) \mid x \in T_i^*/T_i\}$ is diagonally embedded in $T_i^*/T_i \oplus M_i^*/M_i$.

The space of such maps θ is in correspondence with the group $O^+(8, 2)$. We have an action of $O(T_i) \times O(M_i)$ on this set of maps, and for this action, the right factor $O(M_i)$ is transitive. We obtain that $O(T_i) \cong 2^{1+8} \cdot \Omega^+(8, 2)$ acts with kernel of the shape 2^{1+8}. So we have transitivity, and L is unique up to the action of (*). We summarize:

Theorem 11.1.9. *A Leech lattice is unique up to isometry. In more detail, a Leech lattice contains a Leech triple and the set of Leech lattices containing this Leech triple forms one orbit under the isometry group of that Leech triple.*

Theorem 11.1.10. *If M_1, M_2, M_3 is a Leech triple and Λ is a Leech overlattice, then $Stab_{O(\Lambda)}(M_i)$ has shape $(2^{1+8} \times 2) \cdot \Omega^+(8, 2)$ and induces on T_i, M_i the groups $O(T_i) \cong BRW^+(2^3)$, $O(M_i)' \cong O(E_8)'$, respectively.*

The above uniqueness theory for the Leech lattice is shorter than the original version [38] since we have uniqueness results for Barnes-Wall lattices.

11.2 The order of the group $O(\Lambda)$

Next, we prove some transitivity results for the isometry group $O(\Lambda)$ of the Leech lattice Λ and obtain its order $|O(\Lambda)|$. More details may be found in [38].

Lemma 11.2.1. *If $M \leqslant \Lambda, M \cong \sqrt{2}E_8$, then M is part of a Leech triple.*

Proof. Let $x \in M$, $(x, x) = 8$. There is a frame F in Λ such that $x \in F$. This gives the structure of a Golay code $\mathcal{G}$ (11.1.3). The frame F consists of all 48 norm 4 vectors in $x + 2\Lambda$. Since M is a direct summand of Λ, $F \cap M$ gives a set of 16 vectors in M; see (4.2.33), (4.2.34). The set $M \cap F$ and M have the same rational span. Therefore $(z, M) = 0$ if $z \in F \setminus M$. Consequently $M = M(\mathcal{O})$, for some octad $\mathcal{O}$ (see the proof of (11.1.5). □

Lemma 11.2.2. *The isometry group $O(\Lambda)$ is transitive on*
(a) *the set of Leech triples;*
(b) *the set of pairs (M, x), where $M \leqslant \Lambda$, $M \cong \sqrt{2}E_8$, and $x \in M$, $(x, x) = 8$;*
(c) *the set of norm 8 vectors.*

Proof. (a) Suppose $M_1 \perp M_2 \perp M_3$ and $M_1' \perp M_2' \perp M_3'$ are two lattices in Λ. There is an orthogonal transformation g on $\mathbb{Q} \otimes \Lambda$ such that $g(M_i) = M_i'$ for $i = 1, 2, 3$. So Λ and $g(\Lambda)$ are Leech lattices which contain $M_1' \perp M_2' \perp M_3'$.

By (11.1.9), there is $h \in O(M_1') \times O(M_2') \times O(M_3')$ so that $h(g(\Lambda)) = \Lambda$. So, $hg \in O(\Lambda)$ and $hg(M_i) = h(M_i') = M_i'$. This proves (a).

Let $(M, x), (M', y)$ be as in (b). By (a), there is $g \in O(\Lambda)$ so that $g(M) = M'$ and so $g(x), y$ are vectors of norm 8 in $M' \cong \sqrt{2}E_8$. By (4.2.41) and (11.1.10), there is $h \in Stab_{O(\Lambda)}(M')$ so that $h(g(x)) = y$.

(b) In the proof of (11.1.8), we saw that the stabilizer of $M_i \perp T_i$ in $O(\Lambda)$ induces on M_i the group $O(M_i)'$ (index 2 in $O(M_i)$). This group is transitive on vectors of norm 8 in M_i (4.2.41).

(c) All we have to do is show that a norm 8 vector lies in a subgroup isometric to $\sqrt{2}E_8$. Such a vector lies in a frame, F, and we can use F and a the binary code of (11.1.3) to realize Λ as a standard model of the Leech lattice. A point in the index set lies in an octad, so we use the octad to create a sublattice isometric to $\sqrt{2}E_8$. □

Lemma 11.2.3. *Let $M \leqslant \Lambda, M \cong \sqrt{2}E_8$, and $x \in M, (x, x) = 8$. Then the stabilizer group $Stab_{O(\Lambda)}(x, M) \cong 2^{1+8} \cdot 2^6 \cdot Alt_7$ has order $2^{18} \cdot 2^2 \cdot 5 \cdot 7$.*

Proof. We know that $Stab_{O(\Lambda)}(M)$ has the form $(2^{1+8} \times 2) \cdot \Omega^+(8, 2)$ and is embedded in $O(T_i) \times O(M_i)$. It acts on M_i as the commutator group $O(M_i)'$, so is transitive on the set of norm 8 vectors, of which there are $135 \cdot 16$. In it, the stabilizer of x has index $135 \cdot 16$, so has the form above. □

Lemma 11.2.4. *We have a decomposition $Stab_{O(\Lambda)}(F) = DP$, where $D = \{\varepsilon_A \mid A \in \mathcal{G}\}$ and P is the subgroup of Sym_Ω which fixes the Golay code $\mathcal{G}$.*

Proof. Take $g \in Stab_{O(\Lambda)}(F), g = \varepsilon_T p$, where $T \subseteq \Omega$, and p is a permutation matrix. Let O be an octad, we have $g(\frac{1}{2}\alpha_O) = p(\frac{1}{2}\varepsilon_T(\alpha_O)) = p(\frac{1}{2}\alpha_O - \alpha_{T \cap O}) = \frac{1}{2}\alpha_{p(O)} - \alpha_{p(T \cap O)}$. So p is in $Aut(\mathcal{G}) \leqslant O(\Lambda)$. Therefore, $\alpha_{p(T \cap O)} \in \Lambda$. So $|T \cap O|$ is even, for all O. So $T \in \mathcal{G}^\perp = \mathcal{G}$, implying that $\varepsilon_T \in O(\Lambda)$. □

11.2 The order of the group $O(\Lambda)$

Notation 11.2.5. *We let $D \cong 2^{12}$ and P be the diagonal group, group of permutation matrices, respectively, in (11.2.4).*

Lemma 11.2.6. *Let F be a frame, $x \in F$. Then $|Stab_{O(\Lambda)}(F) : Stab_{O(\Lambda)}(F,x)| = 48$.*

Proof. Take $x, y \in F$. There are octads $\mathcal{O}_x, \mathcal{O}_y$ in $\mathcal{G}$ so that $x \in M(\mathcal{O}_x), y \in M(\mathcal{O}_y)$. There is $g \in O(\Lambda)$ so that g takes $(x, M(\mathcal{O}_x))$ to $(y, M(\mathcal{O}_y))$. So $g \in Stab(F)$. $\square$

Lemma 11.2.7. *If $M \leqslant \Lambda, M \cong \sqrt{2}E_8$ and $x \in F \cap M$, then there is an octad $\mathcal{O}$ so that $x \in M(\mathcal{O})$. Also, P is transitive on octads.*

Proof. We use transitivity of $O(\Lambda)$ on pairs (x, M), where $x \in M$ has norm 8 and $M \cong \sqrt{2}E_8$. If M and $M(\mathcal{O})$ contain x, there is $g \in P \leqslant Stab_{O(\Lambda)}(F)$ so that $g(M) = M(\mathcal{O})$. So $M = M(g^{-1}(\mathcal{O}))$. The last statement follows since if $\mathcal{O}$ is an octad, we obtain a pair $(x, M(\mathcal{O}))$, where $x \in M(\mathcal{O})$ and $(x,x) = 8$. $\square$

Corollary 11.2.8. *We have $|Stab_{O(\Lambda)}(x) : Stab_{O(\Lambda)}(x, M)| = 253$.*

Proof. Use the octad triangle (10.3.12). $\square$

Corollary 11.2.9. *Let u_8 denote the number of norm 8 vectors. We have the order $|O(\Lambda)| = u_8 \cdot |Stab_{O(\Lambda)}(x)| = 2^{22} \cdot 3^9 \cdot 5^4 \cdot 7^2 \cdot 11 \cdot 13 \cdot 23$, where $x \in \Lambda$ and $(x,x) = 8$.*

Proof. The first factor 398034000 is obtained from the coefficient of the theta function (10.0.1),(10.5.2). $\square$

Corollary 11.2.10. *The group $Stab_{O(\Lambda)}(F) = DP$ has order $2^{22} \cdot 3^3 \cdot 5 \cdot 7 \cdot 11 \cdot 23$.*

Proof. We have transitivity on the set of norm 8 vectors. $\square$

Corollary 11.2.11. *The order of P is $|P| = 2^{10} \cdot 3^3 \cdot 5 \cdot 7 \cdot 11 \cdot 23$.*

Lemma 11.2.12. *If $g \in P$, g fixes every point of $\mathcal{O}$ and a single point of $\Omega + \mathcal{O}$, then $g = 1$.*

Proof. Such a g is in $Stab(T \perp M)$, where $M \cong \sqrt{2}E_8$ and $T \cong BW_{2^4}$. We know from the discussion of (11.1.9) that g acts on T as an element of $O_2(O(T))$ and as a permutation matrix. This puts g in the translation subgroup (the largest normal 2-subgroup) of the natural $AGL(4,2)$ subgroup of permutation matrices. Thus g operates regularly on the 16 indices represented in T. $\square$

Corollary 11.2.13. *The group P is 5-fold transitive on the index set Ω.*

Proof. Let $a_1, \ldots, a_5$ and $b_1 \ldots, b_5$ be sequences of distinct points in Ω. There is a unique octad $\mathcal{O}_a$ containing $a_1, \ldots, a_5$ and $\mathcal{O}_b$ containing $b_1, \ldots, b_5$. Let $x, y \in F$ correspond to a_1, b_1. So we have pairs $(x, M(\mathcal{O}_a))$ and $(y, M(\mathcal{O}_b))$. There is $g \in O(\Lambda)$ which takes the first pair to the second. So $g = \varepsilon_T p \in DP = Stab_{O(\Lambda)}(F)$. Then p takes $\mathcal{O}_a$ to $\mathcal{O}_b$. So we may assume $\mathcal{O}_a = \mathcal{O}_b = \mathcal{O}$. If $H = Stab_{O(\Lambda)}(F)$ is 5-transitive on $\mathcal{O}$, we are done. This is true since this subgroup acts as the alternating group $Alt_{\mathcal{O}}$ on the octad $\mathcal{O}$ (4.2.38),(11.1.10). We know $|P:H| = 759$. So, $|H| = 2^{10} \cdot 3^2 \cdot 5 \cdot 7$. Note that $|Alt_8| = \frac{1}{2}8!$. If K is the kernel of action of H on $\mathcal{O}$, then $2^4 | |K|$. It suffices to prove $2^4 = |K|$, or to show that if $g \in P$, g fixes every point of $\mathcal{O}$. This follows from (11.2.12). $\square$

Definition 11.2.14. (The Mathieu groups) *We define M_{24} to be the group P, considered as a subgroup of Sym_Ω. For $n \in \{20, 21, 22, 23, 24\}$, we define M_n to be the stabilizer of a set of size $24 - n$ in Ω. Since M_{24} is 5-transitive, this is well-defined up to conjugacy. We furthermore define $M_{12} = Stab_P(\mathcal{D})$, where $\mathcal{D}$ is a 12-set in $\mathcal{G}$ and define, for $n \in \{8, 9, 10, 11, 12\}$, the stabilizer in M_{12} of a subset of D of size 12. conjugacy.*

The Mathieu groups are the groups $M_{24}, M_{23}, M_{22}, M_{12}, M_{11}$ of the above.

Remark 11.2.15. *The other M_n are not usually called Mathieu groups since their composition factors are alternating groups or projective matrix groups. For example, $M_{21} \cong PSL(3,4)$ and $M_{10} \cong Alt_6 \cdot 2$ and the other M_n are solvable.*

We shall prove simplicity for M_{24}. In fact, all five Mathieu groups are simple. Since M_{12} is 5-transitive, the groups M_n are well-defined up to conjugacy.

We point out that M_{24} is not 6-transitive (or else its order would be divisible by $24 \cdot 23 \cdot 22 \cdot 21 \cdot 20 \cdot 19$, which is not the case (11.2.11)).

More details may be read in [37].

11.3 The simplicity of M_{24}

We studied the isometry group $O(\Lambda)$ and found that if F is a frame of norm 8 vectors, then $Stab_{O(\Lambda)}(F) = DP$, where $D = \{\varepsilon_A | A \in \mathcal{G}\}$ and P is the group of permutation matrices with respect to a suitable basis $F' \subseteq F$. See (11.2.4), (11.1.9).

We have $|P| = 2^{10} \cdot 3^3 \cdot 5 \cdot 7 \cdot 11 \cdot 23$ and P is 5-transitive on $\Omega = F/\{\pm 1\}$ (11.2.13).

Theorem 11.3.1. *The group P is simple.*

We shall prove simplicity in a series of steps. Our proof requires introducing theory and techniques and results from basic finite group theory. This should not be surprising since simplicity is a group-theoretic concept.

We assume existence of a minimal normal subgroup K of P, $1 \neq K \neq P$, then derive a contradiction.

Definition 11.3.2. *Suppose that a group G acts transitively on the set X. A system of imprimitivity is a partition of X such that if B is a subset of the partition, then for all g in G, $gB = B$ or $gB \cap B = \emptyset$. This action of G on X is called imprimitive if there exists a system of imprimitivity whose parts are not 1-sets nor X. Otherwise, the action of G is called primitive.*

Trivially, if G acts doubly transitively on X, the action is primitive. Note that our P is doubly transitive (11.2.13)) on Ω.

Lemma 11.3.3. *If the group G acts faithfully and primitively on a set X and K is a nontrivial normal subgroup, then K acts transitively.*

Proof. The orbits of K form a system of imprimitivity. □

Step 1. K acts transitively on Ω, so $24 | |K|$. Prove this with (11.3.3).

Lemma 11.3.4. *Let c be an n-cycle in $G = Sym_n$ or Sym_{n+1}. Then $C_G(c) = \langle c \rangle$.*

Proof. Suppose $G = Sym_n$. We count n-cycles in G. A cycle may start with 1: $(1, a_2, a_3, \ldots, a_n)$. So $|C_G(c)| = \frac{n!}{(n-1)!} = n$. Since $\langle c \rangle \leqslant C_G(c)$, we are done. Now suppose $G = Sym_{n+1}$. An n-cycle fixes one symbol, say $a_1 \in \{1, 2, \ldots, n+1\}$, so it permutes $\{a_i \mid i \neq 1\}$, which may be considered as an element of Sym_n. We deduce that $|G : C_G(c)| = (n+1)(n-1)!$, so $|C_G(c)| = n$. $\square$

Lemma 11.3.5. *Suppose A is a p-group for some prime number p, and F is a p'-group (i.e. $(p, |F|) = 1$). Suppose also that A acts on F as automorphisms. Then, for every prime number $q \neq p$, there is a Sylow q-subgroup S of F so that S is invariant under A.*

Proof. In fact, this action of A fixes the set $Syl_q(F)$ of Sylow subgroups. Note that $|Syl_q(F)|$ divides $|F|$ and is prime to p, by the conjugacy part of the Sylow theorems. On the set $Syl_q(F)$, the group A has orbits of size dividing $|A|$, a power of p. Since $(|F|, p) = 1$, there is an orbit of size 1. Let $\{S\}$ be this orbit. $\square$

Step 2. We have $23 \mid |K|$.

Suppose $23 \nmid |K|$. We have $1 \neq K \lneqq KS \leqslant P \leqslant Sym_\Omega$, where $S \in Syl_{23}(P)$. For every prime number q dividing $|K|$, there is a group $T \in Syl_q(K)$, so $S \leqslant N_P(T)$ (11.3.5). Take $q = 3$. Then $T \neq 1$ and $|T| \mid 3^3$ since $|P| = 2^{10} \cdot 3^3 \cdot 5 \cdot 7 \cdot 11 \cdot 23$. If $|T| < 23$, then $C_P(S) \geqslant TS \gneqq S$, contradict with Step 2. If $|T| = 27$, then $C_T(S)$ has order 4 or 27, so we obtain another contradiction.

Step 3. For $S \in Syl_{23}(P)$, $|N_P(S)| = 23 \cdot 11$, .

Note that $C_P(S) = S$ (11.3.4). Also $N_P(S)/S$ is isomorphic to a subgroup of $Aut(S) \cong \mathbb{Z}_{22}$ since $|Aut(\mathbb{Z}_n)| = \phi(n)$, the Euler number. There is a divisor d of 22 so that $|N_P(S)| = 23 \cdot d$. Since $|P : N_P(S)| \equiv 1 \pmod{23}$, by Sylow's theorem, $d = 11$.

Step 4. We have $P = KN_P(S)$. This follows from the Frattini argument, which we prove next.

Theorem 11.3.6. *(Frattini argument) Let G be a group with a finite normal subgroup $K \triangleleft G$, p be a prime number and $S \in Syl_P(K)$. Then $G = N_G(S) \cdot K$.*

Proof. Let $g \in G$. Then $gSg^{-1} \in Syl_P(K)$. By Sylow's theorems, there exist $y \in K$ such that $y(gSg^{-1})y^{-1} = S$, or $(yg)S(yg)^{-1}$. So $yg \in N_G(S)$, and $g \in y^{-1}N_G(S) \subseteq KN_G(S) = N_G(S) \cdot K$. $\square$

Step 5. $N_P(S) \not\leqslant K$. For otherwise, $P = K$, a contradiction to the choice of K.

Step 6. $N_K(S) = S$. Use Step 5, $23 \mid |N_K(S)|$, and $23 \cdot 11 = |N_P(S)|$.

Step 7. $N_K(S) \neq S$.

Theorem 11.3.7. *(Burnside) Suppose G is a finite group, p is a prime number, $S \in Syl_P(G)$ and $S \leqslant Z(N_G(S))$. Then there is a normal subgroup $K \triangleleft G$ such that $G = KS$ and $K \cap S = 1$.*

Proof. The proof uses the group transfer. See [31, 51]. $\square$

If $N_K(S) = S$, then there is $J \triangleleft K$ so that $K = JS$ and $J \cap S = 1$. If we show $J \triangleleft P$, then we have a contradiction to the choice of K, as a minimal normal subgroup.

Definition 11.3.8. *Let G be a group and H a subgroup. We say that H is a characteristic subgroup if $\alpha(H) = H$, for every $\alpha \in Aut(G)$.*

Lemma 11.3.9. *If $T \leqslant K \triangleleft G$, and T is characteristic in K, then $T \triangleleft G$.*

Proof. Let $g \in G$, we want to show $gTg^{-1} = T$. The function $\alpha : K \to K$ defined by $x \mapsto gxg^{-1}$ is in $Aut(K)$. So $\alpha(T) = T$. $\square$

Lemma 11.3.10. *Suppose $T \leqslant K$ are groups, K finite, $T \triangleleft K$ and $(|T|, |K : T|) = 1$. Then T is characteristic in K.*

Proof. Let $\alpha \in Aut(K)$. Then $T \triangleleft K$ implies $L := T \cdot \alpha(T) \leqslant K$. This subgroup has order $|L| = \frac{|T||\alpha(T)|}{|T \cap \alpha(T)|} = |T|b$, where $b||T|$. Also, $L/T \leqslant K/T$, which has order relatively prime to $|T|$, by hypothesis. So $b = 1$. Therefore, $\alpha(T) \leqslant T$. By finiteness, $\alpha(T) = T$. $\square$

We deduce that the above J is normal in P. This verifies Step 7.

Since Steps 6 and 7 are incompatible, the proof of simplicity for P (11.3.1) is complete. $\square$

11.4 Sublattices of Leech and subgroups of the isometry group

For detailed discussions of subgroups of $O(\Lambda)$, see [18, 34]. New and quicker proofs for several of these cases are given in [38].

Here, we give only a summary of subgroups and their relations to the geometry of the Leech lattice. The two tables below come from [18, 19]. Harder techniques from finite group theory are need for complete justifications of those tables. See [37] for discussion of some cases.

The simple groups which occur as subquotient of $O(\Lambda) = Co_0$ include just 12 of the 26 sporadic simple groups. For context, we refer the reader to the list of the finite simple groups.

Tables of Stabilizers

Table 11.1. The vectors of $\Lambda_2, \Lambda_3, \Lambda_4$

Class	Λ_2^2	Λ_2^3	Λ_2^4	Λ_3^2	Λ_3^3	Λ_3^4	Λ_3^5	
Shape	$(2^8 0^1 6)$	$(3 1^2 3)$	$(4^2 0^{22})$	$(2^{12} 0^{12})$	$(3^3 1^{21})$	$(4\, 2^8 0^{15})$	$(5 1^{23})$	
No.	$2^7 . 759$	$2^{12} . 24$	$2^2 \binom{24}{2}$	$2^{11} . 2576$	$2^{12} \binom{24}{3}$	$2^6 . 759 . 16$	$2^{12} . 24$	
Class	Λ_4^{2+}	Λ_4^{2-}	Λ_4^3	Λ_4^4	Λ_4^{4+}	Λ_4^{4-}	Λ_4^5	Λ_4^6	Λ_4^8
Shape	$(2^{16} 0^8)$	$(2^{16} 0^8)$	$(3^5 1^{19})$	$(4^4 0^{20})$	$(4^2 2^8 0^{14})$	$(4\, 2^{12} 0^{11})$	$(5\, 3^2 1^{21})$	$(6\, 2^7 0^{16})$	$(8\, 0^{23})$
No.	$2^{11} . 759$	$2^{11} . 759 . 15$	$2^{12} \binom{24}{5}$	$2^4 \binom{24}{4}$	$2^9 . 759 . \binom{16}{2}$	$2^{12} . 2576 . 12$	$2^{12} \binom{24}{3} . 3$	$2^7 . 759 . 8$	$2^1 . 24$

11.4 Sublattices of Leech and subgroups of the isometry group

Table 11.2.

Name	Order	Structure	Name	Order	Structure
·0	$2^{22}3^95^47^211.13.23$	New perfect	·222	$2^{15}3^65.7.11$	$PSU_6(2)$
·1	$2^{21}3^95^47^211.13.23$	New simple	·322	$2^73^65^37.11$	M^c
·2	$2^{18}3^65^37.11.23$	New simple	·332	$2^93^25^37.11$	HS
·3	$2^{10}3^75^37.11.23$	New simple	·333	$2^43^75.11$	$3^5.M_{11}$
·4	$2^{18}3^25.7.11.23$	$2^{11}M_{23}$	·422	$2^{17}3^25.7.11$	$2^{10}.M_{22}$
·5	$2^83^65^37.11$	$M^c.2$	·432	$2^73^25.7.11.23$	M_{23}
·6$_{22}$	$2^{16}3^65.7.11$	$PSU_6(2).2$	·433	$2^{10}3^25.7$	$2^4.A_8$
·6$_{32}$	$2^{10}3^35.7.11.23$	M_{24}	·442	$2^{12}3^25.7$	$2^{1+8}A_7$
·7	$2^93^25^37.11$	HS	·443	$2^73^25.7$	$M_{21}.2$
·8$_{22}$	$2^{18}3^65^37.11.23$	·2	·522	$2^73^65^37.11$	M^c
·8$_{32}$	$2^73^65^37.11$	M^c	·532	$2^83^65.7$	$PSU_4(3).2$
·8$_{42}$	$2^{15}2^25.7$	$2^5.2^4.A_8$	·533	$2^43^25^37$	$PSU_3(5)$
·9$_{33}$	$2^53^75.11$	$3^6.M_{11}.2$	·542	$2^73^25.7.11$	M_{22}
·9$_{42}$	$2^73^25.7.11.23$	M_{23}	·633	$2^63^35.11$	M_{12}
·10$_{33}$	$2^{10}3^25^37.11$	$HS.2$	·2 =!2	$2^{19}3^65^37.11.23$	$(\cdot 2)\times 2$
·10$_{42}$	$2^{17}3^25.7.11$	$2^{10}.M_{22}$	·3 =!3	$2^{11}3^75^37.11.23$	$(\cdot 3)\times 2$
·11$_{43}$	$2^{10}3^25.7$	$2^4.A_8$	·4	$2^{19}3^25.7.11.23$	$(\cdot 4)\times 2$
·11$_{52}$	$2^83^65.7$	$PSU_4(3).2$	!4	$2^{22}3^35.7.11.23$	$2^{12}.M_{24}$
			!333	$2^73^95.11$	$3^6.2.M_{12}$
			!442	$2^{15}3^45.7$	$2^{1+8}A_9$

HS, M^c, p^n denote respectively the Higman-Sims group, the McLaughlin group, and the elementary group of order p^n. $A.B$ denotes an extension of the group A by the group B. The notation is otherwise standard.

Notation 11.4.1. *We now explain some notation in those tables. A type a vector in Λ is a lattice vector of norm 2a. A triangle of type abc in Λ is a triple of vectors x, y, z so that $(x, x) = 2a, (y, y) = 2b, (z, z) = 2c$, and $x + y + z = 0$. For a vector of type a, i.e., of norm 2a, there is just one orbit on vectors of types $a = 2, 3, 4$. Its stabilizer is denoted $\cdot a$.*

For certain triples a, b, c, there is just one orbit on triangles of type abc. If so, we call the stabilizer $\cdot abc$. For example, $\cdot 222 \cong PSU(6, 2)$, which has order $2^{15} \cdot 3^6 \cdot 5 \cdot 7$; $\cdot 223 \cong McL$, the group of McLaughlin; $\cdot 233 \cong HS$, the group of Higman-Sims.

The group Co_0 (the big Conway group, commonly denoted by $\cdot 0$) is $O(\Lambda)$, which is a perfect group of order $2^{22} \cdot 3^9 \cdot 5^4 \cdot 7^2 \cdot 11 \cdot 13 \cdot 23$ and whose quotient $Co_1 := O(\Lambda)/\{\pm 1\}$ (denoted by $\cdot 1$) is simple.

The group $O(\Lambda)$ acts transitively on the set of norm 4 vectors. The stabilizer of one is a group Co_2 (also denoted by $\cdot 2$), simple of order $2^{18}3^65^37 \cdot 11 \cdot 23$. It acts transitively on the set of norm 6 vectors. The stabilizer of one is a group Co_3 (denoted by $\cdot 3$), which is a simple group of order $2^{10}3^75^37 \cdot 11 \cdot 23$.

As discussed earlier in this text, $O(\Lambda)$ acts transitively norm 8 vectors and on frames. The stabilizer of a frame is isomorphic to $\{\varepsilon_A | A \in \mathcal{G}\}{:}M_{24} \cong 2^{12} : M_{24}$. Here M_{24} is the Mathieu group; which is simple of order $2^{10}3^35{\cdot}7{\cdot}11{\cdot}23$. The stabilizer of a norm 8 vector is a subgroup of the above group of the form $2^{11}{:}M_{23}$.

Some sporadic groups have a good context in Λ. Here are examples, two of many.

Example 11.4.2. *The Higman-Sims group HS has order $2^93^25^3{\cdot}7{\cdot}11$. It was originally described as automorphisms of a graph Γ with 100 points and valency 22. The graph automorphism group $Aut(\Gamma)$ is the set of all permutations of the 100 vertices which respect edges. The group $Aut(\Gamma)$ has a simple subgroup of index 2 (which is HS). The stabilizer in $Aut(\Gamma)$ of a point is a group of the form $M_{22}{:}2$.*

Higman and Sims defined the graph and proved that $Aut(\Gamma)$ is transitive on the 100 points. This group may be represented on Λ. Take a triangle of type 233, for example this triple of vectors:

$$x = \frac{1}{4}(1,\cdots,1,5,1,\cdots,1,1,1,\cdots,1),$$
$$x' = \frac{1}{4}(-1,\cdots,-1,\cdots,-1,-5,-1,\cdots,-1),$$
$$x'' = \frac{1}{4}(0,\cdots,0,-4,0,\cdots,0,4,0,\cdots,0);$$

then $x + x' + x'' = 0$. We take the set Y of all norm 4 vectors which have inner product -3 with x and inner product 0 with x''. There are only two inner products (y_1, y_2), for distinct $y_1, y_2 \in Y$ and so we get a graph. See [37] for more details.

Example 11.4.3. *We describe the sporadic groups of Suzuki and Hall-Janko. They occurs here as a quotient of the centralizer of an isometry.*

There exists a unique conjugacy class of elements $g_3 \in O(\Lambda)$ so that g_3 has minimal polynomial $\frac{x^3-1}{x-1} = x^2+x+1$ (so g_3 has order 3). The centralizer $C_{O(\Lambda)}(g_3)$ has structure $6{\cdot}Suz$ (central extension of simple group Suz of order $2^{13}{\cdot}3^7{\cdot}5^3{\cdot}7{\cdot}11{\cdot}23$).

The conjugacy class g_5 has minimal polynomial $\frac{x^5-1}{x-1}$ (so g_5 has order 5). Its centralizer $C_{O(\Lambda)}(g_5)$ has structure $5 \times 2{\cdot}HJ$, of order 604800, where HJ is the group of Hall-Janko, a group which acts transitively on a graph with 100 points (this is not the same graph as the one associated to the Higman-Sims group).

11.5 Involutions on the Leech lattice

We can give a good description of involutions on the Leech lattice with the SSD concept (2.6.7) and our theory of the Golay code. There are four conjugacy classes of involutions (see [37] for a proof).

Obviously, -1 is the SSD involution associated to Λ.

The involutions associated to the SSD sublattices isomorphic to $\sqrt{2}E_8$ have been discussed. Their negatives are associated to the SSD sublattices which are orthogonal to the previous type.

One type remains, the ones associated to SSD sublattices which, in the standard model, are supported by a dodecad $D \in \mathcal{G}$. These sublattices are isometric to $\sqrt{2}Q$, where $Q \cong HS_{12}$, an odd, rank 12 unimodular lattice.

References

[1] J. L. Alperin, Local Representation Theory, Modular Representations as an Introduction to the Local Representation Theory of Finite groups. Cambridge Studies in Advanced Mathematics, 11, Cambridge University Press, Cambridge, 1986.

[2] J. L. Alperin, R. B. Bell, Groups and Representations, Graduate Texts in Mathematics, 162, Springer, New York, 1995.

[3] E. Artin, Geometric Algebra, Wiley Interscience, New York, 1957.

[4] E. F. Assmus, Jr., H. F. Mattson, Jr. and R. J. Turyn, Research to Develop the Algebraic Theory of Codes, Report AFCRL-67-0365, Air Force Cambridge Res. Labs., Bedford, Mass, June 1967.
(Remark: SPLAG gives reference as [64], Chap. 18, Sect. 6, Th. 12, p 588.)

[5] E. S. Barnes and G. E. Wall, Some extreme forms defined in terms of abelian groups, J. Aust. Math. Soc. 1 (1959), 47–63.

[6] D. J. Benson, Modular Representation Theory: New Trends and Methods, Second printing of the 1984 original, Lecture Notes in Mathematics, 1081, Springer, Berlin, 2006, xii+231 pp.

[7] B. Bolt, T. G. Room and G. E. Wall, On the Clifford collineations, transform and similarity groups, I, J. Aust. Math. Soc. 2 (1961/1962), 60–79.

[8] B. Bolt, T. G. Room and G. E. Wall, On the Clifford collineations, transform and similarity groups, II, J. Aust. Math. Soc. 2 (1961/1962), 80–96.

[9] B. Bolt, On the Clifford collineations, transform and similarity groups, III; generators and relations, J. Aust. Math. Soc. 2 (1961/1962), 334–344.

[10] R. E. Borcherds, The Leech lattice, Proc. Roy. Soc. London Ser. A 398 (1985), no. 1815, 365–376.

[11] N. Bourbaki, Groupes et Algbres de Lie, Chapitres 4, 5, 6, Masson, Paris, 1981.

[12] M. Broué and M. Enguehard, Sur cetaines rseaux unimodulaires pairs, (French) C. R. Acad. Sci. Paris Sr. A-B 272 (1971), A210–A213. (Reviewer: J. A. Todd) 20.22 (06.00)

[13] M. Broué and M. Enguehard, Polynmes des poids de certains codes et fonctions thta de certains rseaux, (French) Ann. Sci. cole Norm. Sup. (4) 5 (1972), 157–181. (Reviewer: H. F. Mattson, Jr.) 94A10 (10D05)

[14] M. Broué and M. Enguehard, Une famille infinie de formes quadratiques entires; leurs groupes d'automorphismes. (French) C. R. Acad. Sci. Paris Sr. A-B 274 (1972), A19–A22. (Reviewer: J. A. Todd) 20.22 (10.00)

[15] M. Broué and M. Enguehard, Une famille infinie de formes quadratiques entière; leurs groupes d'automorphismes, Ann. scient. Éc. Norm. Sup., 4eme série, t. 6(1973), 17–52.

[16] R. Carter, Simple Groups of Lie Type, Wiley-Interscience, London, 1972.

[17] J. W. S. Cassels, Rational Quadratic Forms, London Mathematical Society Monographs, 13, Academic Press, Inc. [Harcourt Brace Jovanovich, Publishers], London–New York, 1978.

[18] J. H. Conway, A perfect group of order 8, 315, 553, 613, 086, 720, 000 and the sporadic simple groups, PNAS 61 (1968), 398-400.

[19] J. H. Conway, A Group of Order 8, 315, 553, 613, 086, 720, 000. Bull. Lond. Math. Soc. 1 (1969), 79-88.

[20] J. H. Conway, A characterization for Leech's lattice, Invent. Math. 7 (1969), 137–142.

[21] J. H. Conway and N. J. A. Sloane, Grundlehren der mathematischen Wissenschaften 290 Sphere Packings, Lattices and Groups, Springer, 1988.

[22] J. H. Conway and N. J. A. Sloane, Self-dual codes over the integers modulo 4, J. Combin. Theory Ser. A 62 (1993), 30–45.

[23] C. W. Curtis and I. Reiner, Representation Theory of Finite Groups and Associative Algebras, Interscience, New York, 1962.

[24] C. W. Curtis and I. Reiner, Representation theory of finite groups and associative algebras, Reprint of the 1962 original, AMS Chelsea Publishing, Providence, RI, 2006, xiv+689 pp.

[25] C. W. Curtis and I. Reiner, Methods of Representation Theory. Vol. I. With Applications to Finite Groups and Orders, Reprint of the 1981 original. Wiley Classics Library, A Wiley-Interscience Publication, John Wiley & Sons, Inc., New York, 1990, xxiv+819 pp.

[26] C. W. Curtis and I. Reiner, Methods of Representation Theory. Vol. II. With Applications to Finite Groups and Orders, Pure and Applied Mathematics (New York). A Wiley-Interscience Publication, John Wiley & Sons, Inc., New York, 1987, xviii+951 pp.

[27] R. Curtis, A new combinatorial approach to M_{24}, Math. Proc. Cambridge Philos. Soc. 79 (1976), 25–42.

[28] J. Dieudonné, La Gomtrie des Groupes Classiques, Springer, Berlin Heidelberg New York, 1971.

[29] W. Feit, Characters of Finite Groups, W. A. Benjamin, Inc., New York-Amsterdam, 1967

[30] W. Feit, The Representation Theory of Finite Groups, North-Holland Mathematical Library, 25, North-Holland Publishing Co., Amsterdam-New York, 1982.

[31] D. Gorenstein, Finite Groups, Harper and Row, New York, 1968.

[32] R. L. Griess, Jr., Automorphisms of extra special groups and nonvanishing degree 2 cohomology (research announcement for [33]), in Finite Groups 1972: Proceedings of the Gainesville Conference on Finite Groups, (T. Gagen, M. P. Hale and E. E. Shult, eds.), North Holland Publishing Co., Amsterdam, 68-73, 1973.

[33] R. L. Griess, Jr., Automorphisms of extra special groups and nonvanishing degree 2 cohomology, Pacific J. Math. 48 (1973), 403–422.

[34] R. L. Griess, Jr., On a subgroup of order $2^{15}|GL(5,2)|$ in $E_8(C)$, the Dempwolff group and $Aut(D_8 \circ D_8 \circ D_8)$, J. Algebra 40 (1976), 271–279.

[35] R. L. Griess, Jr., The monster and its nonassociative algebra, in Proceedings of the Montreal Conference on Finite Groups, Contemporary Mathematics, 45 (1985), 121–157, American Mathematical Society, Providence, RI.
[36] R. L. Griess, Jr., Sporadic groups, code loops and nonvanishing cohomology, J. Pure Appl. Algebra 44 (1987), 191–214.
[37] R. L. Griess, Jr., Twelve Sporadic Groups, Springer Monographs in Mathematics, Springer, 1998.
[38] R. L. Griess, Jr., Pieces of Eight, Advances in Mathematics 148 (1999), 75–104.
[39] R. L. Griess, Jr., Positive definite lattices of rank at most 8, J. Number Theory 103 (2003), 77–84.
[40] R. L. Griess, Jr., Pieces of 2^d: existence and uniqueness for Barnes-Wall and Ypsilanti lattices. Advances in Mathematics, 196 (2005), 147–192. math.GR/0403480 See also: Corrections and additions to " Pieces of 2^d: existence and uniqueness for Barnes-Wall and Ypsilanti lattices.", Advances in Mathematics 211 (2007), 819–824.
[41] R. L. Griess, Jr., Involutions on the the Barnes-Wall lattices and their fixed point sublattices, I. Pure Appl. Math. 1 (2005), no.4 (Special Issue: In Memory of Armand Borel, Part 3 of 3), 989–1022.
[42] R. L. Griess, Jr. and G. Höhn, Frame stabilizers for the lattice vertex operator algebra of type E_8, J. reine angew. Math. 561 (2003), 1–37.
[43] R. L. Griess, Jr. and C. H. Lam, Rootless pairs of EE_8 lattices, Electron. Res. Announc.Math. Sci. 15 (2008), 52–61.
[44] R. L. Griess, Jr. and C. H. Lam, Dihedral groups and EE_8 lattices, preprint 87 pages. pdf ; arxiv 15 jun 08; to appear in Pare and Applied Math Quarterly.
[45] B. Gross, Group representations and lattices, J. Amer. Math. Soc. 4 (1990), no.3, 929–960.
[46] K. W. Gruenberg, Cohomological Topics in Group Theory, Lecture Notes in Mathematics, Vol. 143, Springer-Verlag, Berlin-New York, 1970, xiv+275 pp.
[47] P. Hall and G. Higman, On the p-length of p-soluble groups and reduction theorems for Burnside's problem. Proc. London Math. Soc. 6 (1956), no.3, 1–42.
[48] B. Hartley and T. O. Hawkes, Rings, Modules and Linear Algebra, A Further Course in Algebra Describing the Structure of Abelian Groups and Canonical Forms of Matrices through the Study of Rings and Modules, A reprinting, Chapman & Hall, London-New York, 1980, xi+210 pp. ISBN: 0-412-09810-5 00A05 (16-01)
[49] P. J. Hilton and U. Stammbach, A Course in Homological Algebra, Second edition, Graduate Texts in Mathematics 4, Springer, New York, 1997, xii+364 pp.
[50] J. Humphreys, Introduction to Lie Algebras and Representation Theory, Graduate Texts in Mathematics 9, Springer, 1972; third printing 1980.
[51] B. Huppert, Endliche Gruppen, I, Springer, 1967.
[52] I. M. Isaacs, Character theory of finite groups, Corrected reprint of the 1976 original [Academic Press, New York; MR0460423], AMS Chelsea Publishing, Providence, RI, 2006.
[53] I. M. Isaacs, Algebra: A Graduate Course Reprint of the 1994 original, Graduate Studies in Mathematics, 100, American Mathematical Society, Providence, RI, 2009. xii+516 pp. ISBN: 978-0-8218-4799-2
[54] D. G. James, On Witt's theorem for unimodular quadratic forms, Pacific J. Math 26 (1868), 303–316.

[55] B. W. Jones, The Arithmetic Theory of Quadratic Forms, Carcus Monograph Series, no. 10, The Mathematical Association of America, Buffalo, N. Y., 1950, x+212 pp.

[56] O. D. King, A mass formula for unimodular lattices with no roots, Math. Comp. 72 (2003), no. 242, 839–863 (electronic). (Reviewer: Hidenori Katsurada) 11H55 (11E41)

[57] Y. Kitaoka, Arithmetic of Quadratic Forms, Cambridge University Press, 1993.

[58] M. Kneser, Quadratische Formen, Springer, 2002.

[59] A. I. Kostrikin and P. H. Tiep, Orthogonal Decompositions and Integral Lattices, De Gruyter Expositions in Mathematics 15, 1994.

[60] A. Kumar and H. Cohn, The densest lattice in twenty-four dimensions, research announcement; Abhinav Kumar and Henry Cohn, Optimality and uniqueness of the Leech lattice among lattices; Henry Cohn and Abinhav Kumar, Universally optimal distribution of points on spheres, to appear in Journal of the American Mathematical Society.

[61] S. Lang, Algebra, Revised third edition, Graduate Texts in Mathematics, 211, Springer, New York, 2002, xvi+914 pp. ISBN: 0-387-95385-X 00A05 (15-02)

[62] J. Leech, Notes on sphere packings. Canad. J. Math. 19 (1967), 251–267.

[63] James Lepowsky and Arne Meurman, J. Lepowsky and A. Meurman, An E_8 approach to the Leech lattice and the Conway group, J. Algebra 77 (1982), 484-504.

[64] J. MacWilliams and N. Sloane, The Theory of Error Correcting Codes, North-Holland, 1977.

[65] J. E. McLaughlin, Some groups generated by transvections. Arch. Math. (Basel) 18, 1967, 364–368.

[66] H. V. Niemeier, Definite Quadratische Formen der Diskriminante 1 und Dimension 24, Doctoral Dissertation, Göttingen, 1968.

[67] V. V. Nikulin, Integral symmetric bilinear forms and their applications, Math. USSR Izv. 14 (1980), 103–167.

[68] V. S. Pless, Introduction to the Theory of Error-correcting Codes, Third edition, Wiley-Interscience Series in Discrete Mathematics and Optimization. A Wiley-Interscience Publication, John Wiley & Sons, Inc., New York, 1998, xiv+207 pp.

[69] V. S. Pless, W. C. Huffman and R. A. Brualdi, Handbook of Coding Theory. Vol. I, II, North-Holland, Amsterdam, 1998. Vol. I: xvi+1138+I-58 pp.; Vol. II: pp. i–xvi, 1139–2169 and I-1–I-58.

[70] J. J. Rotman, An Introduction to Homological Algebra, Second edition, Universitext, Springer, New York, 2009. xiv+709 pp.

[71] J.-P. Serre, A Course in Arithmetic, Graduate Texts in Mathematics 7, Springer, 1973.

[72] G. Shimura, An Introduction to the Arithmetic Theory of Automorphic Functions, Iwanami Shoten Publishers and Princeton University Press, 1971.

[73] S. L. Snover, The Uniqueness of the Nordstrom-Robinson code, Ph. D. Thesis, Department of Mathematics, Michigan State University, 1973.

[74] J. G. Thompson, Nonsolvable finite groups all of whose local subgroups are solvable, Bull. Amer. Math. Soc. 74 (1968), 383–437.

[75] J. Tits, Four Presentations of Leechs lattice, in Finite Simple Groups, II, Proceedings of a London Math. Soc. Research Symposium, Durham, 1978, ed. M. J. Colllins, pp. 306-307, Adademic Press, London, New York, 1980.

[76] J. Todd, A representation of the Mathieu group M_{24} as a collineation group, Annali di Math. Pura ed. Applicata 71 (1966), 199–238.
[77] J. H. van Lint, Introduction to Coding Theory, Third edition, Graduate Texts in Mathematics 86. Springer, Berlin, 1999, xiv+227 pp.
[78] B. B. Venkov, The classification of integral even unimodular 24-dimensional quadratic forms, Trudy Maatematicheskogo Instituta imeni V. A. Steklova 148 (1978), 65–76 (Proceedings of the Steklov Institute of Mathematics (No. 4, 1980).
[79] J. L. Walker, Codes and Curves, Student Mathematical Library, 7, IAS/Park City Mathematical Subseries, American Mathematical Society, Providence, RI; Institute for Advanced Study (IAS), Princeton, NJ, 2000, xii+66 pp.
[80] E. Weiss, Algebraic Number Theory, Dover Publications, 1998.
[81] E. Witt, Theorie der quadratischen Formen in beliebigen Körpern, JRAM 176 (1937), 31–44.
[82] E. Witt, Die 5-fach transitiven Grupen von Maqthieu, Hamburg Universit t Abhandlungen aus dem Mathematischen Seminar 12 (1938), 256–264.

Index

Word [definition], approximate occurrence(s) in text

Symbols, Special Notations

$A \circ B$, central product of groups A and B (7.1.10)
$D_n^\pm, HS_n, HS_n^\pm$ (2.3.6), (10.0.1)
$G', [G,G]$, notations for the commutator subgroup of the group G (7.1.10)
L^{even} (2.2.11)
ν, ν_i (10.2.7)
p_ε^{1+2n}, an extraspecial p-group of type $\varepsilon = \pm$ and order p^{1+2n} (7.2.6)
$\mathcal{P}, \mathcal{PE}$, power set, even subsets   (6.2.1)
Φ, Φ_X, symbol for a root system, root system of type X (4.2.1)
$\Phi(G)$, Frattini subgroup of the group G (7.1.10)
$Tel(L,t)$, the total eigenlattice for t on L (2.6.9)
$Z(G)$, center of the group G (7.1.10)
1, the all-one vector in a code   (6.2.1)

A

affine general linear group, $AGL(n,K)$ (6.1.9)
A_n lattice   (4.2.5)

B

Barnes-Wall lattice   (9.1.1), (2.6.12), (7.3.10)
Bolt, Room, Wall   (7.3.7)
Boolean sum   (6.2.1)
BRW   (7.3.7)
$BRW(p^n, \varepsilon), BRW^0(p^n, \varepsilon)$ (7.3.10)
Burnside   (11.3.7)

C

central product   (7.1.10)

characteristic subgroup   (11.3.8)
commutator   (8.2.1)
commutator density   (8.2.3)
Coxeter number   (10.1.1)

D

degenerate   (2.0.3)
defect, defective   (2.1.4)
diagonal, $Diag(n,K)$ (6.1.4)
Dickson homomorphism   (7.4.7)
discriminant group (of a lattice)   (2.3.1)
D_n-lattice   (4.2.6)
dodecad   (10.3.14)
dual lattice   (2.2.13)

E

eigenlattice   (2.6.9)
E_n-lattice   (4.2.7), (4.2.47), (4.2.50)
equivalence of codes   (6.1.5)
even lattice   (2.2.11)
exponent   (7.1.1)
extraspecial   (7.2.1)
extraspecial group, type of   (7.2.6)

F

fixed points   (8.2.1)
fourvolution   (2.7.1)
frame   (4.2.13), (4.2.34)
frame, lower   (9.1.7)
Frattini argument   (11.3.6)

G

2/4-generation, 3/4-generation   (8.2.5)
Golay code, definition   (10.2.3)
Golay code, existence   (10.2.5)
Golay code, uniqueness   (10.3.30)
Gram matrix   (2.2.8)
group algebra   (7.3.2)
group of a code   (6.1.5)

H

Hall-Janko group   (11.4.3)
Hamming   (6.1.1)
Hermite   (5.0.1), (5.0.3)

I

Higman-Sims group and graph (11.4.2)
holomorph (7.4.15)

I

index (Witt index) (2.0.8), (2.1.6)
index-determinant formula (2.3.3)
induced representation (7.3.2)
inequality, Cauchy-Schwarz (2.0.10)
inequality, triangle (2.0.10)
integral lattice (2.2.10)
irreducible representation (7.3.1)
isometry, isometry group (2.0.2), (3.2.1)

L

lattice (2.0.9)
Leech lattice (10.2.2)
Leech lattice, standard model Λ (10.2.7)
Leech trio (11.1.1)
length (2.0.10)
lower frame (9.1.7)

M

Mathieu groups (11.2.15)
minimal vectors Barnes-Wall (9.1.2)
minimal vectors Leech lattice (10.4.1)
Minkowski (5.0.1)
monomial, $Mon(n, K)$ (6.1.4)
MV, mv (9.1.2)
$MV(\cdot)$, the minimal vectors of a lattice (10.4.1)

N

Niemeier (10.1.1), (10.2.1)
nondefective (2.1.4)
nondegenerate (2.0.3)
nonsingular (2.2.5)
norm (2.0.10)
normal form for extraspecial groups (7.2.6)

O

octad (10.3.7)
order of $GL(n, q)$ (6.2.1)
order of orthogonal and symplectic groups in characteristic 2 (2.1.7)
orthogonal decomposition of lattice (3.5.2), (3.5.3)

P

parameters of a code   (6.1.8)
perfect field   (2.1.2)
perfect group   (7.1.1)
permutation module   (7.4.1)
primitive vector   (2.0.11)

Q

quadratic form   (2.0.3)
quadratic space   (2.0.8)

R

radical (left, right)   (2.2.1)
rectangular lattice   (5.1.3)
reducible representation   (7.3.1)
Reed-Muller code, $R(k,d)$ (6.2.3)
reflection   (2.5.1)
reflection group   (4.2.2)
representation on quotient module   (4.2.24)
$R(k,d)$ (6.2.3)
$RM(1,d)$ (6.2.4)
root   (2.6.4)
root lattice   (3.5.1)
RSSD   (2.6.7)

S

Schur   (7.3.9)
Schur-Zassenhaus theorem   (7.4.9)
self-orthogonal   (6.1.2)
semidirect product   (3.4.1)
septacode   (6.3.1)
Skolem-Nöther   (7.3.8)
Smith invariant and canonical form   (2.1.8)
square lattice   (5.1.3)
SSD   (2.6.7)
Steiner system   (10.3.6)
Suzuki sporadic group   (11.4.3)

T

tensor product of representations   (7.3.13)
tetracode   (6.1.12)
theta series of a lattice   (10.0.1)
theta series Leech lattice   (10.5.2)
total eigenlattice, $Tel(L,t)$ (2.6.9)

transvection (2.5.3), (2.6.2)
triangle of type abc (11.4.1)
trio of octads (10.3.24)
Turyn style construction of Golay code (10.2.5)
Turyn style construction of Leech lattice (10.7.7)
type (of a lattice vector or of a triangle) (11.4.1)

U

uniqueness for Barnes-Wall lattices (9.2.3)
uniqueness for E_8-lattice (5.2.1)
uniqueness for Golay code (10.3.30)
uniqueness for Leech lattice (10.5.4), (11.1.6)

V

vector of type a (11.4.1)

W

weight (6.1.1)
weight distribution (6.1.10)
Weyl group (4.2.1)
Witt classifications of unimodular lattices (10.0.1)
Witt index (2.0.8), (2.1.6), (7.2.6)
Witt theorem (extension of isometries) (2.1.1)
wreath product (3.4.3)

Appendix A
The Finite Simple Groups

The Orders of the Finite Simple Groups

Alternating group

$Alt_n, n \geq 5$	$\frac{1}{2} \cdot n!$

Group G of Lie type	Order of G = expression below divided by d	d
$A_n(q)$	$q^{n(n+1)/2} \prod_{i=1}^{n}(q^{i+1} - 1)$	$(n+1, q-1)$
$B_n(q), n \geq 1$	$q^{n^2} \prod_{i=1}^{n}(q^{2i} - 1)$	$(2, q-1)$
$C_n(q), n > 2$	$q^{n^2} \prod_{i=1}^{n}(q^{2i} - 1)$	$(2, q-1)$
$D_n(q), n > 3$	$q^{n(n-1)}(q^n - 1) \prod_{i=1}^{n-1}(q^{2i} - 1)$	$(4, q^n - 1)$
$G_n(q)$	$q^6(q^6 - 1)(q^2 - 1)$	1
$F_4(q)$	$q^{24}(q^{12} - 1)(q^8 - 1)(q^6 - 1)(q^2 - 1)$	1
$E_6(q)$	$q^{36}(q^{12} - 1)(q^9 - 1)(q^8 - 1)(q^6 - 1)(q^5 - 1)(q^2 - 1)$	$(3, q-1)$
$E_7(q)$	$q^{63}(q^{18} - 1)(q^{14} - 1)(q^{12} - 1)(q^{10} - 1)(q^8 - 1)(q^6 - 1)(q^2 - 1)$	$(2, q-1)$
$E_8(q)$	$q^{120}(q^{30} - 1)(q^{24} - 1)(q^{20} - 1)(q^{18} - 1)(q^{14} - 1)(q^{12} - 1)(q^8 - 1)(q^2 - 1)$	1
$^2A_n(q), n > 1$	$q^{n(n+1)/2} \prod_{i=1}^{n}(q^{i+1} - (-1)^{i+1})$	$(n+1, q+1)$
$^2B_2(q), q = 2^{2m+1}$	$q^2(q^2 + 1)(q - 1)$	1
$^2D_n(q), n > 3$	$q^{n(n-1)}(q^n + 1) \prod_{i=1}^{n-1}(q^{2i} - 1)$	$(4, q^n + 1)$
$^3D_4(q)$	$q^{12}(q^8 + q^4 + 1)(q^6 - 1)(q^2 - 1)$	1
$^2G_2(q), q = 3^{2m+1}$	$q^3(q^3 + 1)(q - 1)$	1
$^2F_4(q), q = 2^{2m+1}$	$q^{12}(q^6 + 1)(q^4 - 1)(q^3 + 1)(q - 1)$	1
$^2E_6(q)$	$q^{36}(q^{12} - 1)(q^9 + 1)(q^8 - 1)(q^6 - 1)(q^5 + 1)(q^2 - 1)$	$(3, q+1)$

Sporadic Groups

Sporadic simple group			
M_{11}	$7920 = 2^4 \cdot 3^2 \cdot 5 \cdot 11$	He	$2^{10} 3^3 5^2 \cdot 7^3 \cdot 17$
M_{12}	$95040 = 2^6 \cdot 3^3 \cdot 5 \cdot 11$	Ly	$2^8 3^7 5^6 7 \cdot 11 \cdot 31 \cdot 37 \cdot 67$
M_{22}	$443520 = 2^7 \cdot 3^2 \cdot 5 \cdot 7 \cdot 11$	ON	$2^9 3^4 5 \cdot 7^3 \cdot 11 \cdot 19 \cdot 31$
M_{23}	$10200960 = 2^7 \cdot 3^2 \cdot 5 \cdot 7 \cdot 11 \cdot 23$	Co_1	$2^{21} 3^9 5^4 7^2 11 \cdot 13 \cdot 23$
M_{24}	$244823040 = 2^{10} \cdot 3^3 \cdot 5 \cdot 7 \cdot 11 \cdot 23$	Co_2	$2^{18} 3^6 5^3 7 \cdot 11 \cdot 23$
J_1	$175560 = 2^3 \cdot 3 \cdot 5 \cdot 7 \cdot 11 \cdot 19$	Co_3	$2^{10} 3^7 5^3 7 \cdot 11 \cdot 23$
$HJ = J_2$	$2^7 \cdot 3^3 \cdot 5^2 \cdot 7$	Fi_{22}	$2^{17} 3^9 5^2 7 \cdot 11 \cdot 23$
J_3	$2^7 \cdot 3^5 \cdot 5 \cdot 17 \cdot 19$	Fi_{23}	$2^{18} 3^{13} 5^2 7 \cdot 11 \cdot 13 \cdot 17 \cdot 23$
J_4	$2^{21} \cdot 3^3 \cdot 5 \cdot 7 \cdot 11^3 \cdot 23 \cdot 29 \cdot 31 \cdot 37 \cdot 43$	Fi_{24}	$2^{21} 3^{16} 5^2 7^3 \cdot 11 \cdot 13 \cdot 17 \cdot 23 \cdot 29$
HS	$2^9 3^2 5^3 \cdot 7 \cdot 11$	F_5	$2^{15} 3^{10} 5^3 7^2 13 \cdot 19 \cdot 31$
McL	$2^7 3^6 5^3 \cdot 7 \cdot 11$	F_3	$2^{14} 3^6 5^6 \cdot 7 \cdot 11 \cdot 19$
Suz	$2^{13} 3^7 5^2 7 \cdot 11 \cdot 23$	F_2	$2^{41} 3^{13} 5^6 7^2 11 \cdot 13 \cdot 17 \cdot 19 \cdot 23 \cdot 31 \cdot 47$
Ru	$2^{14} 3^3 5^3 7 \cdot 13 \cdot 29$	$F_1 = \mathbb{M}$	$2^{46} 3^{20} 5^9 \cdot 7^6 \cdot 11^2 \cdot 13^3 \cdot 17 \cdot 19 \cdot 23 \cdot 29 \cdot 31 \cdot 41 \cdot 47 \cdot 59 \cdot 71$

Appendix B
Reprints of Selected Articles

We include reprints of two articles, plus a rewritten version of [40], taking into account some corrections [40] and some style changes. The text of the present book has referred to [38, 40] many times. The article [41] represents a natural continuation of the Pieces of 2^d theory described in the first half of [40]. The cubi sum visualiation of codewords in $RM(2,d)$ has been quite useful. We have discontinued use of some terminology in [40].

The titles of the three articles are:

B.1 Pieces of Eight: Semiselfdual Lattices and a New Foundation for the Theory of Conway and Mathieu Groups

B.2 Pieces of 2^d: Existence and Uniqueness for Barnes-Wall and Ypsilanti Lattices

B.3 Involutions on the Barnes-Wall Lattices and Their Fixed Point Sublattices, I.

B.1 Pieces of Eight: Semiselfdual Lattices and a New Foundation for the Theory of Conway and Mathieu Groups

Robert L. Griess, Jr.*

Abstract

We give a new theory of uniqueness of the Leech lattice, based on sublattices which are orthogonal sums of sublattices isometric to $\sqrt{2}$ times the E_8-lattice. We obtain new proofs of existence and properties of the groups of Mathieu and Conway. Our theory has logical advantages.

Keywords and Phrases: Leech lattice, E_8-lattice, Mathieu groups, Conway groups, uniqueness, semiselfdual lattice, semiselfdual involutions.

1 Introduction

We give a new and easy construction of the largest groups of Mathieu and Conway, as the full automorphism groups of the Golay code and Leech lattice, respectively. In addition we get a new uniqueness proof for the Leech lattice and the Golay code. The main technique is intensive use of semiselfdual sublattices: instead of rank 1 lattices as the basis for coordinate concepts, we use scaled versions of L_{E_8}, the E_8 root lattice; the semiselfdual lattices we use most of the time are isometric to $\sqrt{2}L_{E_8}$. While it has been recognized for decades that one can use copies of L_{E_8} to describe the Leech lattice (see [23][33][34]), our uses of it to *create the theory* of Conway and Mathieu groups are new. Using these "smarter coordinates", properties of their automorphism groups and appropriate uniqueness theorems, we get a compact foundation of this theory (see Section 3, esp. (3.7) and (3.19)).

The logical order of our steps is new. We first take a Leech lattice (*any* rank 24 even integral unimodular lattice without vectors of squared length 2) then deduce its uniqueness and properties of its automorphism group, the large Conway group. We use our uniqueness theorems to get transitivity of $\Lambda_2, \Lambda_3, \Lambda_4$ and other configurations of vectors and sublattices *without listing members of these sets or even knowing* $|Aut(\Lambda)|$, in contrast with earlier treatments (some of our transitivity results may not be in the literature). Next, we deduce existence and uniqueness of the Golay code, then existence and order of the Mathieu group, M_{24}. Simple

---

*Department of Mathematics, University of Michigan, Ann Arbor, MI 48109 U.S.A. E-mail: rlg@umich.edu

observations of the Golay code then give immediate results about permutation representations of M_{24}. Only minimal examination of particular codes is ever done, in contrast with [4][7][8][18]. Our logical sequence of *first* obtaining the Conway group, *then* the Mathieu groups, is in defiance of the classical theories for these groups. Our characterization of the Leech lattice is a logical improvement over that of [5], which *depends on* the characterization of the Golay code because we *deduce* its characterization.

The construction of the Leech lattice we give (3.3.i) is not really new (see [20][33][34]), but our treatment of uniqueness and analysis of the automorphism group is different, notably in avoidance of Conway's characterization [5] and avoidance of displaying explicit "extra automorphisms" with respect to a frame basis [4][6][18]. Furthermore, our foundation of the theory of Conway and Mathieu groups is "elementary", if one takes the structure of the E_8 lattice, its automorphism group and basic lattice management (2.1) for granted. We emphasize that the present article contains a complete proof of this foundation (modulo standard background material about lattices and finite groups in Section 2 and the Appendices), a fact which should be taken into account in making comparisons.

Our break with the past is not complete since we still rely on the theta series of an E_8 lattice and a Leech lattice and in (3.4), (3.7), (3.17), (3.18), we have to use a few elementary properties of a code associated to a frame (A.4)(A.8) (this code turns out to be the Golay code, but we do not need to quote its characterization). Possibly, reliance on the theta series can be lessened; for instance, one would like a direct, elementary proof of the fact that in a Leech lattice, Λ, the orthogonal of a sublattice isometric to $\sqrt{2}L_{E_8}$ contains a copy of $\sqrt{2}L_{E_8}$ (and even $\sqrt{2}L_{E_8} \perp \sqrt{2}L_{E_8}$). Even the weaker statement that a vector x in the Leech lattice contains a copy of $\sqrt{2}L_{E_8}$ in the sublattice $\{y \in \Lambda \mid (x,y) \in 2\mathbb{Z}\}$ (the "annihilator of x mod 2") would be useful (and seems hard to prove without using theta series).

We introduce *semiselfdual involutions* in (2.5), a concept with potential for wider applications. In case one wishes to follow the spirit of an earlier construction, one can choose an extra automorphism from our family of semiselfdual involutions (2.5)(3.6)(5.3).

We thank Dan Frohardt, George Glauberman and Jacques Tits for useful comments. For financial support during work on this article, we acknowledge NSF Grant DMS 9623038 and University of Michigan Faculty Recognition Grant (1993-96).

2 Basic notations and definitions

Throughout this article, $(\cdot,\cdot)$ denotes a positive definite bilinear form on a lattice of finite rank or a finite dimensional rational vector space. If L is a lattice, we write $\mathbb{Q}L$ for the ambient rational vector space, which may be identified with $\mathbb{Q} \otimes L$. Groups act on the right. The notation $A.B$ stands for a group extension, with extension kernel A and quotient B, with $A{:}B$ and $A{\cdot}B$ denoting split and nonsplit extensions, respectively. If m is an integer, m^n denotes the direct product of n cyclic groups of order n and $m^{n+\cdots+q+r}$ denotes a compound extension corresponding copies of $W_i := Aut(M_i) \cong W_{E_8}$. The choices for θ form an orbit under the natural action of W_1 or of W_2; its stabilizer is a diagonal subgroup W_{12} of $W_1 \times W_2$.

The subgroup S_{12} of W_{12} stabilizing the sublattice T is isomorphic to a subgroup of W_{E_8} of shape $2^{1+6}GL(4,2)$, (A.3); S_{12} acts on T^*/T as $GL(4,2)$ (since $[M_i, O_2(S_{12})] = N_i$ and $[N_i, O_2(S_{12})] = 2M_i$, for $i = 1, 2$; see (3.2)). The set of isometries $T^*/T \to M_3^*/M_3$ forms a single orbit under the natural action of W_3. It follows that the isometry of (ii) has stabilizer S contained in $S_{12} \times W_3$ and is isomorphic to $[2 \times 2_+^{1+6}]GL(4,2)$; the projection of S to $W_1 \times W_2$ is S_{12} and the projection to W_3 is isomorphic to $2.GL(4,2)$ since the normal subgroup of shape 2^{1+6} acts trivially on T^*/T. □

(3.4) Corollary. If L is a Leech lattice and M is a sublattice isometric to $\sqrt{2}L_{E_8}$, then isomorphism type of $M^\perp$ is determined.

Proof. Since L is unimodular and M is a direct summand, $\det(M^\perp) = \det(M) = 2^8$. We finish with (3.3.ii) if we find a sublattice of $M^\perp$ isometric to $\sqrt{2}L_{E_8} \perp \sqrt{2}L_{E_8}$. This follows from (A.6,7,8). □

(3.5) Proposition. Choose a quadratic form on $\mathbb{Q}^{24}$ so that $\mathbb{Q}^{24}$, endowed with this form, contains a Leech lattice, say Λ. Then $\mathbb{Q}^{24} = \mathbb{Q}\Lambda$ and $O(\mathbb{Q}^{24})$ acts transitively on the following sets:

(i) pairs (L, M), where L is a Leech lattice in $\mathbb{Q}\Lambda$ and M is a sublattice of L isometric to $\sqrt{2}L_{E_8}$;

(ii) triples (L, M_1, M_2), where L is a Leech lattice in $\mathbb{Q}\Lambda$ and M_1, M_2 is a pair of orthogonal sublattices, each isometric to $\sqrt{2}L_{E_8}$;

(iii) quadruples (L, M_1, M_2, M_3), where L is a Leech lattice in $\mathbb{Q}\Lambda$ and (M_1, M_2, M_3) is a Leech trio in L.

Proof. (iii) Use (3.3.iii).

(i) A proof may be obtained from the ideas in the proofs of (3.3.ii,iii) and (3.4).

(ii) It suffices, by (iii), to prove that this ordered pair is part of a Leech trio. Define $Q := M_1 \perp M_2$, $M_3 := Q^\perp \cap L$. Since the invariants of Q are (2^{16}), the invariants of both M_3 and $R := M_3^\perp \cap L$ are (2^a), for some $a \leq 16$. Since rank$(M_3) = 8$, $a \leq 8$. We have $|R:Q| = 2^{8-a/2}$ (2.1). If $a < 8$, then for some $k \in \{1,2\}$, $R_k := R \cap \mathbb{Q}M_k > M_k$. Since L is an even lattice, R_k/M_k must be a totally singular subspace with respect to the natural nonsingular bilinear form on M_k^*/M_k; by (A.2.i) (applied to $M_k^* \cong \frac{1}{\sqrt{2}}L_{E_8}$), $R_k \setminus M_k$, hence L, contains a root, a contradiction. So, $a = 8$, whence $M_3 = 2M_3^*$ is doubly even and so $M_3 \cong \sqrt{2}L_{E_8}$ (A.9). □

(3.6) Corollary. $Aut(\Lambda)$ acts transitively on the set of sublattices isometric to $\sqrt{2}L_{E_8}$. We also have a conjugacy class of involutions in $Aut(\Lambda)$, the SSD involutions (2.5) associated to these sublattices.

(3.7) Corollary. Any two Leech lattices are isometric.

Proof. Any Leech lattice has a Leech trio (A.7). Now use (3.5). □

At this point, we know little about $Aut(L)$ beyond some transitivity properties. We need to study the sublattices which occur as $M^\perp$, for a sublattice

$M \cong \sqrt{2}L_{E_8}$.

(3.8) Theorem. (i) In a Leech lattice, Λ, let M be a sublattice isometric to $\sqrt{2}L_{E_8}$ and set $T := M^\perp$. Then $Aut(T)$ is an extension $2^{1+8}\Omega^+(8,2)$.

(ii) The noncentral involutions of $Aut(T)$ which lie in $O_2(Aut(T))$ form a single $Aut(T)$ conjugacy class. For such an involution, the sum of the fixed point sublattice and the negated sublattice is isometric to $\sqrt{2}L_{E_8} \perp \sqrt{2}L_{E_8}$.

(iii) There is a bijection between the involutions of (ii) and ordered pairs of orthogonal sublattices of T, each isometric to $\sqrt{2}L_{E_8}$. If $M_1 \cong M_2 \cong \sqrt{2}L_{E_8}$ are orthogonal sublattices of T, then $Stab_{Aut(T)}(M_1) \cap Stab_{Aut(T)}(M_2) \cong 2 \times 2^{1+6}GL(4,2)$ and the image of this group in $Aut(M_i)$, for $i = 1, 2$, is of the form $2^{1+6}GL(4,2)$.

(iv) The set of unordered pairs of sublattices as in (ii) is in bijection with the set of maximal totally singular subspaces of $T/2T$ (all of which have dimension 12 and contain the 8-dimensional radical).

(v) $T \cap 2\Lambda = 2M_1 + 2M_2 + N_{12} = 2T$ and $T/T \cap 2\Lambda \cong 2^{16}$; also $T + M + 2\Lambda = T + 2\Lambda$.

(vi) The actions of $Stab(M)$ on T^*/T and $M^*/M = \frac{1}{2}M/M \cong M/2M$ are equivalent and may be identified with the action of $\Omega^+(8,2)$ on the space $\mathbb{F}_2^8$ stabilizing a nondegenerate quadratic form. (We remark that these actions are equivalent to the irreducible action on $U/U \cap 2\Lambda \cong U + 2\Lambda/2\Lambda$ but not to T/U, where U is the radical modulo 2 of T.)

Proof. First, we show that $B := \{a \in Aut(T) \mid [T,a] \leq U\} \cong 2_+^{1+8}$.

Take the sublattice $Q := M_1 \perp M_2$ of T, with $M_j \cong \sqrt{2}L_{E_8}$, for $j = 1, 2$, as in (3.2). Note that $U = N_1 + N_2 + M_{12}$. Let A be the stabilizer in $Aut(T)$ of Q. Note that A lies in a wreath product $[Aut(M_1) \times Aut(M_2)]2$ and that A has the form $[2_+^{1+6} \times 2_+^{1+6}].[GL(4,2) \times 2]$; see (A.3) and consider the subgroup of the above wreath product which stabilizes N_1 and N_2. In more detail, let $R_i \cong 2_+^{1+6}$, for $i = 1, 2$ be the normal subgroup of $Stab_{Aut(M_i)}(N_i)$ as in (A.3); then $B \leq A$ has the form $R_{12} \circ E$, a central product, where R_{12} is diagonally embedded in $R_1 \times R_2$, E is dihedral of order 8, $E \cap O_2(A) \cong 2 \times 2$ and E contains an involution which interchanges R_1 and R_2 under conjugation. The statement about B follows.

Since B acts absolutely irreducibly on T, $C(B)$ consists of scalar matrices and so the quotient $Aut(T)/BC(B)$ embeds in $Out(B) \cong O^+(8,2)$ (C.3). Our subgroup A contains B and maps onto a parabolic subgroup P of $Out(B)'$ of the form $2^6 \colon GL(4,2)$.

We may do the above for any sublattice of T which is isometric to $M_1 \perp M_2$. One such sublattice is $\frac{1}{2}N_{12;\theta} \perp \frac{1}{2}N_{12;-\theta}$, and we thereby get a subgroup isomorphic to P and distinct from P. Since P is a maximal parabolic subgroup of $Out(B)' = \Omega^+(8,2)$ [3], these two subgroups generate $Out(B)'$. It follows that $Aut(T)/B \cong O^+(8,2)$ or $\Omega^+(8,2)$.

There are several ways to see that $Aut(T)/B \not\cong O^+(8,2)$: (a) a subgroup H of $GL(\mathbb{C} \otimes T)$ which contains B as a normal subgroup and with quotient $O^+(8,2)$ has the property that certain elements of $H \setminus H'$ have traces of the form $\varepsilon 2^{c/2}$, where ε is a root of unity and c is an odd integer (C.4); since the representation of $Aut(T)$ on T is rational, this does not happen; (b) study the centralizer in

A of B_1, a 2_+^{1+6} subgroup of B; $C(B_1)$ lies in $GL(2,\mathbb{Q})$ and contains a copy of $C_B(B_1) \cong Dih_8$, which is a maximal finite 2-subgroup in $GL(2,\mathbb{Q})$; (c) in case $Aut(T)/B$ were $O^+(8,2)$, we would get a contradiction when we examine the structure of the frame group; see (A.5).

We now have (i). For (ii), it is clear since $Aut(T)$ has index 2 in a holomorph of an extraspecial group (Appendix C), that the noncentral involutions in $O_2(Aut(T))$ form a single conjugacy class. Since two such involutions are the central involutions of R_1 and R_2, the connection with the stated sublattices follows. For (iii), note that such a decomposition (with ordered summands) leads to a SSD involution t (-1 on M_1, 1 on M_2) which preserves T. It remains to show that t satisfies $[T,t] \leq U$. It is clear from the proof of (3.3.ii) that there are appropriate sublattices $N_k \leq M_k$ and exactly two isometries $\pm\theta : M_1 \to M_2$ so that $T = M_1 + M_2 + \frac{1}{2}N_{12}$ as in (3.2). Then we get $[T,t] = N_1 \leq U = N_1 + N_2 + M_{12}$, as required.

For (iv), note that $Aut(T)$ induces $\Omega^+(8,2)$ on $T/U \cong 2^8$ and that $M_1 \perp M_2/U$ represents a maximal totally singular subspace.

The first part of (v) follows since T is a direct summand of Λ. From (2.3), we get that $T + M \geq 2\Lambda$ and $T + M/2\Lambda$ has codimension 8 in $\Lambda/2\Lambda$. Since the image of T in $\Lambda/2\Lambda$ has dimension 16, $T + 2\Lambda = T + M + 2\Lambda$.

For the first part of (vi), just note that Λ projects onto each of M^*/M and T^*/T. Since $T^* = \frac{1}{2}U$ and $U \cap 2\Lambda = T \cap \Lambda = 2T$, $U/2T \cong T^*/T$ as modules for $Stab(M)$; these modules are self-dual. Clearly, $\Lambda/2\Lambda$ has a composition series with factors $U/2T \cong U + 2\Lambda/2\Lambda$, T/U ($T + 2\Lambda/2\Lambda$ is the annihilator of $U + 2\Lambda/2\Lambda$) and finally Λ/T (isomorphic to $\Lambda/2\Lambda$ modulo the annihilator of $U + 2\Lambda/2\Lambda$). The first and third are dual, hence isomorphic. The middle factor turns out to be not isomorphic to these, but the proof is perhaps not easy (this fact is not a necessary part of our theory; anyway, here is a nonelementary proof: we take an element x of order 3 in $Stab(M)$ for which 1 occurs as an eigenvalue in the first and third factor with multiplicity 6; so, on Λ, the multiplicity of 1 is at least 12; since we know the classes of elements of order 3 in $Aut(\Lambda)$ [7] [18]), we deduce that x has 1 with multiplicity exactly 12 and so on the middle composition factor, x does not have 1 as an eigenvalue). $\square$

(3.9) Corollary. Notation as in (3.8.i). The group $Stab_{Aut(\Lambda)}(M) = Stab_{Aut(\Lambda)}(M^\perp)$ is of the form $2_+^{1+8}.W'_{E_8}$ (though we do not need it, we mention that this is a nonsplit extension (C.3)) and it induces W'_{E_8} on M and $Aut(T) \cong 2_+^{1+8}\Omega^+(8,2)$ on $M^\perp$.

Proof. Because of the decompositon (3.2), it is clear that $Stab(M) = Stab(M^\perp)$ induces exactly W'_{E_8} on M (3.8). The normal extraspecial group is generated by its involutions. By (3.8.ii) and how the involutions may be interpreted as SSD maps that act trivially on M, the normal subgroup of $Stab(M)$ of shape 2_+^{1+8} acts trivially on M. $\square$

(3.10) Corollary. If (M_1, M_2, M_3) is a Leech trio in Λ, its stabilizer in $Aut(\Lambda)$ is of the form $2^{3+12}GL(4,2)$, of order $2^{21}3^25.7$. The stabilizer of the unordered trio has the form $2^{3+12}[GL(4,2) \times Sym_3]$, a group of order $2^{22}3^35.7$. This subgroup determines the Leech trio by the three nontrivial linear characters of the normal eights group which occur in $\mathbb{Q}^{24} = \mathbb{Q}\Lambda$ (each with multiplicity 8).

Proof. Let B and A be these respective subgroups. We have $A/B \cong Sym_3$ (3.5.iii) and $C_B(M_3) \cong 2 \times 2^{1+6}$ which acts on M_1, M_2 with respective kernels the two direct factor of order 2; see the proofs of (3.5.iii) and (3.8). The result follows since A embeds in the subgroup $H \cong [2_+^{1+6}GL(4,2)] \wr Sym_3$ of $Aut(M_1 \perp M_2 \perp M_3)$, where the wreathing is done with the natural degree 3 action; in other words, B is forced to have shape 2^{3+12} and be the unique normal subgroup of A of index 2^6 in $O_2(H)$. The last statement is trivial. $\square$

(3.11) Definition. Write $\Lambda_n := \{x \in \Lambda \mid (x,x) = 2n\}$, the set of lattice vectors of *type n*.

(3.12) Lemma. (i) $Aut(\Lambda)$ acts transitively on the set of pairs (x, M), where M is a sublattice isometric to $\sqrt{2}L_{E_8}$ and $x \in M$ has type 4;
(ii) $Stab((x,M)) \cong [2^{1+8+6}]Alt_7 (\leq 2^{1+1+8+6}\Omega^+(6,2))$ and its order is $2^{18}3^25.7$;
(iii) $\Omega^+(6,2) \cong Alt_8$.

Proof. By (3.6), it suffices for (i) to prove transitivity of $Stab(M)$ on $M \cap \Lambda_4$. Define an equivalence relation on $M \cap \Lambda_4$ by congruence modulo $2M$. Each class consists of 16 vectors, two of which are orthogonal or opposite (A.2). Now, $Stab(M)$ induces W'_{E_8} on M (3.9), which is transitive on the set of equivalence classes since they correspond to singular points in $M/2M$. Let K be a class and a, b, c, d linearly independent vectors in K. Then $\frac{1}{2}(a-b)$ and $\frac{1}{2}(c-d)$ are roots and the product of the corresponding reflections interchanges a and b. Transitivity follows since all "even" transformations on M come from the action of $Stab(M)$ (3.9). Note that the group $Stab(M)$ has a permutation representation of degree 8 on $K/\{\pm 1\}$ and that the simple group $\Omega^+(6,2)$ has order $8!/2$. $\square$

(3.13) Proposition. Let $T_n := \{x \in T \mid (x,x) = 2n\}$. Then
(i) $Aut(T)$ has one orbit on T_2 (length $4320 = 2^5 3^3 5$, stabilizer of shape $2^{4+6}\Omega^+(6,2)$) and it has two orbits on T_4, size 522720; one orbit in U, the radical modulo 2 (length $4320 = 2^5 3^3 5$, stabilizer of shape $2^{4+6}\Omega^+(6,2)$), and the second orbit outside U (length $518400 = 2^8 3^4 5^2$, stabilizer of shape $2^{4+3+3}GL(3,2)$).
(ii) $Aut(T)$ has one orbit on T_3 (length $61440 = 2^{12}3.5$). A stabilizer is isomorphic to $Sp(6,2)$.
(iii) The theta series for T begins $1 + 4320q^2 + 61440q^3 + 522720q^4 + \cdots$.

Proof. (i) Since $Aut(T)$ induces $\Omega^+(8,2)$ on T/U (3.8.vi), any singular coset in T/U lies in a natural sublattice of the form $M_1 \perp M_2$, with $M_i \cong \sqrt{2}L_{E_8}$. Since $U \cap M_i \cong 2L_{E_8}$, (3.5.iii), (3.10) and (A.3.ii) imply the statement about elements of type 2 and show that the elements x of type 4 in $M_i \setminus N_i$ lie in one $Aut(T)$-orbit. To get the cardinality of the first orbit, count ordered triples (x, M_1, M_2), where $x \in T_2 \cap M_1$, $M_1 \cong M_2 \cong \sqrt{2}L_{E_8}$, $(M_1, M_2) = 0$. Such ordered pairs M_1, M_2 correspond to the $2 \cdot 135 = 2.3^3 5$ noncentral involutions of $O_2(Aut(T))$. Given M_1, there are $240 = 2^4 3.5$ such x, hence $2^5 3^4 5^2$ such triples. In T/U, there are 15 totally singular subspaces containing the singular vector $x + U$, and such subspaces have the form $M_1 + M_2/U$. So the number of such x is $2^5 3^4 5^2/15 = 2^5 3^3 5 = 4320$. Since $U \cong \sqrt{2}T$, we get the same count for the orbit $T_4 \cap U$. Now, let S be a stabilizer for this orbit, $|S| = 2^{21}3^5 5^2 7/2^5 3^3 5 = 2^{16} 3^2 5.7$. Since $-1_T \notin S$, $S \cap O_2(Aut(T))$

is elementary abelian of rank at most 4. Since $S/S \cap O_2(Aut(T))$ embeds in a subgroup of shape $2^6{:}\Omega^+(6,2)$, this embedding is onto and $S \cap O_2(Aut(T)) \cong 2^4$.

For $T_4 \setminus U$, we have transitivity if we show that every element is in the $Aut(T)$-orbit of an element of $M_i \setminus N_i$. Clearly every element of type 4 in $M_1 \perp M_2$ has the form $x = x_1 + x_2$, $x_i \in M_i$, where (a) one of the summands is 0 (whence x has the desired form); or (b) each x_i has type 2. Assume the latter for x and, by way of contradiction assume that (*) x is not in the orbit of an element from (a). Then by (3.8.ii), no involution of $O_2(Aut(T))$ fixes x, so the orbit of x under $O_2(Aut(T))$ has length 2^9 and lies in a single coset of U. Because U is the radical modulo 2, if L is the span of the orbit of $\frac{1}{2}x$, L is an even integral lattice containing at least 512 roots. Since the action of $Aut(T)$ on $\mathbb{Q}T$ is an absolutely irreducible representation, the action of $Aut(Q)$ on the orthogonally indecomposable summands of L is transitive, so all have the same isometry type and are generated by roots. So, the roots of L form a system of type D_m^n, A_m^n or E_8^2, for $mn = 16$. No such lattice has 512 roots, a contradiction. We count pairs $(x, M_1 + M_2)$ with M_1, M_2 as above, with x a type 4 vector in $M_1 + M_2 \setminus U$; in each M_i, there are $2160 - 240 = 1920 = 2^7 3.5$ such x, and the remaining x have the form $x_1 + x_2$, where $x_i \in M_i$ has type 2. The number of such pairs is 240^2, but the requirement $x \notin U$ requires us to remove 240.2^4 such pairs. Given $M_1 + M_2$, there are $2^7 3.5 + 2^7 3.5 + 2^8 3^2 5^2 - 2^8 3.5 = 2^8 3^2 5^2$ such x. The number of such $M_1 + M_2$ is $135 = 3^3 5$, so the number of such pairs is $2^8 3^5 5^3$. Since these pairs form an orbit under $Aut(T)$ and there are 15 such $M_1 + M_2$ containing a given x, the cardinality of $T_4 \setminus U$ is $2^8 3^5 5^3 / 15 = 2^8 3^4 5^2$.

A stabilizer S for x in this orbit has order $2^{21} 3^5 5^2 7 / 2^8 3^4 5^2 = 2^{13} 3.7$. Since -1_T does not stabilize S, $S \cap Z(Aut(T)) = 1$ and $S \cap O_2(Aut(T))$ is an elementary abelian group of rank at most 4. We may take x to have the form (b) as above. Then, since $S/S \cap O_2(Aut(T))$ embeds in a subgroup of shape $Sp(6,2)$, the order forces $S/S \cap O_2(Aut(T)) \cong 2^{3+3} GL(3,2)$ because the index prime to 2 means the subgroup is a parabolic [3], and the order allows one parabolic, up to conjugacy.

(ii) An element x of T_3 maps to a nonsingular vector in T/U, where $Aut(T)$ acts as $\Omega^+(8,2)$, so is transitive on such vectors with the stabilizer of $x + U$ an $Sp(6,2)$-subgroup. We claim that the only element of $O_2(Aut(T)) \cong 2^{1+8}_+$ which stabilizes $\{x, -x\}$, for $x \in T_3$, is $\{\pm 1\}$. But this is clear from (3.8.ii) where it is shown that the noncentral involutions of $O_2(Aut(T))$ have, for their action on T, fixed point sublattices isometric to $\sqrt{2}L_{E_8}$; these contain no vectors of odd type. We next claim that $O_2(Aut(T))$ acts transitively on $\Lambda_3 \cap (x + U)$. Since two elements of type 3 are congruent modulo 2Λ if and only if they are equal or negatives (an easy exercise), for any $y \in \Lambda_3 \cap (x+U)$, the map $O_2(Aut(T))/Z(O_2(Aut(T))) \to U/2T$ derived from $g \mapsto y - y^g + 2T$ is injective, and this proves the claim since $U/2T \cong 2^8$. Therefore, $|T_3| = 2^9.120 = 2^{12} 3.5 = 61440$. A stabilizer for this orbit meets $O_2(Aut(T))$ trivially and embeds in $Sp(6,2)$ since these vectors have odd type. Since this $|\Omega^+(8,2)|/|Sp(6,2)| = 120$, we deduce that this embedding is onto.

(iii) This follows from (i) and (ii). Also, see (A.10) and (A.11). □

(3.14) Lemma. If $x \in \Lambda_2 \cup \Lambda_3$, $A(x) := \{y \in \Lambda \mid (x,y) \in 2\mathbb{Z}\}$, the "annihilator mod 2", contains a sublattice isometric to $\sqrt{2}L_{E_8}$.

Proof. Let (M_1, M_2, M_3) be a Leech triple and let $x = x_1 + x_2 + x_3$, where x_i is the projection of x to the rational span of M_i. Then each $x_i \in M_i^* = \frac{1}{2}M_i$, whence $n_i := (x_i, x_i)$ is a nonnegative integer. If some n_i is 0, we are done, so assume all are positive.

Since Λ is even, we may reindex to assume $n_3 > 0$ is even. Let $y := x_1 + x_2$. Each x_i is nonzero. If x has type 2, $(y, y) = 2$ and $y \in T^* \cong \frac{1}{\sqrt{2}}T$; we quote (3.13.i) to transform x by $Stab(M_3)$ to an element where some n_i is 0. If x has type 3, y has type 1 or 2 and a similar use of transitivity (3.13) works (there are cases: $y \in T$ implies $x_3 \in M_3$ and this is impossible since minimum squared lengths in T and M_3 are 4; if $y \notin T$, then x_3 has type 1 or 2 and y has type 2 or 1). $\square$

(3.15) Lemma. (i) Suppose that $x \in \Lambda_n$ and $M \leq \Lambda$, $M \cong \sqrt{2}L_{E_8}$ satisfies $(M, x) \leq 2\mathbb{Z}$ (such M exist if $n \leq 3$, by (3.14)). Then $x \in M \perp (M^\perp \cap \Lambda)$, so we write $x = u + v$, $u \in M$, $v \in M^\perp$.

(ii) If $n = 2$, $u = 0$ or $v = 0$. In case $u = 0$, there is a sublattice $M_1 \perp M_2$ of $M^\perp$ as in (3.2) containing x.

(iii) If $n = 3$, $x = v \in M^\perp$.

Proof. For (i), use (2.3.ii). The remaining statements follow from (3.13).

(3.16) Theorem. $Aut(\Lambda)$ is transitive on Λ_2.

Proof. From (3.13.ii), a type 2 vector lies in a $\sqrt{2}L_{E_8}$ sublattice, say M, of Λ. From (3.9), we know that $Stab(M)$ induces W'_{E_8} on M, so all vectors of type 2 in M lie in a single orbit under $Stab(M)$. $\square$

(3.17) Theorem. $Aut(\Lambda)$ is transitive on
(i) pairs (M, x), where $M \cong L_{E_8}$ is a sublattice and $x \in \Lambda_3 \cap M^\perp$;
(ii) Λ_3;
(iii) quadruples (M_1, M_2, M_3, x), where (M_1, M_2, M_3) is a Leech trio and $x \in \Lambda_3 \cap M_3^\perp$.

The stabilizer of pair as in (i) has the form $2.Sp(6,2)$, order $2^{10}3^45.7$, and the stabilizer of a quadruple as in (iii) has the form $2.2^{3+3}.GL(3,2)$, order $2^{10}3.7$.

Proof. (i) follows from (3.13). Note that (3.15) tells us that any type 3 vector is part of a pair as in (i), so (ii) follows from (i). For (iii), use Witt's theorem to see that H, the stabilizer in $Aut(T)$ of a maximal totally singular subspace in T/U, is transitive on nonsingular vectors. As in the proof of (3.13), we know that if $x \in T$ has type 3, $O_2(Aut(T))$ is transitive on the type 3 vectors in $x + U$, so we get the stabilizer for (i). The stabilizer for (iii) has index 135 in the stabilizer for (i), so by surveying the maximal parabolics for $Sp(6,2)$, we get the indicated subgroup. $\square$

(3.18) Lemma. Let $x \in \Lambda_4$. Then the set of sublattices $M \cong \sqrt{2}L_{E_8}$ which contain x form an orbit of length 253 under $Stab_{Aut(\Lambda)}(x)$.

Proof. (A.4), (A.7), (3.12.i). $\square$

(3.19) Theorem. $Aut(\Lambda)$ is transitive on frames (A.4) and on Λ_n, for $n = 0, 2, 3, 4$; $|Aut(\Lambda)| = 2^{22}3^95^47^211.13.23$.

Proof. By (3.16) and (3.17), it suffices to prove transitivity on Λ_4. Let

$G = Aut(\Lambda)$. Any type 4 element lies in a frame and any frame gives rise to a Golay code (A.4) and so we may take our element x of type 4 and embed it in a pair (M, x) as in (A.7)(3.12.i). Transitivities on Λ_4 and frames follow (3.12). The number of such pairs containing x is 253 (3.18). Then, we have $|G : G_{(M,x)}| = |G : G_x||G_x : G_{(M,x)}|$. The first factor on the right side is $u_4 = 2^4 3^7 5^3 7.13$ (A.1.ii) and the second is 253 (A.6). We conclude that $|G : G_{(M,x)}| = 2^4 3^7 5^3 7.11.13.23$, whence (3.12.ii) $|G| = 2^{22} 3^9 5^4 7^2 11.13.23$. □

(3.20) Theorem. Let Λ be any Leech lattice. Then, for any frame Σ, there is an ordered basis $\Sigma^+ \subset \Sigma$ and a code so that Λ is as described in (A.4). The stabilizer Σ in $Aut(\Lambda)$ has the form $D : P$ (A.4.1), (A.5).

(3.21) Remark. The preceeding statement is not trivial. Moving from a frame plus Golay code to a containing Leech lattice involves work over $\mathbb{Z}/4\mathbb{Z}$, where the sign problems are harder to deal with than over $\mathbb{Z}/2\mathbb{Z}$, e.g. [18], Appendix 9A.

(3.22) Remark. (The TU-tower) We call attention to the chain $\cdots < 2T < U < T < \frac{1}{2}U < \cdots$ whose every member is a scaled copy of T by a power of $\sqrt{2}$ (3.2). This is an analogue of (2.9).

We get another nice uniqueness result, which makes the Barnes-Wall lattice a member of the Broué-Enguehard series; see (A.10) and (A.11).

(3.23) Theorem. There is a unique even integer lattice which has rank 16, theta series which begins like $1 + 4320q^2 + \cdots$ and has no vectors of type 2 in its radical modulo 2.

Proof. Let T be such a lattice and $U := \{x \in T \mid (x, T) \leq 2\mathbb{Z}\}$, the radical modulo 2. As usual, for an even integral lattice, L, we write L_n for $\{x \in L \mid (x, x) = 2n\}$. Our assumption may be expressed: $T_1 = \emptyset = U_2$ and $\text{rank}(T) = 16$.

The main thing we have to establish is that the relation "congruence modulo U" is an equivalence relation on T_2 whose classes consist of "frames", that is, 16-sets of type 2 vectors, two of which are proportional or orthogonal. So, let x, y be nonproportional members of T_2 such that $x + U = y + U$. Then, $x + y = u \in U$ and so $(x, x + y) \in 2\mathbb{Z}$, whence $(x, y) \in 2\mathbb{Z}$. We want $(x, y) = 0$. If nonzero, we may replace y by $-y$ if necessary to arrange $(x, y) < 0$. Then $(x, y) \leq -2$ and $0 < (x + y, x + y) \leq 4 + 4 - 4 = 4$, whence $x + y \in U_2 = \emptyset$, a contradiction. So, each class has at most 16 vectors.

Since $4320/16 = 135 > 2^7$, it follows that the finite abelian 2-group T/U is elementary abelian and that the nonsingular quadratic form on it inherited from T has maximal Witt index (the other possibility, a quadratic form of nonmaximal Witt index, would have exactly 119 singular points). It follows that the 4320 elements of T_2 are distributed among these 135 singular cosets of U in T, whence each class has exactly 16 members.

In case $\frac{1}{2}U/T$ has maximal Witt index (for its form taking values in $\frac{1}{2}\mathbb{Z}/\mathbb{Z}$), we follow an idea from Section 3. Let $Q \cong \sqrt{2}L_{E_8}$. There is an isometry of $\frac{1}{2}Q/Q$ and $\frac{1}{2}U/T$, so can take Λ to be the preimage in $\frac{1}{2}Q \perp \frac{1}{2}U$ of the diagonal subgroup of $\frac{1}{2}Q/Q \times \frac{1}{2}U/T$. Then, Λ is an even unimodular lattice with no roots (since the minimum norm in T is 4), so L is isometric to the standard Leech lattice (3.7).

From (3.4), we deduce that our T is isometric to the lattice T of (3.2), and we are done.

The isometry type of $\frac{1}{2}U/T$ has not been established. However, we did prove that T/U has maximal Witt index, so if we use $\frac{1}{\sqrt{2}}T, \frac{1}{\sqrt{2}}U$ for $\frac{1}{2}U, T$, respectively, in the argument of the last paragraph, we get a rootless even unimodular lattice since the minimum norm in T is 4 and we deduce from (3.4) that the isometry type of $\frac{1}{\sqrt{2}}U$ is determined. Uniqueness of T then follows from (3.22). □

4 The Mathieu group, with the Golay code (finally!)

(4.1) Definition. A *Golay code*, $\mathcal{G}$, is a binary code of length 24, dimension 12 and minimum weight *at least* 8. A group M_{24} is the automorphism group of a binary Golay code. We use the notations of (A.4), (A.5).

(4.2) Theorem. (Uniqueness of the binary Golay code) There is a unique (up to equivalence, i.e., coordinate permutations) binary code of length 24, dimension 12 and minimum weight at least 8 (hence equal to 8).

Proof. Existence of such codes comes from (A.4.3) and existence of a Leech lattice. To prove uniqueness, we suppose that $\mathcal{C}$ is any Golay code. We define a lattice, L with $\mathcal{C}$ following the recipe in (A.4). Let $\Omega := \{1,\ldots,24\}$ be an index set and $\{x_i \mid i \in \Omega\}$ be a basis of Euclidean space $\mathbb{R}^\Omega$ such that $(x_i, x_j) = 2\delta_{i,j}$. For $A \subseteq \Omega$, define $x_A := \sum_{i \in A} x_i$, $\nu_i := -x_i + \frac{1}{4}x_\Omega$. Let L be the span of all $2x_i$, $\pm x_i \pm x_j$, ν_i and all $\frac{1}{2}x_A$, $A \in \mathcal{C}$. Then L is an EUL without vectors of type 2, hence is isometric to Λ (3.7). By transitivity of $Aut(\Lambda)$ on frames (3.19), we may assume that $\{\pm 2x_i \mid i \in \Omega\}$ corresponds by our isometry to the standard frame (A.4), whence $\mathcal{C} \cong \mathcal{G}$. □

(4.3) Proposition. $Aut(\mathcal{G}) = P$ and $|P| = 2^{10}3^35.7.11.23$ (see (A.5) for the definition of P; the big Mathieu group M_{24} is *defined* to be $Aut(\mathcal{G})$, so we get $|M_{24}| = 2^{10}3^35.7.11.23$).

Proof. (A.5), (3.19). □

(4.4) Proposition. P acts transitively (i) on the set of octads; (ii) and on Ω.

Proof. (i) If $\mathcal{O}$ and $\mathcal{O}'$ are octads, then $M(\mathcal{O})$ (A.6) contains some $x \in \Sigma$, the standard frame (A.4), and $M(\mathcal{O}')$ contains some $x' \in \Sigma$. Now use (3.12.i) and note that an element of $Aut(\Lambda)$ which carries the first pair to the second stabilizes Σ.

(ii) Given $i, j \in \Omega$, expand each to an octad to create pairs $(M(\mathcal{O}), 2\alpha_i)$ and $(M(\mathcal{O}'), 2\alpha_j)$. Now use (3.12.i) and (3.19). □

(4.5) Proposition. The stabilizer in P of an octad is isomorphic to $AGL(4,2)$, the affine general linear group, which is a semidirect product of the general linear group $GL(4,2)$ by the group of translations, isomorphic to 2^4.

Proof. Let H be the stabilizer in P of the octad $\mathcal{O}$; form the sublattice

$M(\mathcal{O})$ (A.6) and use (3.9) and (A.5) to conclude that the stabilizer of $(M(\mathcal{O}), F)$, where F is a frame in the sense of (A.2.i), has the form DH, order $2^{16}|GL(4,2)|$, whence $|D| = 2^{12}$ implies that $|H| = 2^4|GL(4,2)| = |AGL(4,2)|$. It is clear from (3.3.iii) that H has a normal 2-subgroup R of order 16 whose quotient is $\Omega^+(6,2)$. We now display $R \cong 2^4$.

The involutions of R can be associated to semiselfdual sublattices (2.4) as follows: the space W of Golay sets disjoint from $\mathcal{O}$ consist of 30 octads, $\emptyset$ and $\mathcal{O} + \Omega$ (A.7). Given a codimension 1 subspace, W_0 of W which contains $\langle \Omega \rangle$, we define an involution $t \in P$ as follows. Let $\mathcal{O}_i$, $i = 1, 2, 3$ be octads in W_0 which are linearly independent modulo $\mathcal{O} + \Omega$. Define an associated TI (*triple intersection*) as a set of the form $\mathcal{O}'_1 \cap \mathcal{O}'_2 \cap \mathcal{O}'_3$, where $\mathcal{O}'_i$ represents $\mathcal{O}_i$ or its complement $\mathcal{O}_i + [\mathcal{O} + \Omega]$ in $\mathcal{O} + \Omega$; a TI is a 2-set. Similarly, call a set a DI, a *double intersection*, if it has the form $\mathcal{O}'_1 \cap \mathcal{O}'_2$, notation as above; a DI is a 4-set. The intersection of any two octads disjoint from $\mathcal{O}$ is a 0-, 4- or 8-set. Define $M := M(\mathcal{O}_1, \mathcal{O}_2, \mathcal{O}_3)$ as the span of all $\alpha_i - \alpha_j$, where $\{i, j\}$ is an associated TI and all $\frac{1}{2}[\alpha_i - \alpha_j + \alpha_k - \alpha_\ell]$, where $\{i, j\}$ and $\{k, \ell\}$ are TIs whose union forms a DI. (It is easy to check that the definition of M depends just on W_0.) For such an M, we have a SSD involution, $t = t_M \in P$. Its effect on the α_i is to interchange two whose indices form a TI. If W_0, W_1 and W_2 are three such subspaces of W with associated involutions t_0, t_1, t_2, then $t_0 t_1 t_2 = 1$ if $W_0 \cap W_1 = W_1 \cap W_2 = W_2 \cap W_0$.

It is clear that we have 15 involutions which have cycle shape $1^8 2^8$ and which, with the identity, form an elementary abelian group, R, which acts regularly on $\mathcal{O} + \Omega$. At once, H splits over R. Let K be a complement. Since the action of K on T is faithful (because the action of $Stab(M)$ on $M^\perp \cap \Lambda = \Lambda \cap [\mathbb{Q}\Lambda, R]$ has kernel $\{\pm 1\}$ (3.8)), we are done. □

(4.6) Remark. (A trio of sporadic isomorphisms) We note that the sporadic isomorphisms $\Omega^+(6,2) \cong GL(4,2) \cong Alt_8$ follow from our analysis (C.3), (A.3), (3.12), (4.5); this says more than $GL(4,2) \cong Alt_8$, which is observed in the traditional course of studying M_{24} and the octad stabilizer [8] [18].

(4.7) Remark. It follows from (3.13) that if $x = 2\alpha_i \in \Lambda_4$ is in the standard frame, there are two orbits of $Stab(x)$ on the set of $M \cong \sqrt{2}L_{E_8}$ in $x^\perp \cap \Lambda$. One is the set of all $M(\mathcal{O})$ (A.4)(A.6), where $\mathcal{O}$ is an octad avoiding i, and a sublattice from the other orbit contains no vectors from the standard frame and is a sublattice of $\Lambda \cap \sum_{i \notin \mathcal{O}'} \mathbb{Q}\alpha_i$, where $\mathcal{O}'$ is an octad containing i.

Next, we will give a uniqueness argument for the ternary Golay code. We use existence of a ternary Golay code to prove uniqueness. If $\mathcal{TG}$ is such a code, namely, a $[12, 6, 6]$ code over $\mathbb{F}_3$, then we use a lattice M which is the orthogonal direct sum of 12 lattices isometric to L_{A_2}. Using rational linear combinations of elements of the M_i related to the elements of $\mathcal{TG}$, we will get a rootless even unimodular rank 24 lattice, Λ, which by our characterization (3.7), must be unique up to isomorphism. We give our version here (cf. [8][18]).

(4.8) Ternary Construction of the Leech lattice. Let $\mathcal{TG}$ be a ternary Golay code, a [12,6,6] code. One may imitate the analogous construction of the E_8-lattice, described in [18], (8.22). For $i = 1, \ldots, 12$, take a base α_i, β_i of the A_2 root system in $M_i \cong L_{A_2}$, define $\nu_i := \frac{1}{3}[\alpha_i - \beta_i]$. Then $M_i^* = M_i + \mathbb{Z}\nu_i$. Define M_0 as the span of all $3M_i^*$ and all $\sum c_i \alpha_i$, where $c_i \in \mathbb{Z}$ and $\sum_i c_i \in 3\mathbb{Z}$. For $c = (c_i) \in \mathbb{F}_3^6$ and x one of the symbols α, β, ν, define $x_S := \sum_{i \in S} c_i x_i$, where we think of $\mathbb{F}_3$ as the subset $\{-1, 0, 1\}$ of $\mathbb{Z}$. Finally, define $\gamma_i := \frac{1}{3}[4\alpha_i + \sum_{j \neq i} \alpha_j]$. We define the lattice $\Lambda_0 := M_0 + \sum_{c \in \mathcal{TG}} \mathbb{Z}\nu_c$; then $L_0/M_0 \cong \mathcal{TG}$. We take $\Lambda := L_0 + \mathbb{Z}\gamma_i$ (the definition of Λ is independent of i since $\gamma_i - \gamma_j = \alpha_i - \alpha_j \in M_0$); see (2.7). Since $\det(M_i) = 3$ and $\det(M) = 3^{12}$, we get $\det(M_0) = 3^{14}$, $\det(L_0) = 3^2$ and $\det(\Lambda) = 1$ by repeated use of the formula $\det(K) = \det(J)|J:K|^2$, for lattices $K \leq J$ (2.1). The rootless property of Λ is easy to verify. Though the code is not described explicitly, we may still construct a group of automorphisms of Λ which is isomorphic to $\mathcal{TG}$ and preserves M, namely the maps $\varepsilon_s := \prod_i \varepsilon_i^{s_i}$, where $s = (s_i) \in \mathcal{TG}$ is a codeword and ε_i is the identity on M_j if $j \neq i$ and

$$\varepsilon_i : \begin{cases} \alpha_i \\ \beta_i \end{cases} \mapsto \begin{cases} \beta_i \\ -\alpha_i - \beta_i \end{cases}, \text{ a rotation by } 2\pi/3 \text{ on } M_i. \text{ Note that } \alpha_i^{\varepsilon_i} = \alpha_i - 3\nu_i,$$

$\beta_i^{\varepsilon_i} = \beta_i - 3\beta_i - 3\nu_i$ and $\nu_i^{\varepsilon_i} = \nu_i - \beta_i$, whence ε_i satisfies $|M_i : M_i(\varepsilon_i - 1)^k| = 3^k$. Proof that the ε_c preserve Λ is routine (one must use the property that all inner products in $\mathcal{TG}$ are zero mod 3). Obviously, $Aut(\mathcal{TG})$ is a group of lattice automorphisms (by coordinate permutations) and it normalizes the above group of order 3^6.

From this particular way of constructing Λ, we can see an important property of the Sylow 3-group of $Aut(\Lambda)$, which has order 3^9.

(4.9) Lemma. A Sylow 3-group contains exactly one elementary abelian subgroup of order 3^6. Such a subgroup is therefore weakly closed in a Sylow 3-subgroup.

Proof. This follows from the fact that an element of order 3 in $Aut(\mathcal{TG}) \cong 2 \cdot M_{12}$ acts with Jordan canonical form $2J_3$ or $J_1 J_2 J_3$. Proof of this fact is an exercise using Section 7 of [18], (7.37). □

This Lemma can be proved with much less work than a general analysis of $\mathcal{TG}$ and $Aut(\mathcal{TG})$. Since we deduce that $|Aut(\mathcal{TG})|_3 = 3^3$ from our knowledge of $|Aut(\Lambda)|$, it would suffice to display a group of automorphisms of order 3^3 (this Sylow 3-group is nonabelian) and just observe the Jordan canonical forms of its elements; this may be done by selective use of Section 7 of [18].

(4.10) Theorem. There is a unique ternary code with parameters of the form $[12, 6, 6]$.

Proof. Any such code may be used with a lattice $M := M_1 \perp \cdots \perp M_{12}$, where $M_i \cong L_{A_2}$, to construct a Leech lattice, Λ, as above.

This group of shape 3^6 is isomorphic to $\mathcal{TG}$ by the inverse of $c \mapsto \varepsilon_c$ (4.8) and may be used to recover the lattices $M_0 \cap M_i$ (in the notation of (4.8)) as the sublattices affording 12 distinct rational characters of 3^6; one gets M_i from $M_0 \cap M_i$ as $M_i = 3[M_0 \cap M_i]^*$.

The weak closure property of such a group in a Sylow 3-subgroup of $Aut(\Lambda)$ (4.9) then implies that any two such lattices M are in a single orbit under the action of $Aut(\Lambda)$. From this, it follows that the associated code in M^*/M is unique up to equivalence.□

5 Other consequences for the Leech lattice and its automorphism group

We make no attempt to systematically derive the standard results about Mathieu groups and Conway groups using SSD theory, but merely give a sample to illustrate use of our theory. The next result analyzes an entry from the list of triangle stabilizers in $Aut(\Lambda)$ [4] [18].

(5.1) Proposition. G is transitive on triangles of type 222. Let a, b, c form a triangle of type 222, that is, a triple of vectors of type 2 which sum to 0. Let $H := Stab_G(a, b, c)$. Then, H contains 2-central involutions with centralizers of shapes $2^{1+6}_{-}\Omega^{-}(6, 2)$.

Proof. Note that the image mod 2Λ of any triangle with edges of even type lies in a maximal totally singular subspace. By (3.16), we may assume that $a \in M_3$, where (M_1, M_2, M_3) is a Leech triple. If $b \in M_1 \perp M_2 \perp M_3$, we are done since $(b, b) = 4$, $(b, a) \neq 0$ and $M_i \cong \sqrt{2}L_{E_8}$ imply that $b \in M_3$. It is therefore enough, by (3.6) to show that any triangle of type 222 lies in a sublattice isometric to $\sqrt{2}L_{E_8} \perp \sqrt{2}L_{E_8} \perp \sqrt{2}L_{E_8}$ since $Stab(M_3)$ induces W'_{E_8} on M_3.

For an index k, define $T = T_k := \Lambda \cap M_k^\perp$ and let $U = U_k/2T$ be the radical mod 2 of T (2.3). For an index k and $x \in \Lambda$, define $A_k(x) := \{y \in T_k \mid (x, y) \in 2\mathbb{Z}\}$, a sublattice of T_k of index 1 or 2.

If there is an index $i \neq 3$ so that $(M_i, b) = 0$, (3.8.ii) and the first paragraph imply that we are also done because the images of a, b in $T_i/2T_i$ lie in a common maximal totally singular subspace, we are done. So, we may assume that there is no such i.

We let b_i be the projection of b to $\mathbb{Q}M_i$; each (b_i, b_i) is a positive integer. Since $4 = (b, b) = (b_1, b_1) + (b_2, b_2) + (b_3, b_3)$, we conclude that the three summands on the right side are $(1, 1, 2)$, in some order.

If $(b_3, b_3) = 2$, $b' := b_1 + b_2 \in T_3^* \setminus T^3$. By transitivity (3.13.i), (3.8.ii) and the singularity of $b + T_3 \in T_3^*/T_3$, we may assume that $b' \in \frac{1}{2}[N_1 + N_2]$, whence b' lies in one of the $\frac{1}{2}N_j$. If i is the remaining index, we have $\{a, b\} \in M_i^\perp \cap \Lambda = T_i$, and we are done by another use of (3.8) and the first paragraph of this proof.

Finally, we assume that $(b_3, b_3) = 1$; let j be the index so that $(b_j, b_j) = 2$ and set $b' := b - b_j \in T_j^* \setminus T_j$; $b' + T_j$ is a singular vector in T_j^*/T_j. Then the image in T_j^*/T_j lies in a totally singular subspace, say R/T_j, where $R = R' \perp R'' \cong L_{E_8} \perp L_{E_8}$. By transitivity (3.13.i), we may assume that a lies in one summand, say R'. Since $(b', b') = 2$, b' lies in one summand, and since $(a, b') = (a, b) = -2 \neq 0$, this summand must be R'. Therefore, $\{a, b\}^\perp \geq R'' \cap \Lambda \cong \sqrt{2}L_{E_8}$. Now, $Stab(R'' \cap \Lambda)$ will transform the set $\{a, b\}$ into a sublattice of $(R'' \cap \Lambda)^\perp$ which is isometric to $\sqrt{2}L_{E_8} \perp \sqrt{2}L_{E_8}$, and finally we are done.

The structure of the centralizer of an involution follows from making the choice of triangle of type 222 in the sublattice M of (3.9). □

(5.2) Remark. The proof of (5.1) can be adapted to show that the group

Co_2 has an involution with centralizer of the form $2_+^{1+8}Sp(6,2)$ and by looking in $\frac{1}{2}N_{12}$ (2.3), we can find another with centralizer of the form $2^{1+6}2^4.GL(4,2)$. The stabilizer H of a triangle of type 222 should be isomorphic to $PSU(6,2)$, but I am not aware of any proof in the literature; indeed, in [4], the table of stabilizers had a question mark at 222 (removed in later versions). Centralizer of involution characterizations of $PSU(6,2)$ in the literature seem to require more than the centralizer shape in (5.1). One can identify H by verifying that it has the 3-transposition property [12][13].

(5.3) Remark. There are four classes of involutions in $Aut(\Lambda)$, [7][18]. All may be interpreted as SSD involutions. We know already (3.6) about involutions associated to $M \cong \sqrt{2}L_{E_8}$; their negatives are also SSD, associated to $M^\perp \cap \Lambda$. The involutions of trace 0 are associated to the sublattice of Λ consisting of vectors supported at a dodecad; it is isometric to the halfspin lattice for D_{12} [18] and has invariants (2^{12}). Finally, the involution -1 is the SSD involution associated to Λ itself.

(5.4) Remark. Something like SSD theory should work for elements of order greater than 2. We would expect a theory of existence and uniqueness of other codes used in other descriptions of the Leech lattice [18], e.g the ternary Golay code which is associated to the subgroup $3^{12}2 \cdot M_{12}$.

(5.5) Proposition. A Leech trio stabilizer is a 2-local in $Aut(\Lambda)$ of shape $2^{1+2+12}[Sym_3 \times GL(4,2)]$ (3.10) and it acts absolutely irreducibly on $\Lambda/p\Lambda$, for all odd primes p.

Proof. If M is any sublattice isomorphic to $\sqrt{2}L_{E_8}$, $O_2(Stab_{Aut(\Lambda)}(M))$ acts on $\Lambda/p\Lambda$ with an irreducible direct summand of dimension 16. Also, $O_2(Stab_{Aut(\Lambda)}(M))$ stabilizes any Leech trio containing M (the other two sublattices of a trio are just the fixed points of x and xz, where z is the SSD involution associated to M and x is a noncentral involution in $O_2(Stab_{Aut(\Lambda)}(M))$ (3.8.ii). If we let M range over members of a Leech trio, it follows that the trio stabilizer acts irreducibly on $\Lambda/p\Lambda$. □

(5.6) Corollary. $\Lambda/p\Lambda$ is an absolutely irreducible G-module, for all prime numbers p.

Proof. If p is odd, we use (5.5). Suppose that $p = 2$. Let $Q/2\Lambda$ be a proper submodule of $\Lambda/p\Lambda$, of dimension $d > 0$. By transitivity of G on vectors of type 3 (3.17) and $d < 24$, we have $Q \cap \Lambda_3 = \emptyset$ (A.2.ii). Then, $Q/2\Lambda$ is totally singular, whence $d \le 12$. But then $2^d < \min\{u_2/2, u_4/24\}$, a contradiction to transitivity on Λ_2 and Λ_4. So, we have irreducibility. Suppose that we do not have absolute irreducibility. Then we have a nontrivial centralizer algebra and an integer $e > 1$ so that when $\Lambda/2\Lambda$ is extended to a splitting field, every irreducible for $Aut(\Lambda)$ occurs with multiplicity divisible by e. If we take an element x of order 23 in $Aut(\Lambda)$ and note that $x^{23} - 1$ has a single nontrivial irreducible factor, we see that $e > 1$ is impossible, a contradiction.

(5.7) Remark. We have absolutely irreducible action modulo all primes for W_{E_8} on L_{E_8} (easy to prove) and also that of the sporadic simple group F_3 on a 248 dimensional lattice. For other examples, see [10][11][27][28][29][30][31][32].

Appendices background

Appendix A Elementary lattice theory

(A.1) Proposition. Theta functions: for an even integral lattice L, the theta function is $\sum_{k\geq 0} u_k q^k$, where u_k is the number of lattice vectors of squared length $2k$.

(i) The theta series for L_{E_8} begins $1 + 240q + 2160q^2 + \cdots$.

(ii) The theta series for a Leech lattice, Λ, begins $1 + 196560q^2 + 16773120q^3 + 398034000q^4 + \cdots$.

Proof. [24], for instance; or [8], p.135. □

(A.2) Proposition. (The E_8 lattice and Leech lattice mod 2) Let L be L_{E_8} or a Leech lattice Λ and let $L_n := \{x \in L \mid (x,x) = 2n\}$.

(i) For $L = L_{E_8}$, a coset of $2L$ meets L_n for exactly one value of $n \in \{0,1,2\}$ and such a nonempty intersection has the form $\{\pm x\}$, except for $n = 2$ for which it is a "frame", a set of 16 vectors, two of which are equal, opposite or orthogonal.

(ii) For $L = \Lambda$, a coset of $2L$ meets L_n for exactly one value of $n \in \{0,2,3,4\}$ and such a nonempty intersection has the form $\{\pm x\}$, except for $n = 4$ for which it is a "frame", a set of 48 vectors, two of which are equal, opposite or orthogonal.

Proof. (i) is trivial. For (ii), which is almost as trivial, see [4] or [18]. □

(A.3) Lemma. (i) In W_{E_8}, the stabilizer P of a maximal totally singular subspace in L_{E_8} mod 2 has the form $2^{1+6}_+ GL(4,2) \cong 2^{1+6}_+ \Omega^+(6,2)$; P splits over $O_2(P)/Z(P)$ but not over $O_2(P)$, i.e., P contains a perfect group of the form $2.GL(4,2)$ but does not contain $GL(4,2)$. Also, the nontrivial cosets of this subspace each contain 16 roots.

(ii) W_{E_8} acts transitively on

(a) pairs (M,x) where $M \cong \sqrt{2}L_{E_8}$ is a sublattice of $L = L_{E_8}$ and $x \in L$, $(x,x) = 2$;

(b) pairs (M,x) where $M \cong \sqrt{2}L_{E_8}$ is a sublattice of $L = L_{E_8}$ and $x \in L \setminus M$, $(x,x) = 4$;

(c) pairs (M,x) where $M \cong \sqrt{2}L_{E_8}$ is a sublattice of $L = L_{E_8}$ and $x \in M$, $(x,x) = 4$.

Proof. (i) Since all such stabilizers are conjugate, it suffices to examine a convenient one. The chief factors of such a group P are clear; the only issue is the structure of the maximal normal 2-subgroup. Use the following description of L_{E_8}. Let $v_1, \ldots, v_8$ be a basis of 8-dimensional Euclidean space such that $(v_i, v_j) = 2\delta_{i,j}$ and let $\mathcal{H}$ be a dimension 4, length 8, minimum weight 4 extended binary Hamming code, e.g. the span in $\mathbb{F}_2^8$ of $(11110000), (11001100), (10101010), (11111111)$. Its group is a subgroup $S \cong AGL(3,2)$ of the full group Sym_8 of coordinate permutations. We may and do identify the index set with the normal eights group $E := O_2(S)$ of S; the codewords of weight 4 may be identified with affine subspaces of dimension 3 in E. For $A \subseteq E$, define $v_A := \sum_{i \in A} v_i$. Define the lattice L to be the $\mathbb{Z}$-span of all $\pm v_i \pm v_j$, $\frac{1}{2}v_A$ for $A \in \mathcal{H}$ and all $-v_i + \frac{1}{4}v_E$. This lattice is even and unimodular so is isometric to L_{E_8}. Denote by M the sublattice spanned by all $\pm v_i \pm v_j$ and $\frac{1}{2}v_E$. Both L and M admit the group S by coordinate permutations and admit the group of sign changes consisting of maps $\varepsilon_A, A \subseteq E$, which are defined by $v_i \mapsto \begin{cases} -v_i & i \in A; \\ v_i & i \notin A. \end{cases}$ The group $\langle E, F \rangle$ is extraspecial (C.1) and acts trivially on both L/M and $M/2L$. For a discussion of a related family of lattices, see [2],(A.11). For the statement about splittings in P, we observe that P has a subgroup of the form $2.GL(4,2)$ which occurs as the stabilizer of a maximal totally singular subspace complementary to the one stablized by P.

To show that any subgroup $2.GL(4,2)$ of P is nonsplit, we note that, if split, the ambient 8 dimensional complex representation would have irreducible constituents of degrees 1 and 7 [7]; on any invariant lattice taken modulo 2 therefore we would have the trivial module, which is not the case (we have two 4-dimensional irreducible constituents).

(ii) Let $L_n := \{x \in L \mid (x,x) = 2n\}$. By Witt's Theorem and the fact that W_{E_8} induces the full orthogonal group on $L/2L$, there is a single orbit of W_{E_8} on singular vectors outside the subspace $M/2L$ of $L/2L$ and a single orbit on nonsingular vectors outside this subspace. These nonsingular vectors are the images in $L/2L$ of L_1 and a nonsingular vector corresponds to a pair $\{x, -x\}$ in L_1. Since $-1 \in P$, the stabilizer in W_{E_8} of M, we clearly have one orbit, whence (a).

(b) Now, let $X := M \cap L_2$. A singular vector outside $M/2L$ corresponds via $L \to L/2L$ to a subset R of X of 16 vectors, two of which are equal, negatives or orthogonal. It suffices to show that $Stab_P(R)$ is transitive on R. Since P induces $GL(4,2)$ on L/M, it suffices to take R to contain an "odd" vector, say $-v_i + \frac{1}{4}v_E$, and show R is contained in a single K-orbit. Since membership in R is determined by congruence modulo $2L$, every vector in R is odd. Finally, it suffices to show that K has a single orbit on Y, the set of all odd vectors of squared length 4. It is easy to see that Y consists of $[-v_i + \frac{1}{4}v_E]^{\varepsilon_A}$, for all indices $i \in E$ and all even subsets $A \subseteq E$. Since S is transitive on E and K contains D, the group of all ε_A, transitivity is clear.

(c) We can view the action of P as that of the stabilizer in $Aut(M)$ of $2L$, so the preceeding argument applies here since $M \cong \sqrt{2}L_{E_8}$ and $2L/2M$ is a maximal totally isotropic in the sense of the associated nonsingular quadratic form (3.1) on $M/2M$. We deduce (c) from (b). $\square$

(A.4) Frames, codes and a description of the Leech lattice

(A.4.1) (Frame and code concepts) Given a Leech lattice, Λ, a *frame* is a set of 48 vectors in Λ of squared length 8, two of which are opposite or orthogonal; a frame is an equivalence class of sets of vectors of type 4 in which x and y are equivalent if and only if $x - y \in 2\Lambda$. An *oriented frame* or a *frame basis* is a subset of a frame which is a basis of $\mathbb{Q}\Lambda$.

(A.4.2) (Standard frame and standard description of a Leech lattice) We now assume the existence of a Golay code, $\mathcal{G}$, and its use in describing a Leech lattice, Λ, in the standard way with a basis consisting of an orthogonal set of roots $\{\alpha_i \mid i \in \Omega\}$. The *standard frame* is $\Sigma := \{\pm 2\alpha_i \mid i \in \Omega\}$ and the *standard oriented frame* is $\Sigma^+ := \{2\alpha_i \mid i \in \Omega\}$. We define $\Lambda := span_{\mathbb{Z}}\{2\alpha_i, \pm\alpha_i \pm \alpha_j, \frac{1}{2}\alpha_S, \nu_i \mid i, j \in \Omega, S \in \mathcal{G}\}$, where, for $S \subseteq \Omega$, $\alpha_S := \sum_{i \in S} \alpha_i$ and $\nu_i := \frac{1}{4}\alpha_\Omega - \alpha_i\}$ (this set of generators is unnecessarily large, but shows symmetry). The *frame group* is the stabilizer of a given frame in $Aut(\Lambda)$. Clearly, in the standard frame group $Stab(\Sigma)$, there is a natural subgroup of the form $D : P$, where D acts diagonally with respect to the frame and where P is the group of permutation matrices identified with the automorphism group of the code $\mathcal{G}$. The orthogonal transformations which stabilize each 1-space spanned by elements of Σ have the form $\varepsilon_A, A \in \Omega$, which are defined by $\alpha_i \mapsto \begin{cases} -\alpha_i & i \in A; \\ \alpha_i & i \notin A. \end{cases}$

(A.4.3) (Deduction of a code from a Leech lattice) Conversely, given a Leech lattice, Λ, and a frame, F (which exists, by (A.1.ii)), we find that a code occurs naturally. We define $\Lambda(4) := span\{\frac{1}{2}(x-y) \mid x, y \in F\}$, $\Lambda(2) := \{x \in L \mid 2x \in \Lambda(4)\}$. Then, by using the 24 coordinate spaces $\frac{1}{2}\mathbb{Z}x$ mod $\mathbb{Z}F, x \in F$, $\Lambda(2)/\Lambda(4)$ gives a binary code $\mathcal{C}$ and, since Λ is a Leech lattice, it is straightforward to see that the code is doubly even of dimension 12 (whence the universe set is in $\mathcal{C}$, making it closed under complementation), the code has minimum weight at least eight and that $|\Lambda : \Lambda(2)| = 2$.

We now prove that the minimum weight is eight. Let $S \in \mathcal{C}$ have minimum weight, say $w \geq 8$. Since $\mathcal{C}$ is closed under complementation, we may assume that $w \leq 12$. We suppose that that $w \geq 9$ and obtain a contradiction. Consider the map $\psi : \mathcal{C} \to P(S)$, the power set, defined by $A \mapsto A \cap S$. Since Λ is integral, all intersections of pairs of sets in $\mathcal{C}$ are even, so $\dim(\text{Im}(\psi)) \leq w - 1$ and $\dim(Ker(\psi)) \geq 13 - w$. Assuming that $\dim(Ker(\psi)) \geq 2$, we have $A \in Ker(\psi)$, $A \neq 0, \Omega + S$. Since $|A| \geq 9$, $18 \leq |A + S| < 24$, whence $0 < |A + S + \Omega| \leq 6$, a contradiction. We have $\dim(Ker(\psi)) = 1$ and $w = 12$ and so every set in $\mathcal{C}$ is $0, \Omega$ or a 12-set. Since $\text{Im}(\psi)$ is the codimension 1 space of even sets in $P(S)$, there is $A \in \mathcal{C}$ so that $A \cap S$ is a 2-set. Since A is a 12-set, $A + S$ is a 20-set, our final contradiction.

(A.4.4) (The Steiner system, octads and dodecads.) Let $S(5, 8, 24)$ denote a Steiner system with parameters $(5, 8, 24)$, that is, a family of 8-sets in a fixed 24-set such that any 5-set is contained in a unique member of this family. Sets of weight 8 in $\mathcal{C}$ are called *octads* (4.4). The octads in a Golay code as above form such a Steiner system. Sets of weight 12 in $\mathcal{C}$ are called *dodecads*; their stabilizers are, by definition, the group M_{12}.

(A.5) Proposition. $Stab_{Aut(\Lambda)}(\Sigma) = D : P$ and $D := \{\varepsilon_A \mid A \in \mathcal{G}\} \cong 2^{12}$ (see (A.4.2)).

Proof. Since the code $\mathcal{G}$ is its annihilator in the power set $P(\Omega)$ (with addition the symmetric difference and bilinear form $(A, B) \mapsto |A \cap B|$ mod 2), application of ε_A to lattice elements of the form $\frac{1}{2}\alpha_S, S \in \mathcal{G}$, shows that if it stabilizes Λ, then $A \in \mathcal{G}$. Trivially, $\varepsilon_A \in Aut(\Lambda)$ if $A \in \mathcal{G}$.

Now, let $g \in Stab(\Sigma)$; then g is a product dp, where d is diagonal and p is a permutation matrix. Applying g to a vector of the form $\frac{1}{2}\alpha_S, S \in \mathcal{G}$ gives a vector of the form $\frac{1}{2}\sum_{i \in T} \pm \alpha_i$ in Λ, for some $T \subseteq \Omega$. Since its inner product with every $\frac{1}{2}\alpha_S, S \in \mathcal{G}$ is an integer, we get $|T \cap S|$ even, for every $S \in \mathcal{G}$, which implies that $T \in \mathcal{G}$. This means that p is in the group of the code, whence both p and d

stabilize Σ. $\square$

(A.6) Notation. $M(\mathcal{O})$ is the set of lattice vectors supported at the octad $\mathcal{O}$ (A.4). From (A.4.3), it is clear that $\Lambda \cap M(\mathcal{O})$ is just the 112 $\pm\alpha_i \pm \alpha_j$ and all 128 $\frac{1}{2}\alpha_{\mathcal{O}}\varepsilon_A$, for all $A \in \mathcal{G}$.

(A.7) Lemma. For $x \in \Sigma$ (A.4.2), the set of $M \cong \sqrt{2}L_{E_8}$ containing x is just the set of lattices of the form $M(\mathcal{O})$ where $\mathcal{O}$ is an octad containing the index i, where $x = \pm 2\alpha_i$. (There are 253=11.23 such.)

Proof. It is clear from (3.6) that the stabilizer of x is transitive on the set of such M which contain it. Since $Stab(x) \leq Stab(\Sigma)$, which acts monomially with respect to the double basis Σ, it follows that every such M has the form $M(\mathcal{O})$. (The count $253 = \binom{23}{4}/\binom{7}{4}$ follows from (A.4.4)). $\square$

(A.8) Proposition. The Golay sets disjoint from an octad consist of 30 octads, the empty set and the octad complement. Any octad is part of a trio (a partition of Ω into three octads.

Proof. Let $\mathcal{O}$ be an octad. Study the map $\mathcal{G} \to P(\mathcal{O})$, as in (A.4). $\square$

(A.9) Theorem. (Classifications of EULs) An EUL has rank divisible by 8. In rank 8, there is, up to isometry, just one EUL, L_{E_8}[36]. In rank 16, there are two, the halfspin lattice of rank 16 and the direct sum of two copies of L_{E_8}[36]. In rank 24, there are 24, and they are distinguished by the systems formed by their sets of roots; the empty root system corresponds to the Leech lattice [23]. See also [21] [24] [35].

(A.10) [2] (The lattices of Broué-Enguehard.) Given an integer $n \geq 3$, there is a lattice L_n of rank 2^n with the following properties:

(i) $\det(L_n) = \begin{cases} 1 & n \text{ odd}; \\ 2^8 & n \text{ even}; \end{cases}$

(ii) $Aut(L_3) \cong W_{E_8} \cong 2\dot{}O^+(8,2)$; $Aut(L_n) \cong 2_+^{1+2n}\Omega^+(2n,2)$ if $n \geq 4$;

(iii) the minimum squared length is $2^{[n/2]}$.

(iv) $Aut(L_n)$ is transitive on minimal vectors.

(A.11) Remark. (i) These lattices are beautifully described by an error correcting code of length 2^n based on the action of $GL(n,2)$ on $\mathbb{F}_2^n$ and the minimal vectors are listed. In case $n = 3$, the expected group of automorphisms $2_+^{1+6}\Omega^+(6,2)$ is proper in the full group of automorphisms, the Weyl group for E_8.

(ii) The Barnes-Wall lattice [1] is a rank 16 lattice with determinant 256 and minimum squared length 4; the theta series begins $1 + 4320q^2 + 61440q^3 + 522720q^4 + \cdots$. In [8], p.131, its occurrence as our lattice T (3.2) and certain other properties are asserted but no reference or proof is given. This lattice turns out to be a member of the Broué-Enguehard series, by (3.23); in [2][8] is mentioned but there seems to be no explanation of its relationship to the Barnes-Wall lattice.

Appendix B Orthogonal groups in characteristic 2

For basic theory, see [7][9]. We summarize what we need.

(B.1) The groups. We are in even dimension $2n \geq 2$ over the field F, which is perfect of characteristic 2 (this includes finite fields). There are two equivalence classes of quadratic forms, according to the Witt index, the dimension of a maximal totally singular subspace; the possibilities for the Witt index are n (plus type) and $n-1$ (minus type) and the isomorphism types of the isometry groups are denoted $O^\varepsilon(2n, F)$, $\varepsilon = \pm$. This group is not simple since it has a normal subgroup $\Omega^\varepsilon(2n, F)$ of index 2, the kernel of the Dickson invariant [9], which is a simple group except in the case $(n, \varepsilon) = (2, +)$.

(B.2) Permutation representations. By Witt's theorem, which says that in V, a finite dimensional vector space with a nonsingular quadratic form, an isometry between subspaces extends to an isometry on V, the groups $O^\varepsilon(2n, F)$ are transitive on the totally singular subspaces of a given dimension $d \leq n$; when restricted to $\Omega^\varepsilon(2n, F)$ we still have transitivity, except for $d = n$, where we have two orbits. For one of these subspaces, the stabilizer in $\Omega^\varepsilon(2n, F)$ or $O^\varepsilon(2n, F)$ is a parabolic subgroup of the form $F^{\binom{n}{2}} : GL(n, F)$. The stabilizer in $\Omega^\varepsilon(2n, F)$ of a singular, resp. nonsingular vector, has the shape $F^{2n-2}\Omega^\varepsilon(2n-2, F)$, $\Omega^\varepsilon(2n-1, F) \cong Sp(2n-2, F)$.

(B.3) Weyl group of E_8. The Weyl group $W := W_{E_8}$ satisfies $Z(W) = \{\pm 1\}$, $W/Z(W) \cong O^+(8, 2)$ and $|W| = 2^{14}3^5 5^2 7$.

Appendix C Extraspecial p-groups

(C.1) Definition. [14][19] Given a prime number p, an *extraspecial p-group* is a finite p-group P such that $Z(P) = P'$ has order p. It follows that $\Phi(P) = Z(P)$ and that $P/Z(P)$ is a vector space of dimension $2n$ over $\mathbb{F}_p$, for some integer $n \geq 1$ and that the map $P/Z(P) \times P/Z(P) \to Z(P)$ based on commutation may be interpreted as a nonsingular alternating bilinear form. When $p = 2$, the squaring map induces a map $P/Z(P) \to Z(P)$ which may be interpreted as a nonsingular quadratic form.

(C.2) Theorem. Given P as in (C.1), the irreducible representations consist of p^{2n} linear characters and an algebraically conjugate family of $p-1$ irreducibles of dimension p^n; the latter are faithful.

(C.3) Definition. A *holomorph* of P, as in (C.1), is a group G so that $P \triangleleft G$ and the natural map $G \to Aut(P)$ has kernel $Z(P)$ and image $C_{Aut(P)}(Z(P))$; if the image is a proper subgroup, the G is a *partial holomorph*. A holomorph is *standard* if it exists as a subgroup of $GL(2^m, \mathbb{C})$; otherwise it is a *twisted holomorph*. For $p = 2$ extensions are generally nonsplit [15][16] but are split for p odd. In any holomorph, G, the conjugacy classes within P consist of the p elements of $Z(P)$, plus one or two further ones, distinguished by their orders, p and p^2 (either one or both orders may occur).

(C.4) Theorem. Let G be a standard holomorph and χ the character of G in the p^n dimensional irreducible representation. For $g \in G$, define P_g by $P_g/Z(P) := C_{P/Z(P)}(g)$ and $d_g := \dim(P_g/Z(P))$. We say that g acts *cleanly* on P

if and only if $P_g = C_P(g)$ (the other possibility, left unnamed, is $|P_g : C_P(g)| = p$). Then $\chi(g) = \varepsilon p^{d_g/2}$ for some root of unity ε, if g acts cleanly; and $\chi(g) = 0$ otherwise.

Proof. [17]. □

References

[1] E. S. Barnes and G. E. Wall, Some extreme forms defined in terms of Abelian groups, J. Aust. Math. Soc. 1 (1959), 47-63.

[2] M. Broué and M. Enguehard, Une famillie infinie de formes quadratiques entière; leurs groupes d'automorphismes, C. R. Acad. Sc. Paris, t. 274 (3 Janvier 1972), Série A-19.

[3] R. Carter, Simple Groups of Lie Type, Wiley-Interscience, London, 1972.

[4] J. H. Conway, A Group of Order 8, 315, 553, 613, 086, 720, 000. Bull. Lond. Math. Soc. 1 (1969), 79-88.

[5] J. H. Conway, A characterization of Leech's lattice, Invent. Math. 7 (1969), 137-142.

[6] J. H. Conway, Three Lectures on Exceptional Groups in Higman-Powell, pp. 215–247. Finite Simple Groups. Academic Press, London, 1971.

[7] J. H. Conway, R. T. Curtis, S. P. Norton, R.A. Parker and R. A. Wilson, An Atlas of Finite Groups, Oxford, Clarendon, 1985.

[8] J. H. Conway and N. J. A. Sloane, Sphere Packings, Lattices and Groups, Springer, Berlin, 1988.

[9] J. Dieudonné, La Gomtrie des Groupes Classiques, Springer, Berlin Heidelberg New York, 1971.

[10] N. Dummigan and P. H. Tiep, Congruences for certain theta series, J. Number Theory 71 (1998), 86-105.

[11] N. Dummigan and P. H. Tiep, Lower bounds for certain symplectic group and unitary group lattices, Amer. J. Math. (to appear, joint with N. Dummigan)

[12] B. Fischer, Finite groups generated by 3-transpositions, University of Warwick notes, 1969.

[13] B. Fischer, Finite groups generated by 3-transpositions, Invent. Math. 13 (1971), 232-246.

[14] D. Gorenstein, Finite Groups, Harper and Row, New York, 1968; 2nd ed. Chelsea, New York 1980.

[15] R. L. Griess, Jr., Automorphisms of Extra Special Groups and Nonvanishing Degree 2 Cohomology. Pacific Jour. of Math. 48(1973), 403-422.

[16] R. L. Griess, Jr., On a subgroup of order $2^{15}|GL(5,2)|$ in $E_8(\mathbb{C})$, the Dempwolff group and $Aut(D_8 \circ D_8 \circ D_8)$, J. Algebra 40 (1976), 271-279.

[17] R. L. Griess, Jr., The monster and its nonassociative algebra, article in Finite Groups - Coming of Age, Contemp. Math. 45 (1985), edited by John McKay.

[18] R. L. Griess, Jr., Twelve Sporadic Groups, Springer Monographs in Mathematics, Springer, 1998.

[19] B. Huppert, Endliche Gruppen, I, Springer, 1967.

[20] J. Lepowsky and A. Meurman, An E_8 approach to the Leech lattice and the Conway group, J. Algebra 77 (1982), 484-504.

[21] J. MacWilliams and N. Sloane, The Theory of Error Correcting Codes, North-Holland, 1977.

[22] J. Milnor and D. Husemoller, Symmetric Bilinear Forms, Springer, Berlin Heidelberg New York, 1973.

[23] H. V. Niemeyer, Definite Quadratische Formen der Diskriminante 1 und Dimension 24, Doctoral Dissertation, Göttingen, 1968.

[24] J.-P. Serre, A Course in Arithmetic, Graduate Texts in Mathematics 7, Springer, 1973.

[25] N. J. A. Sloane, review of [20], Mathematical Reviews.

[26] J. G. Thompson, A Simple Subgroup of E_8 (3), In: N. Iwahori, Finite Groups Symposium, Japan Soc. for Promotion of Sci., Tokyo, 1976, 113-116.

[27] P. H. Tiep, Globally irreducible representations of the finite symplectic group $Sp_4(q)$, Comm. Algebra 22 (1994), 6439-6457.

[28] P. H. Tiep, Basic spin representations of $2\mathbb{S}_n$ and $2\mathbb{A}_n$ as globally irreducible representations, Arch. Math. (Basel) 64 (1995), 103-112.

[29] P. H. Tiep, Weil representations as globally irreducible representations, Math. Nachr. 184 (1997), 313-327.

[30] P. H. Tiep, Globally irreducible representations of finite groups and integral lattices, Geometriae Dedicata 64 (1997), 85-123.

[31] P. H. Tiep, Globally irreducible representations of $SL_3(q)$ and $SU_3(q)$, Israel J. Math. 109 (1999), no. 1, 225-251.

[32] P. H. Tiep, and A. E. Zalesskii, Reduction modulo p of unramified representations(in preparation).

[33] J. Tits, Four Presentations of Leech's lattice, in Finite Simple Groups, II, Proceedings of a London Math. Soc. Research Symposium, Durham, 1978, ed. M. J. Colllins, pp, 306-307, Adademic Press, London, New York, 1980.

[34] J. Tits, Quaternions over $\mathbb{Q}(\sqrt{5})$, Leech's lattice and the sporadic group of Hall-Janko, J. Algebra 63 (1980), 56-75.

[35] B. B. Venkov, The classification of integral even unimodular 24-dimensional quadratic forms, TMIS 148 (1978), 65-76 = PIMS (no. 4, 1980), 1967-1974.

[36] E. Witt, Theorie der quadratischedn Formen in beliebigen Körpern, JRAM 176 (1937), 31-44.

B.2 Pieces of 2^d: Existence and Uniqueness for Barnes-Wall and Ypsilanti Lattices

Robert L. Griess Jr.[*][†]

Abstract

We give a new existence proof for the rank 2^d even lattices usually called the Barnes-Wall lattices, and establish new results on uniqueness, structure and transitivity of the automorphism group on certain kinds of sublattices. Our proofs are relatively free of calculations, matrix work and counting, due to the uniqueness viewpoint. We deduce the labeling of coordinates on which earlier constructions depend.

Extending these ideas, we construct in dimensions 2^d, for $d >> 0$, *the Ypsilanti lattices*, which are families of indecomposable even unimodular lattices which resemble the Barnes-Wall lattices. The number $\Upsilon(2^d)$ of isometry types here is large: $\log_2(\Upsilon(2^d))$ has dominant term at least $\frac{r}{4}d\,2^{2d}$, for any $r \in [0, \frac{1}{2})$. Our lattices may be the first explicitly given families whose sizes are asymptotically comparable to the Siegel mass formula estimate ($\log_2(mass(n))$) which has dominant term $\frac{1}{4}\log_2(n)n^2$.

This work continues our general uniqueness program for lattices, begun in Pieces of Eight [19]. See also our new uniqueness proof for the E_8-lattice [20].

Keywords and Phrases: Barnes-Wall lattices, Bolt-Room-Wall group, uniqueness, semiselfdual lattice, Siegel mass formula, Ypsilanti lattice.

1 Notation and terminology

admissible	14.10
A_{ij} and other diagonal notation	5.2
ancestors and generations, ancestral	13.5, 13.7
annihilator, self annihilating	Section 4
$BRW^0(2^d, \pm)$	Bolt, Room and Wall group, A.2
BW_{2^d}, the Barnes-Wall lattice in dimension 2^d	3.4
classification	10.2
coelementary abelian subgroup, p-coelementary abelian	a subgroup $B \leq A$ so A/B is

---

[*]2009 revision.

[†]Department of Mathematics University of Michigan Ann Arbor, MI 48109, USA. rlg@umich.edu *Dedicated to Donald G. Higman.*

	p-elementary abelian
D, a lower dihedral group	7.2
defect of an involution	6.10
density, commutator density	6.16
determinant of a lattice, L	$|\mathcal{D}(L)|$
d-invariant	12.1
$\mathcal{D}(L)$, discriminant group of a lattice L	$\mathcal{D}(L) = L^*/L$
double basis	5.3
DT, DTL	14.15
duality level	6.7
eigenlattice, total eigenlattice, Tel	6.8
f, f_i, f_{12}; various fourvolutions	7.2
F_i	7.2
fourvolution	6.1
frame, plain frame PF, sultry frame SF	6.15, 8.7
generation, 2/4- and 3/4	6.19
G_{2^d}	8.2
Hamming codes	4.2, 4.4
$I(d, p, q)$	11.12
labeling	11.16
lattice of BW-type	3.4
L, an integral lattice of rank n	Section 5
$L_i, L_i[k]$	7.2
lower	8.3
L^*, the dual of the lattice L	Section 5
mass formula, $mass(n)$	14.30
$M_i, M_i[k]$	7.2
minimal vectors, $MinVec(L)$, $\mu(L)$	7.7
nextbw	7.2
power set, even sets	Section 4
Q_i	7.2
r-modular	5.10
R	8.2, 7.2
R_{2^d}	8.2
R_{2^i}	7.2
sBW, ssBW	7.11
$Scalar(G, M)$, the scalar subgroup	6.14
SSD, semiselfdual, RSSD, relatively semiselfdual	5.11
sultry frame	8.7
sultry transformation, twist	6.3
$t_i, t_{ij}, t_{ij'}$	7.2
upper	8.3
Condition $X(2^d)$	3.3
$\mathfrak{X}$	10.1
$\mathfrak{Y}$	14.11
Ypsilanti lattices, cousins, etc.	Section 14

zop2, zoop2	11.4
ε_A, sign changes, monomial group	5.5
ψ_i	7.2
Ω, index set (often identified with an affine space $\mathbb{F}_2^d$)	Section 5
Ω, universe, **1**, the "all ones vector" $(1, 1, \ldots, 1)$ in $\mathbb{F}_2^\Omega$	Section 4

Conventions. Our groups and most endomorphisms act on the right, often with exponential notation. Group theory notation is mostly consistent with [12, 18, 23]. The commutator of x and y means $[x, y] = x^{-1}y^{-1}xy$ and the conjugate of of x by y means $x^y := y^{-1}xy = x[x, y]$. These notations extend to actions of a group on an additive group; see 6.16, ff.

Here are some fairly standard notations used for particular extensions of groups: p^k means an elementary abelian p-group; $A.B$ means a group extension with normal subgroup A and quotient B; $p^{a+b+\cdots}$ means an iterated group extension, with factors $p^a, p^b, \ldots$ (listed in upward sense); $A{:}B$, $A{\cdot}B$ mean, respectively, a split extension, nonsplit extension.

2 Introduction

All lattices in this article are positive definite. A sublattice is simply an additive subgroup of a lattice (no requirement on the rank).

We prove existence and uniqueness of the Barnes-Wall lattices of rank 2^d by induction and establish properties of them and their automorphism groups, including some new ones. In particular, the uniqueness theorem seems to be new. With future classifications (and discoveries!) of lattices in mind, we promote systematic study of uniqueness for important lattices. In [19], we used scaled unimodular lattices and SSD involutions to give a new uniqueness proof of the Leech lattice and revise the basic theory of the Leech lattice, Conway groups and Mathieu groups. There is a new and elementary uniqueness proof for the E_8 lattice in [20].

The Barnes-Wall lattices BW_{2^d} are even lattices in Euclidean space of dimension 2^d. They have minimum norm $2^{\lfloor \frac{d}{2} \rfloor}$ and remarkable automorphism groups [4] isomorphic to $BRW^0(2^d, +) \cong 2_+^{1+2d}\Omega^+(2d, 2)$, $d \geq 4$.

Various terms have been applied to these abstract groups and their analogues over finite fields in general. We think that *BRW group* for the groups which occur here would be most appropriate since Bolt, Room and Wall seem to have been the first to determine their structure [4]. Compare the later articles [7, 13, 14, 16]. See Appendix A2.

These lattices (and related ones) were defined in [2]. Independently, these lattices were rediscovered and their groups analyzed by Broué and Enguehard in [7]. This coincidence does not seem well recognized in the literature. We first noticed [7], then [2] only years later. The beautiful and definitive analysis of Broué and Enguehard [7] was the main inspiration for this article.

We shall abbreviate Barnes-Wall by BW.

For ranks $2^d \leq 16$, the BW lattices are well-known in several contexts. For

$d = 1$, we have a square lattice, and, depending on scaling, BW_{2^2} is the D_4 or F_4 root lattice. We have $BW_{2^3} \cong L_{E_8}$, though in [2], we find $\sqrt{2}L_{E_8}$. As sublattices of many of the Niemeier lattices, there are scaled copies of BW_{2^3} and BW_{2^4}. See also [19]. About the BRW groups, there are further details in Appendix A2.

We prove existence and uniqueness of the BW lattices of rank 2^d by induction and establish basic properties of them and their automorphism groups. We start not with a frame (a double orthogonal basis) but an orthogonal sum of two scaled BW lattices of rank 2^{d-1}, then show how, by choosing overlattices, to enlarge this to a BW lattice of rank 2^d. Analysis of choices and induction give suitable existence and uniqueness theorems, structure of the set of minimum norm vectors, properties of automorphism groups, transitivity on certain sublattices, etc. The uniqueness and transitivity theorems are new. (The referee called our attention to a characterization of Barnes-Wall lattices in [21].)

Our program emphasizes elementary algebra and involves very little of special calculations, matrix work and combinatorial arguments. We heavily exploit commutator density and equivalent properties, like 3/4-generation and 2/4-generation, which are quite useful for manipulating sublattices and lessening computations. As far as we know, these properties are new.

Reflections on the uniqueness theory led us naturally to the *Ypsilanti lattices*, a very large family of BW-like lattices. The Ypsilanti lattices are fairly explicit and represent a nontrivial share of all the even unimodular lattices of dimension 2^d. Their existence also clarifies the need for some hypothesis like (e) in 3.3, as we now explain.

Let $n > 0$ be an integer divisible by 8. If L is a rank n even, unimodular lattice, the theta function of L lies in a vector space of dimension roughly $\frac{n}{24}$ (see [31], p. 88). For $n = 8$ or 16, the dimension is 1, so the condition constant term 1 determines the theta function. For $n = 24$, the two conditions constant term 1 and no roots determines the theta function. In these cases, one can use arithmetic information about norms to determine structure.

Now take n to be 2^d for d large and L a BW lattice. The condition minimum norm $\mu(L) = 2^{\lfloor \frac{d}{2} \rfloor}$ represents $2^{\lfloor \frac{d}{2} \rfloor}$ linear demands on the theta function. This number is much less than $\frac{2^d}{24}$. It is unclear how knowledge of some higher coefficients can be used effectively to determine structure. The family of Ypsilanti lattices shows that many isometry types in a given dimension have the same minimum norm. To characterize these, or ones like them, we probably need more than hypotheses about their theta functions. We guess that for the Ypsilanti lattices, given theta functions may be shared by large sets of isometry types, and similarly for automorphism groups.

The numerous Ypsilanti lattices have nontrivial normal 2-subgroups (products of extraspecial 2-groups) in their automorphism groups. Their enumeration should be compared with [1], which shows that the portion of even unimodular lattices in rank divisible by 8 which have nontrivial automorphism group goes to 0 as the rank goes to infinity. See also the Mathematical Review article for [1] by Scharlau.

We acknowledge helpful conversations with Alex Ryba, Leonard Scott, Jean-Pierre Serre and Kannan Soundarajan.

The author has been supported by NSA grant USDOD-MDA904-03-1-0098.

3 Statement of results

First, we give some notation, then state the main results.

Definition 3.1. Given a lattice, L, define $\mu(L) := \min\{(x,x)|x \in L, x \neq 0\}$.

Definition 3.2. Given a lattice L, we define the *dual lattice* to be $L^* := \{x \in \mathbb{Q} \otimes L | (x,L) \leq \mathbb{Z}\}$. Given an integral lattice, L, we define the *discriminant group* of L to be $\mathcal{D}(L) := L^*/L$, a finite abelian group. A set of *invariants* of an integral lattice are the orders of the cyclic summands in a direct product decomposition of $\mathcal{D}(L)$. (This depends on choice of decomposition.)

Definition 3.3. Condition $X(2^d)$: This is defined for integers $d \geq 2$. Let $s \in \{0,1\}$ be the remainder of $d+1$ modulo 2.

We say that the quadruple (L, L_1, L_2, t) is a an X-quadruple if it satisfies *condition $X(2^d)$* (or, more simply, *condition X*), listed below:

(a) L is a rank 2^d even integral lattice containing $L_1 \perp L_2$, the orthogonal direct sum of sublattices $L_1 \cong L_2$ of rank 2^{d-1};

(b) When $d = 2$, $L \cong L_{D_4} \cong BW_4$ and $L_1 \cong L_2 \cong L_{A_1^2}$; when $d \geq 3$, $2^{-\frac{s}{2}}L_1$ and $2^{-\frac{s}{2}}L_2$ are initial entries of quadruples which satisfy condition $X(2^{d-1})$.

(c) $\mu(L) = 2^{\lfloor \frac{d}{2} \rfloor}$.

(d) $\mathcal{D}(L) \cong 2^{2^{d-1}}, 1$ as d is even, odd, respectively.

(e) There is an isometry t of order 2 on L which interchanges L_1 and L_2 and satisfies $[L,t] \leq L_1 \perp L_2$, i.e., acts trivially on $L/[L_1 \perp L_2]$.

(f) The projection of L to each V_i are rank 2^{d-1} BRW-sublattices, i.e., are stable under a natural $BRW^0(2^{d-1}, +)$-subgroup of $Aut(L)$ which centralizes t and stabilizes each L_i.

Definition 3.4. Also, we say that the lattice L is a *lattice of Barnes-Wall type* or a *Barnes-Wall type lattice* if there exist sublattices L_1, L_2 of L and an involution $t \in Aut(L)$ so that (L, L_1, L_2, t) satisfies condition $X(2^d)$.

Theorem 3.5. *Let $d \geq 2$. A Barnes-Wall type lattice of rank 2^d exists and is unique up to isometry.*

Corollary 3.6. (i) *For every integer $d \geq 2$, there is an integral even lattice L, unique up to scaled isometry, such that*

(a) *the rank is 2^d;*

(b) *$Aut(L)$ contains a group $G_{2^d} \cong 2_+^{1+2d}\Omega^+(2d,2)$;*

(ii) *For such a lattice, the group of isometries is isomorphic to W_{E_8} if $d = 3$ and is just G_{2^d} for $d \geq 4$. Also, $\mathcal{D}(L) \cong 1$ or $2^{2^{d-1}}$, as d is odd, even, respectively. Also, $\mu(L) = 2^{\lfloor \frac{d}{2} \rfloor}$.*

We mention that the much-studied lattice L_{E_8} is the case $d = 3$ of the above. The author has recently given an elementary uniqueness proof for L_{E_8}. See [20], where previous uniqueness proofs are discussed. Also, a uniqueness proof for BW_4 was given in [19].

In addition we prove transitivity results for certain types of sublattices made of scaled Barnes-Wall lattices, including frames. See 12.4, 13.1, 13.3.

A final application of our theory is the construction of *the Ypsilanti lattices* or the *Ypsilanti cousins*, built in a similar style. (Their definition is a special case of $\mathfrak{Y}$ 14.11, which is in turn a natural extension of the notation $\mathfrak{X}$ 3.3; the idea came during a pleasant moment in Ypsilanti, Michigan.)

Let $j \geq 1, d = 5 + 3j$. The Ypsilanti lattices are indecomposable, even, unimodular in dimension 2^d, and BW-like in the sense of minimum norm. For large dimensions, they become quite numerous. The following easily stated results give a sample of what we proved.

Theorem 3.7. *For $c \in [0, \frac{1}{8})$ and integer $j > 0$ so that $\frac{1}{16}(2 - 2^{1-j} + 3 \cdot 2^{-1-2j}) > c$, there is a family $Ypsi(2^d, j)$ of rank 2^d indecomposable, even unimodular lattices, defined for all $d >> 0$, so that $\log_2$ of the number of isometry types in $Ypsi(2^d, j)$ has dominant term at least $c\, d\, 2^d$ (in other language, at least $(\frac{1}{8} + o(1))d\, 2^d$).*

  (i) *There is an integer m so that for $d >> 0$ and $L \in Ypsi(2^d, j)$, $\mu(L) = 2^m$;*
  (ii) *the minimal vectors of L span a proper sublattice of finite index in L;*
  (iii) *$Aut(L)$ has a normal 2-subgroup U of order divisible by 2^{1+2d}.*

The quotient $Aut(L)/U$ is generally small. The integer m in (iii) is roughly $\lfloor \frac{d-j}{2} \rfloor$. Like the BW lattices, the minimum norms go to infinity roughly like the square root of the dimension.

Corollary 3.8. *Let $b \in [0, \frac{1}{8})$. The number $\Upsilon(n)$ of isometry types of even unimodular lattices of dimension $n \in 8\mathbb{Z}$ which contain a Ypsilanti lattice as an orthogonal direct summand satisfies: $\log_2(\Upsilon(n))$ is asymptotically at least $b \cdot \log_2(mass(n))$ for $n >> 0$, where $mass(n)$ is the number provided by the Siegel mass formula.*

4 Background on codes

Definition 4.1. *An $(n-k) \times n$ matrix of the form $H = (A|I_{n-k})$, where A is an $(n-k) \times k$ matrix, is a parity check matrix for the code C if C is defined as the set of row vectors $x \in F^n$ which satisfy $Hx^{tr} = 0$ [26], p.2.*

Definition 4.2. *The Hamming code $\mathcal{H}_r$ is defined (up to coordinate permutations) by the parity check matrix H_r which is the $r \times (2^r - 1)$ matrix consisting of the $2^r - 1$ nonzero column vectors of height r over $\mathbb{F}_2$. The binary simplex code $\mathcal{S}_r$ is the annihilator of the Hamming code $\mathcal{H}_r$.*

Remark 4.3. The code $\mathcal{H}_r$ can be interpreted as the subsets of nonzero vectors in $\mathbb{F}_2^r$ which sum to zero. It has parameters $[2^r - 1, 2^r - 1 - r, 3]$ [26], p. 23. The minimum weight elements of $\mathcal{H}_r$ are simply the nonzero elements of a 2-dimensional subspace. Therefore, a nonzero codeword A in the annihilator meets every such 3-set in 0 or 2 elements. Equivalently, the complement A' of A meets every such 3-set in a 1-set or the whole 3-set. It is clear that A' with the zero vector is a codimension 1 linear subspace of $\mathbb{F}_2^r$, whence A is an affine codimension 1 subspace. It follows that every nonzero element of $\mathcal{S}_r$ has weight 2^{r-1}, so $\mathcal{S}_r$ has parameters $[2^r - 1, r, 2^{r-1}]$. Note that for $r \geq 2$, $\mathcal{H}_r \geq \mathcal{S}_r$ and that $\mathcal{S}_r$ contains **1**, the all-ones vector, an odd set. Also, $\mathcal{H}_r$ is spanned by the affine planes with 0 removed.

Definition 4.4. The *extended Hamming code* is obtained by appending an overall parity check, so has parameters $[2^r, 2^r - r - 1, 4]$. For $r \geq 2$, it contains the all-ones vector. It is denoted $\mathcal{H}_r^e$. Its annihilator is the *extended simplex code* $\mathcal{S}_r^e$, which has parameters $[2^r, r+1, 2^{r-1}]$. We have for $r \geq 2$, that $\mathcal{H}_r^e \geq \mathcal{S}_r^e$ contains $\mathbf{1}^{2^r}$. Also, $\mathcal{H}_r$ is spanned by the affine planes.

Proposition 4.5. *If $r \geq 1$ is an integer, the Hamming code and simplex code of length 2^r have automorphism group isomorphic to $AGL(r, 2)$.*

Proof. This is well known. Since these two codes are mutual annihilators, they have a common group. A recent proof was given in an appendix of [11]. $\square$

Lemma 4.6. *If S is a subset of $\mathbb{F}_2^d$ of cardinality $2^r > 1$ so that for every affine hyperplane H of $\mathbb{F}_2^d$, $|H \cap S| = 0, 2^r$ or 2^{r-1}, then S is an affine subspace.*

Proof. This is a result of Rothschild and Van Lint, [30]; it is given in [26], Chapter 13, Section 4, Lemma 6, page 379. $\square$

Remark 4.7. The Reed-Muller codes are present in our analysis (the codes $\mathcal{C}_X$ in 11.16) but play a small role.

Definition 4.8. A code $0 \neq C \leq F^X$ is *decomposable* if there is a nontrivial partition $X = Y \cup Z$ of the index set, so that $C = C_Y \oplus C_Z$ is a nontrivial direct sum, where C_W means the set of vectors in C with support contained in $W \subseteq X$. If a code $C \neq 0$ is not decomposable, it is *indecomposable*.

Lemma 4.9. *For all $t \geq 3$, there is a length 2^t indecomposable doubly even self orthogonal binary code.*

Proof. For $t = 3$, take the extended Hamming code. Suppose $t \geq 4$ and set $u = t - 3$. Take a partition of an index set S of size 2^t into 2^u parts S_i of size 8, for $i = 1, \ldots, 2^u$. Let H_i be an extended Hamming code on S_i. Take a vector v_i of weight 2 with support A_i in S_i and define $v = \sum_i v_i$. Form the code C spanned by v_i and the codimension 1 subspace of $\sum_i H_i$ which annihilates v. Then $wt(v) = 2 \cdot 2^u \in 4\mathbb{Z}$, whence C is even. $\square$

5 Background on lattices

Lemma 5.1. *Let L be a positive definite integral lattice. Then L has a unique orthogonal decomposition into indecomposable summands. More precisely, let $X(L)$ be the set of nonzero vectors of L which are not expressible as the orthogonal sum of two nonzero vectors of L. Generate an equivalence relation on $X(L)$ by relating two elements if their inner product is nonzero. An orthogonally indecomposable summand of L is the sublattice spanned by an equivalence class in $X(L)$. In fact, an orthogonal direct summand is a sum of a subset of this set of sublattices.*

Proof. (See [25] and [27], which credits [25].) Let X_i, $i = 1, \ldots, t$ be the equivalence classes in $X = X(L)$ and let L_i be the sublattice spanned by X_i. Positive definiteness implies that L is the sum of the L_i. Also by taking inner products, we deduce $L_i \cap L_j = 0$ for $i \neq j$. So, we have an orthogonal direct sum.

Let M be an arbitrary orthogonal direct summand. Let N be the annihilator of M in L. We show for each i that $L_i \leq M$ or $L_i \leq N$. For $x \in X$, write $x = x_M + x_N$, where $x_M \in M, x_N \in N$. Indecomposability implies that one of these components is 0.

Now, suppose that $M \cap L_i$ is nonempty. Then, there exists $x \in X_i$ so that $(x, M) \neq 0$. The last paragraph implies that $x \in M$. We then deduce $X_i \subset M$ and $L_i \leq M$. If $M \cap L_i = \emptyset$, then $L_i \subset N$. □

Notation 5.2. Given a lattice L, the ambient vector space is $\mathbb{Q} \otimes L$, with natural extension of the symmetric bilinear form on L.

Take isometries $\psi_i : \mathbb{Q} \otimes L \to V_i$ of rational vector spaces. From these, we get isometries $\psi_{ij} = \psi_i^{-1} \psi_j$ from V_i to V_j. Priming on an index means replacement of the corresponding map by its negative.

For a subset $A \subseteq \mathbb{Q} \otimes L$, define the following subsets of $V := V_1 \perp V_2$:
$A_{ij} := \{x^{\psi_i} + x^{\psi_j} | x \in A\}$,
$A_{ij'} := \{x^{\psi_i} - x^{\psi_j} | x \in A\}$,
$A_{i'j} := \{-x^{\psi_i} + x^{\psi_j} | x \in A\}$,
$A_{i'j'} := \{-x^{\psi_i} - x^{\psi_j} | x \in A\}$.

Notation 5.3. Given a basis $\mathcal{B}$ of Euclidean space and binary code $C \leq \mathbb{F}_2^{\mathcal{B}}$, we define $L_{\mathcal{B},C} := \{\sum_{b \in \mathcal{B}} \frac{1}{2} a_b b | a_b \in \mathbb{Z}, (a_b + 2\mathbb{Z})_{b \in \mathcal{B}} \in C\}$. (This lattice is sometimes integral.) Note that $L_{\mathcal{B},C} = \{\sum_{b \in \mathcal{B}} \frac{1}{2} a_b b | a_b \in \mathbb{Z}, \sum_{b \in \mathcal{B}} (a_b + 2\mathbb{Z}) c_b = 0 + 2\mathbb{Z}$ for all $(c_b)_{b \in \mathcal{B}} \in C^\perp\}$.

Notation 5.4. Let α_i, $i \in \Omega = \mathbb{F}_2^3$ be vectors in $\mathbb{R}^\Omega$ which satisfy $(\alpha_i, \alpha_j) = 2\delta_{ij}$. Let $\mathcal{H}_8^e$ be the extended Hamming code 4.4.

Define the A_1^8-description of L_{E_8} or the 2-twisted version of L_{E_8} to be the $\mathbb{Z}$-span of all α_i and all $\frac{1}{2}\alpha_c$, for $c \in \mathcal{H}_8^e$. In the Notation 5.3, this is $L_{\{\alpha_i | i=1,\ldots,8\}, \mathcal{H}_8^e}$.

Notation 5.5. Suppose that Ω is an index set and $\{v_i | i \in \Omega\}$ is a basis of a vector space. For a subset A of Ω, define $v_A := \sum_{i \in A} v_i$. The linear transformation ε_A sends v_i to $-v_i$ if $i \in A$ and to v_i if $i \notin A$. The group of such maps is $\mathcal{E}_{P(\Omega)}$. If C is a subset of $P(\Omega)$, $\mathcal{E}_C$ denotes the set of maps ε_A for $A \in C$. This is a subgroup if C is a subspace of the vector space $P(\Omega)$.

Proposition 5.6. *For an integer $d \geq 2$, define $m := \lfloor \frac{d}{2} \rfloor$. Let Ω be an index set, identified with $\mathbb{F}_2^d$. Take a basis $\mathcal{B} := \{v_i | i \in \Omega\}$ where $(v_i, v_j) = 2^m \delta_{ij}$ of $\mathbb{R}^\Omega$. Form $L_{\mathcal{B}, \mathcal{H}^e}$, as in 5.3; it is integral for $d \geq 3$. Then, if $d \geq 4$, $Aut(L_{\mathcal{B}, \mathcal{H}_d^e})$ is in the monomial group on $\mathcal{B}$ and in fact $Aut(L_{\mathcal{B}, \mathcal{H}_d^e}) = \mathcal{E}_\Omega F$, where F is a natural $AGL(d, 2)$ subgroup of the group of permutation matrices. If $d = 3$, $Aut(L_{\mathcal{B}, \mathcal{H}_d^e}) \cong W_{E_8}$.*

Proof. For $d = 3$, we have the lattice 5.4 and for $d = 2$, we have the F_4 lattice, spanned by vectors of shape $(\pm 1, 0, 0, 0), (\pm \frac{1}{2}, \pm \frac{1}{2}, \pm \frac{1}{2}, \pm \frac{1}{2})$. These automorphism groups are well known to be W_{E_8} and W_{F_4}, respectively.

For any $d \geq 2$, the set of minimal vectors of $L := L_{\mathcal{B},\mathcal{H}_d^e}$ is just $\pm v_i$, $i \in \Omega$ and $\frac{1}{2}v_S\varepsilon_T$, for $S \in \mathcal{H}_d^e$ an affine plane and T a subset of S. These span L since affine planes span $\mathcal{H}_d^e$. All these minimal vectors have norm 2^m, $m \geq 1$.

We now assume $d \geq 4$. The set of these which are in $2^{m-1}L^*$ is exactly $\pm v_i$, $i \in \Omega$, for if A is an affine plane there exists an affine plane B so that $A \cap B$ is a 1-set (because $d \geq 4$), whence $(\frac{1}{2}v_A\varepsilon_S, \frac{1}{2}v_B\varepsilon_T) = \pm 2^{m-2}$. It follows that $Aut(L)$ is contained in the monomial group based on $\mathcal{B}$. Clearly it contains $\mathcal{E}_\Omega F$, described above, and maps to the stabilizer of the code L/Q in $\frac{1}{2}Q/Q \cong \mathbb{F}_2^\Omega$, where Q is the square lattice with basis $\mathcal{B}$. Since $Aut(\mathcal{H}_d^e)$ is a natural $AGL(d, 2)$ subgroup of the symmetric group (4.5), we are done. □

Definition 5.7. Let $c = (c_i) \in \mathbb{F}_2^n$. The *Euclidean lift* of c is the vector in $\{0,1\}^n \subset \mathbb{Z}^n$ which reduces modulo 2 to c. When p is an odd prime and $c = (c_i) \in \mathbb{F}_p^n$, we have a similar definition of lift, using the subset $\{-\frac{p-1}{2}, -\frac{p-3}{2}, \ldots, -1, 0, 1, \ldots, \frac{p-3}{2}, \frac{p-1}{2}\}^n \subset \mathbb{Z}^n$.

Lemma 5.8. *Let L be a lattice with sublattice of finite index M which is a coelementary abelian p-group for some prime p. Let $F := \mathbb{F}_p$. Suppose that $\mathcal{C}$ is an error correcting code in F^n with minimum weight w. Suppose that J is the lattice between M^n and L^n corresponding to $\mathcal{C}$, i.e. spanned by all $(c_1 x, c_2 x, \ldots, c_n x)$ for $x \in L$ and (c_i) is the Euclidean lift of a codeword in $\mathcal{C}$.*

(i) If $(y_1, \ldots, y_n) \in J \setminus M^n$, the weight of $(y_1, \ldots, y_n)(\mathrm{mod}\ M^n)$ is at least w.

(ii) Suppose that M is indecomposable and $\mathcal{C}$ is indecomposable. Then $Aut(J) \cap Aut(M \perp \cdots \perp M) \cap Aut(L \perp \cdots \perp L)$ factorizes as the product of subgroups $A_1 A_2$, where A_1 is the subgroup which fixes each direct summand isometric to M and acts diagonally on $(L/M)^n$, and where A_2 is the subgroup of the natural group of block permutation matrices of degree n which fixes the code $\mathcal{C}$.

Proof. (i) We take a basis $v(1), \ldots, v(d)$ of $\mathcal{C}$. For a codeword $v = (v_i)$ and $x \in L/M$, let vx be the vector in $(L/M)^n$ whose i^{th} component is the Euclidean lift of v_i times $x + M$. So, $J/M^n = \sum_{i=1,\ldots,d; x \in L} v(i)x + M^n = \bigoplus_{i=1,\ldots,d} v(i)(L/M)$.

Suppose that $y = (y_1, \ldots, y_n)(\mathrm{mod}\ M^n)$ represents the minimal weight k in J/M^n. Write it as a linear combination $y = \sum_{i=1}^d v(i)x(i)$, where $x(1), \ldots, x(d)$ is a sequence of elements of L/M and the product $v(i)x(i)$ is as in the previous paragraph.

Take any linear functional $f : L/K \to F$ and extend it componentwise to $g : (L/M)^n \to F^n$. Then $g(J/M) = \mathcal{C}$. Given a nonzero $x(i)$, take an f so that $f(x(i)) = 1$. Then $g(y) \in \mathcal{C}$ has nonzero coefficient 1 at $v(i)$, whence $0 \neq g(y) = (f(y_1), \ldots, f(y_n)) \in \mathcal{C}$ has at least w nonzero entries, whence so does $y = (y_1, \ldots, y_n)$. Therefore, $k \geq w$.

(ii) First, suppose that $h \in Aut(J) \cap Aut(M \perp \cdots \perp M) \cap Aut(L \perp \cdots \perp L)$. We claim that h determines a unique element of the code group, up to scalars. For any $v \in \mathcal{C}, x \in L \setminus M$, h takes vx to an element of J/M. This means that for linear functionals f, g as in (i) where $f(x) = 1$, we have that $g(h(vx))$ is a codeword. Since h takes vx to another element of similar form $v'x'$, it follows that there is a block permutation matrix b so that, for all codewords x, the action of hb stabilizes each direct summand isometric to M and takes vx to an element of the form vx', for all $x \in L/M$. Since the code is indecomposable, we use the property that for any $v, w \in \mathcal{C}$, if hb takes vx to vx', then hb takes wx to wx'. In other words, the actions of hb on the summands of $(L/M)^n$ are identified. $\square$

Lemma 5.9. *Suppose that J is a lattice and that $SMV(J)$, the sublattice spanned by the minimal vectors, has finite index. Suppose that $SMV(J) = J_1 \perp \cdots \perp J_n$, where the J_i are indecomposable lattices.*

Let K satisfy $SMV(J) \leq K \leq J$ and K is homogeneous with respect to the rational vector spaces spanned by the summands, i.e., $K = \sum_{i=1}^n K_i$ where $K_i := K \cap (\mathbb{Q} \otimes J_i)$.

Suppose that J/K corresponds to an indecomposable code 4.8. Then J is orthogonally indecomposable.

Proof. (See 5.1.) Let x be an indecomposable vector of J which is not in K and let S be the indecomposable summand of J which contains it. Let A be the support of $x + K$ in J/K, i.e., those indices where $x + K$ projects nontrivially to $\mathbb{Q} \otimes K_i / K_i$. For $i \in A$, there exists a minimal vector $y \in K_i$ so that $(x, y) \neq 0$. Therefore, $J_i \leq S$, for all $i \in A$. The indecomposability assumption on the code implies that all $J_i, i = 1, \ldots, n$ are in S and since $SMV(J)$ has finite index in J and S is a summand of J, $S = J$. $\square$

Definition 5.10. Let $r > 0$ be an integer. An integral lattice L is *r-modular* if $L \cong \sqrt{r} L^*$.

Definition 5.11. The SSD concepts were established in [19]. Call a lattice M *semiselfdual (SSD)* if $2M^* \leq M \leq M^*$. If the sublattice M of the integral lattice L is semiselfdual, we define the orthogonal transformation t_M on $\mathbb{Q} \otimes L$ by -1 on M and 1 on $M^\perp$. Then t_M leaves L invariant and so gives an isometry of L of order 1 or 2; it has order 2 on L if $M \neq 0$.

A more general notion is that of *relatively SSD (RSSD)*: this is the condtion that the sublattice M of the integral lattice L satisfies the weaker condition $2L \leq M + M^\perp$. In this case, the orthogonal involution defined as above preserves L.

6 Actions of 2-groups and endomorphisms on lattices

We gather an assortment of results on this topic.

Definition 6.1. A *fourvolution* is a linear transformation whose square is -1. A *fourvolution on a lattice* is a lattice isometry whose square is -1. In case we have a lower group as in 8.3, we use the terms *lower and upper fourvolution*. We may call an element in a group a fourvolution with respect to a representation, and even with respect to more than one representation, for example by restriction of one action to a submodule.

Lemma 6.2. *If f is a fourvolution of the lattice L, then the adjoint of $1\pm f$ is $1\mp f$, $(1\pm f)^2 = \pm 2f$, $1\pm f$ is an isometry scaled by $\sqrt{2}$ and we have $L \geq L(1+f) \geq 2L$ and $|L : L(1+f)| = |L(1+f) : 2L| = |L/2L|^{\frac{1}{2}}$. In particular, $\mathrm{rank}(L)$ is even.*

Proof. For $x, y \in L$, we have $(x(1 \pm f), y(1 \pm f)) = (x, y) \pm (x, yf) \pm (xf, y) \pm (xf, yf) = 2(x,y) \pm (x,yf) \pm (xf^2, yf) = 2(x,y)$. For adjointness, just compute $(x(1 \pm f), y) = (x,y) \pm (xf, y) = (x,y) \pm (xf^2, yf) = (x,y) \mp (x, yf)$. The other statements are easy to prove. $\square$

Notation 6.3. Let L be a lattice and f a fourvolution in $Aut(L)$. Define $S[k] := S(1-f)^k$, for $S \subseteq \mathbb{Q} \otimes L$ and $k \in \mathbb{Z}$. Note that this makes sense since the linear map $1-f$ is invertible, with inverse $\frac{1}{2}(1+f)$. Call a transformation of the form $1-f$ a *sultry tranformation* and call $S[k]$ the kth *sultry $(1-f)$-twist of S* or the kth *sultry twist of S*. (The terminology is explained in 9.1.)

We have $S[0] = S$. Note that for all k, $(Sp)[k] = (S[k])p$, where p is any polynomial expression in f. Also, $S[k][\ell] = S[k+\ell]$.

Lemma 6.4. *If S is an f-invariant lattice in $\mathbb{Q} \otimes L$, then for $k \leq \ell$, $|S[k] : S[\ell]| = 2^{\frac{1}{2}\mathrm{rank}(S)(\ell-k)}$.*

Proof. This follows since $(1-f)^2 = -2f$ and because for all integers p, q and all integers $r \geq 0$, $S(1-f)^p/S(1-f)^{p+r} \cong S(1-f)^q/S(1-f)^{q+r}$. $\square$

Lemma 6.5. *Let S, T be subsets of $\mathbb{C} \otimes L$. Then*

(i) $(S[1], T) = -(S, Tf[1])$.

Now assume that S and T are f-invariant. Then $S = Sf = -S$, $T = Tf = -T$ and the following hold.

(ii) For all integers k, ℓ, we have $(S[k], T[\ell]) = 2(S[k-1], T[\ell-1])$ and $(S[k], T[\ell]) = 2(S[k-2], T[\ell]) = 2(S[k], T[\ell-2])$.

(iii) $S^*[k] = S[k]^* f^{-k}$.

(iv) $(S[k], T[\ell]) = (S[k'], T[\ell'])$, for all integers k, k', ℓ, ℓ' such that $k + \ell = k' + \ell'$; and

(v) Assume that the integer ℓ satisfies $S^* = S[\ell]$. Then $S^*[k] = S[k+\ell]$.

Proof. (i) and (v) are clear.

(ii) follows since $1 - f$ is an isometry scaled by $\sqrt{2}$.

(iii) We have $x \in S[k]^*$ if and only if $(x, S[k]) \in \mathbb{Z}$ if and only if $x(1+f)^k = (-1)^k x f^k (1-f)^k \in S^*$ if and only if $x \in (-1)^k S^*[-k] f^{-k} = S^*[-k] f^{-k}$.

(iv) is trivial for $k = 0$ and for $k \geq 1$ it follows from (i) and easy induction. If k is negative, use (ii) and the case $k \geq 0$.

Example 6.6. If $L \cong L_{D_4}$ then $L[-1] \cong L_{F_4}$, where we take the latter to be the span of a standard version of the F_4 root system: $(\pm 1 0^3), (\pm 1^2 0^2), (\pm \frac{1}{2}^4)$.

Definition 6.7. Let L be a lattice with fourvolution f. Suppose that there is an integer r such that $L^* = L[-r]$ (see 6.3). We call r the *duality level* of L. Such a modular lattice (see 5.10 and 6.4) is called an *r-sultrified dual* and is 2^r-modular (see 5.10).

Definition 6.8. Given a group E acting on a lattice L and character $\lambda \in Hom(E, \{\pm 1\})$, define the *eigenlattice* L^λ to be $\{a \in L | ay = \lambda(y)a, \text{ for all } y \in E\}$. Define the *total eigenlattice* to be $Tel(E, L) := \sum_{\lambda \in Hom(E, \{\pm 1\})} L^\lambda$. The notation extends naturally a set of automorphisms. When t has order 1 or 2, define L^+, L^- to be the lattice of fixed, negated points, respectively. To denote dependence on t, we write $L(\pm, t)$ or $L^\pm(t)$ for $L^\pm$.

Remark 6.9. In case E is 2-elementary abelian, $L/Tel(E, L)$ is finite, and is in fact a 2-group, but in general is not elementary abelian. For an example, let E be a fourgroup and $L = \mathbb{Z}[E]$, the regular representation. Then, $L/Tel(E, L) \cong 2 \times 2 \times 4$.

Lemma 6.10. *Suppose that the involution t acts on the additive group A. Let $A^\varepsilon := \{a \in A | a^t = \varepsilon a\}$. Suppose furthermore that the minimal number of generators of A as an abelian group is $r < \infty$. Define integers k, ℓ by $2^k := |A : A^- + A^+|$ and $2^\ell := |A : B|$, where $B := \{x \in A | x(1-t) \in 2A\}$.*

Then: (i) $2A \leq A^- + A^+ \leq B$, whence $\ell \leq k$;

(ii) $\ell \leq r/2$.

(iii) $A^- \geq A(1-t) \geq 2A^-$ and $|A(1-t)/2A^-| = 2^k$, whence $\operatorname{rank}(A^-) \geq k$.

(iv) $A^+ \geq A(1+t) \geq 2A^+$ and $|A(1+t)/2A^+| = 2^k$, whence $\operatorname{rank}(A^+) \geq k$.

(v) If A is free abelian, A^ε is a direct summand of A and $A^+ + A^- = A^+ \oplus A^-$.

(vi) If A is free abelian, $k = \ell$ (whence $k \leq r/2$).

(vii) Suppose that multiplication by 2 is a monomorphism of A (e.g., A is free abelian). If $k = 0$ (i.e., if t is trivial on $A/2A$), $A = A^+ + A^-$.

Proof. (i) The proof follows from the equation $2a = (a + a^t) + (a - a^t)$.

(ii) Let $B := \{x \in A | x(1-t) \in 2A\}$. Then the map $(1-t)$ induces an injection of $A/B \cong 2^\ell$ into $B/2A$, so in particular $\ell \leq k$. If $x_1, x_2, \ldots, x_\ell \in A$ form a basis modulo B, then $x_1, x_1^t, x_2, x_2^t, \ldots, x_\ell, x_\ell^t$ are independent modulo $2A$. Therefore, $2\ell \leq r$.

(iii) For the first statement, notice that the kernel of the map $\phi : A \to A^-/2A^-$, $x \mapsto x(1-t)$ is $A^+ \oplus A^-$ and then use $\operatorname{Im}(\phi) \cong A/Ker(\phi)$, which has rank k.

(iv) This follows from (iii) by replacing t with $-t$.

(v) Clearly, A/A^ε is torsionfree. The second statement follows from $A^+ \cap A^- = 0$.

(vi) This follows from the general classification of free abelian groups which are modules for cyclic groups of prime order, e.g., A.10; (74.3) in [10]. (The result for a cyclic group of order 2 is easy to prove directly.) It states that such a module A has the form $F_1 \oplus \cdots \oplus F_p \oplus E_1 \oplus \cdots \oplus E_q$, where each F_i is a copy of the regular representation $\mathbb{Z}\langle t \rangle$ and where each E_j is infinite cyclic. By reducing such a decomposition modulo 2, one deduces that $k = \ell$.

(vii) This is easy to prove directly (of course it is a consequence of the nontrivial result mentioned in (vi)). Suppose that the involution t is trivial on $A/2A$. Then $t = 1 + 2S$ for some $S \in End(A)$. From $1 = t^2 = 1 + 4(S + S^2)$, we deduce that $S + S^2 = 0$. For $a \in A$, $a = a(1 + S) - aS$. One checks that $aS \in A^-$ and $a(1 + S) \in A^+$. $\square$

Definition 6.11. The *defect* of the involution t acting on the free abelian group A is the integer $k = \ell$, as in 6.10. It is the number of nontrivial Jordan blocks for the action of t on $A/2A$.

Lemma 6.12. *Let L be a unimodular lattice and t an involution acting on L. Then the eigenlattices $L^\varepsilon := \{x \in L | x^t = \varepsilon x\}$ satisfy $\mathcal{D}(L^+) \cong \mathcal{D}(L^-) \cong 2^k$, where k is the defect of t in the sense of Definition 6.11.*

Proof. Since each L^ε is a direct summand of L, which is unimodular, the orthogonal projection takes L onto $[L^\varepsilon]^*$. The kernel of the map from L to $[L^\varepsilon]^*/L^\varepsilon$ is $L^\varepsilon + L^{-\varepsilon}$, so from Lemma 6.10, we deduce that the image is elementary abelian, of order 2^k. □

Remark 6.13. The notion of RSSD involution is essentially the same as that of an involution on a lattice. Let L be a lattice. An involution $t \in Aut(L)$ creates a pair of eigenlattices, $L^\pm$. Since $L^+ \perp L^-$ is 2-coelementary abelian in L and t acts trivially on $\frac{1}{2}[L^+ \perp L^-]/[L^+ \perp L^-]$, t is a RsSSD involution which preserves L (see 6.10, 5.11 and [19]).

Definition 6.14. For a group G acting on the RG-module M, where R is a commutative ring, the *scalar subgroup* is

$Scalar(G, M) := \{g \in G | g$ acts on M as multiplication by an element of $R^\times\}$.

When M is a free abelian group, this is just the subgroup of group elements which act as ± 1.

Definition 6.15. A *frame* or *plain frame* in a rank n lattice is a set of $2n$ vectors of common norm, two of which are linearly dependent or orthogonal.

Later in 8.7, we work with a special case of this.

6.1 Commutator density, $3/4$-generation and $2/4$-generation

In this subsection, "lattice" means just a free abelian group since the bilinear form is irrelevant to the arguments.

Definition 6.16. Let G be a group and M a $\mathbb{Z}G$-module. We define the *commutator submodule* to be $[G, M] := \sum_{x \in G} M(x-1)$. The subset S of G is *commutator dense* on M if $[M, G] = \sum_{x \in S} M(x-1)$.

Notation 6.17. Let L be a lattice with involution s acting on L. Let $L^\varepsilon(s) := \{v \in L | vs = \varepsilon v\}$ be the ε-eigenlattice for the involution s and let $Tel(s) := L^+(s) \oplus L^-(s)$ be the total eigenlattice for s.

Notation 6.18. Let t, u be involutions which generate the dihedral group D of order 8 and let f be an element of order 4 in D.

Definition 6.19. *We use the notation of 6.18. Let M be a module for $\mathbb{Z}D$ such that $[t, u]$ acts as -1 on M. For an involution $r \in D$, define $M^\varepsilon(r) := \{v \in$

$M|vr = \varepsilon v\}$, for $\varepsilon = \pm$. We say that M has 2/4-generation if $M = M^+(s) + M^+(s')$, where s, s' is a pair of generating involutions for D. We say that M has 3/4-generation if $M = M^+(s) + M^+(s') + M^+(s'')$, where $\{s, s', s''\}$ is a set of three involutions from $\{t, -t, u, -u\}$ (so contains a pair of generating involutions for D).

Notation 6.20. Let L be a free abelian group of rank $2n$ which admits an action of D in which the central involution $[t, u]$ acts as -1 and let $L_1 := L^-(t), L_2 := L^+(t)$.

We extend the action of D to the ambient vector space $\mathbb{Q} \otimes L$. For integers $\ell \leq m$, let $Q(\ell, m) := 2^\ell Tel(t)/2^m Tel(t)$.

Define integers d, e to be the number of Jordan blocks of size 2, 1 respectively, for the action of u on the elementary abelian group $L/Tel(t)$.

Lemma 6.21. $|L : Tel(t)| = 2^{2d+e}$ and $|L : 2Tel(t)| = 2^{2n+2d+e}$.

Notation 6.22. We have a chain of D-invariant abelian groups $2Tel(t) \leq L \leq \frac{1}{2}Tel(t)$. For $g \in D$, we denote by $A(g)$ and $B(g)$ the commutator modules $Q(-1, 1)(g-1) = \frac{1}{2}Tel(t)(g-1) + 2Tel(t)/2Tel(t)$ and $L(g-1) + 2Tel(t)/2Tel(t)$, respectively.

Lemma 6.23. $A(u) \cap Q(0, 1) = B(u) \cap Q(0, 1) = Q(0, 1)(u - 1)$.

Proof. Clearly, $A(u) \cap Q(0, 1) \geq B(u) \cap Q(0, 1) \geq Q(0, 1)(u - 1)$. Now to prove the opposite containment. Since $A(u) \cap Q(0, 1)$ consists of elements inverted by u, hence fixed by u, it is contained in the subgroup $Q(0, 1)(u - 1)$ of the free $\mathbb{F}_2\langle u \rangle$ module $Q(0, 1)$. $\square$

Lemma 6.24. $A(t) \cap A(u) = 0$.

Proof. Since $Tel(t)(t-1) = 2L_1$, the image of $A(t)$ is just $L_1 + 2Tel(t)/2Tel(t) \leq Q(0, 1)$. Also, $A(u) \cap Q(0, 1)$ is exactly the image of the diagonal sublattice $\{(x, xu)|x \in L_1\}$ of $L_1 \oplus L_2$ in $Q(0, 1)$. The result follows. $\square$

Lemma 6.25. *A coset of $Tel(t)$ fixed by u contains an element fixed by u.*

Proof. Let $x + Tel(t)$ be such a coset. Since $Tel(t)$ is a free $\mathbb{Z}\langle u \rangle$-module, every element of $Tel(t)$ negated by u is a commutator. Therefore, there exists $v \in Tel(t)$ so that $x(u - 1) = v(u - 1)$. Then $x - v$ is fixed by u and is in $x + Tel(t)$. $\square$

Lemma 6.26. $|B(u)| = 2^{n+d}$.

Proof. The right side is the product of $|Q(0, 1)|^{\frac{1}{2}} = 2^n$ with $2^d = |(L/Tel(t))(u - 1)|$. To evaluate the left side, use 6.23, 6.25. $\square$

Lemma 6.27. *The kernel of the endomorphism induced by $t - 1$ on $L/2Tel(t)$ is just $Q(0, 1)$.*

Proof. If the kernel were larger, there would be $x \in L \setminus Tel(t)$ so that $x(t-1) \in 2Tel(t)$. Then there would be a unique $y \in L^-(t)$ so that $2y = x(t - 1)$, whence $yt = -y$ and $(x+y)t = x+y$ and so $x+y \in L^+(t)$, a contradiction to $x \notin Tel(t)$. $\square$

Corollary 6.28. $|B(t)| = 2^{2d+e}$.

Proof. Since $Q(0,1)$ has index 2^{2d+e} in $L/2Tel(t)$, the result follows from 6.27. □

Lemma 6.29. $B(t) \cap B(u) = 0$ and $|B(t) + B(u)| = 2^{n+3d+e}$.

Proof. For $B(t) \cap B(u) = 0$, use 6.24. The second statement follows from the formula $|B(t) + B(u)| = |B(t)||B(u)|/|B(t) \cap B(u)| = |B(t)||B(u)|$ and 6.26, 6.28. □

Lemma 6.30. $L(t-1) + L(u-1) = [L, D] \geq L(f-1) \geq 2L$.

Proof. First, $L(t-1) + L(u-1) = [L, D]$ holds because t and u generate D. The containment $[L, D] \geq L(f-1)$ is obvious. Since $[L, D]$ contains $L(f-1)$ and $L(f-1)^2 = 2L$, the final containment holds. □

Lemma 6.31. We have $|B(t) + B(u) : (L(f-1)/2Tel(t))| = 2^{n+3d+e-(2d+e+n)} = 2^d$. Therefore, $L(f-1) = L(t-1) + L(u-1)$ if and only if $d = 0$. In other language, commutator density is equivalent to 2/4-generation.

Proof. Observe that if M is any f-invariant subgroup of L, then $M(f-1)^2 = 2M$ and that for any integer j, $(*)$ $|M(f-1)^j : M(f-1)^{j+1}| = 2^n$. Since $Tel(t)(f-1) \leq Tel(t)$, we have $2Tel(t) = Tel(t)(f-1)^2 \leq L(f-1) \leq [L, D]$. Both $L(f-1)$ and $L(t-1) + L(u-1) = [L, D]$ contain $2Tel(t)$, whence a basic isomorphism theorem implies that $|L(t-1) + L(u-1) : L(f-1)| = |B(t) + B(u) : (L(f-1)/2Tel(t))|$. The statements follow from 6.21, 6.29 and $(*)$. □

Lemma 6.32. The properties 2/4-generation and 3/4-generation are equivalent.

Proof. Obviously, 2/4-generation implies 3/4-generation. Conversely, assume that $L = L^+(t) + L^-(t) + L^+(u)$. Using 6.30, we have $L^+(t) + L^+(u) \geq L(t+1) + L(u+1) = [L, D] \geq 2L$. Since $L^+(t) + 2L = L^-(t) + 2L = L^+(t) + L^-(t)$, $L = L^+(t) + L^-(t) + L^+(u) = L^+(t) + L^+(u) + 2L = L^+(t) + L^+(u)$, whence 2/4-generation. □

Theorem 6.33. For the action of a dihedral group of order 8 on a lattice, where the central involution acts as -1, the properties of commutator density, 2/4-generation and 3/4-generation are equivalent.

Proof. Lemmas 6.31 and 6.32. □

Remark 6.34. Note that the present version avoids bilinear forms, so does not use the language of determinants. If a D-invariant bilinear form is present, we may replace statements about finite indices of sublattices with ones about determinants. Note that, for a noncentral involution s of D, each $L^{\pm}(s)$ has the same determinant.

7 Sultry twists and the next BW procedure

We discuss some procedures for proving the main theorem. We continue to let BW abbreviate "Barnes-Wall".

We first show how to start from a BW-type lattice of rank 2^{d-1} and create one of rank 2^d. Later, in 10.2 we show how a BW-type lattice of rank 2^d is uniquely determined by an ancestor of rank 2^{d-1}. Eventually, we use an induction argument which will show that a BW-type lattice is unique, so is the same (up to rescaling) as the lattices constructed in [2, 7].

An important technique here is to use the commutator density enjoyed by these lattices. The twisting by sultry transformations helps control the analysis.

Notation 7.1. Let M be a BW lattice of rank $2^{d-1} \geq 3$. Let Q be a lower group (see A.2) in $Aut(M)$, i.e. in some $BRW^0(2^{d-1}, +)$ subgroup of $Aut(M)$, which by induction is isomorphic to $BRW^0(2^{d-1}, +)$ or $d - 1 = 3$ and $Aut(M) \cong W_{E_8}$. Also, let $f \in Q$ be a fourvolution, $F := N_{Aut(M)}(Q) \cong BRW^0(2^{d-1}, +)$; see A.2. Now let r be duality level of M (see 6.7). Then $r \in \{0, 1\}$ and $r \equiv d \pmod{2}$.

Definition 7.2. *The Next BW Procedure.* We use notations M, F, Q, f as in 7.1.

Form $M_1 \perp M_2$, two orthogonal copies of M based on the isometries $\psi_i : M \to M_i$ and let $V_i := \mathbb{Q} \otimes M_i$ be their ambient rational vector spaces. Set $V := V_1 \perp V_2$. Also, we use $\psi_{ij} := \psi_i^{-1}\psi_j$, the natural isometry from M_i to M_j, extended to $V_i \to V_j$. See Notation 5.2.

Define Q_i, F_i and f_i and the groups and element in $End(V_i)$ corresponding to Q, F and f under ψ_i. Extend their actions to V in the natural way. Also, define the group Q_{12} as the natural diagonal subgroup of $Q_1 \times Q_2$ and element $f_{12} := f_1 f_2 \in Q_{12}$ acting on V (see 5.2).

For the SSD sublattices $M_i[1-r], M_{ij}[1-r], M_{ij'}[-r]$, we denote the respective SSD involutions by $t_i, t_{ij}, t_{ij'}$. Observe that $-1 = t_1 t_2 = t_{12} t_{12'}$. For convenience and symmetry, we define $t_{i'} := -t_i, t_{i'j} := t_{ij'}, t_{i'j'} := t_{ij}$. Finally, we define $D := \langle t_1, t_2, t_{12}, t_{12'} \rangle \cong Dih_8$ and $R := \langle Q_{12}, D \rangle \cong 2^{1+2d}_+$. So, $R = Q_{12}D$, a central product.

Define $L_d := M_1[1-r] + M_2[1-r] + M_{12}[-r]$, and R-invariant lattice. We call L_d the *type BW-successor* to M. From 6.5 and $M^* = M[-r]$, we deduce that L_d is an integral lattice and since elements of the the above generating set have even norms, L_d is even.

Lemma 7.3. $\mathcal{D}(L_d) \cong 1, 2^{2^{d-1}}$ as d is odd, even. Therefore, the duality level is the remainder of $d + 1$ modulo 2.

Proof. When d is even, $L := L_d$ is the kernel of the epimorphism $M_1 \perp M_2 \to M/M[1]$, defined by $(x^{\psi_1}, y^{\psi_2}) \mapsto x + y + M[1]$. Since $M_1 \perp M_2$ is unimodular, $\mathcal{D}(L) \cong 2^{2^{d-1}}$.

When d is odd, this is the same as the kernel of the epimorphism $M_1^* \perp M_2^* \to M^*/M$ defined by $(x^{\psi_1}, y^{\psi_2}) \mapsto x + y + M$. Since $M_1^* \perp M_2^*$ has determinant 2^{-2^d}, L is unimodular.

The statement about duality level follows from 6.4. $\square$

Lemma 7.4. *We take* $L := L_d$ *(in the notation 7.2). Here,* $r \in \{0,1\}$, $d = \mathrm{rank}(L)$ *and* $r \equiv d \pmod{2}$. *Then:*
   (i) *L is the sum of any three of the four lattices*
$$M_1[1-r], M_2[1-r], M_{12}[-r], M_{12'}[-r].$$
   (ii) *L^* is the sum of any three of the four lattices*
$$M_1[1-r], M_2[1-r], M_{12}[-1], M_{12'}[-1].$$

Proof. This follows from 6.33. Here is a different proof. (i) Let $i = 1$ or 2. Define j by: $\{1,2\} = \{i,j\}$.
   Now, we observe that for any integer k,
   (a) $M_i[k] \leq M_j[k] + M_{12}[k] \leq M_j[k] + M_{12}[k-1]$;
   (b) $M_{12'}[k] \leq M_{12'}[k] + M_{12}[k] = M_{12}[k] + M_i[2+k] \leq M_{12}[k] + M_i[1+k]$.
   At once, (i) follows.
   For (ii), (a) and (b) prove equality of $N = M_1[1-r] \perp M_2[1-r] + M_{12}[-1]$ and $N' = M_{12}[-1] + M_{12'}[-1] + M_i[1-r]$. It is clear by taking dot products that $N = N'$ is in L^*. If $r = 0$, $L = N$ and if $r = 1$, $N/L \cong M_{12}[-1]/M_{12}[0] \cong 2^{2^{d-1}}$, whence $N = L^*$ (see 7.3). □

Corollary 7.5. *In the notation of 6.16 and 7.2, the R-module L_d is commutator dense with respect to any fourvolution in R.*

Proof. Since Q_{12} acts diagonally on V_i, we deduce $[M_i[k], Q_{12}] = [M_i[k], f] = M_i[k+1]$ for all k and $i = 1, 2$. Also, $[M_{ij}[k], Q_{12}] = M_{ij}[k+1]$. Note also, that $[M_i[k], t] = M_{ij}[k+1]$ and $[M_{ij}[k], t] = 0$ or $2M_{ij}[k] = M_{ij}[k+2]$ for $t = t_{ij}$ or $t_{ij'}$. Similar statements hold for the $M_{ij'}[k]$. Since R is generated by Q_{12} and $\langle t_1, t_{ij} \rangle$, $[L_d, Q_{12}] = [L_d, R]$.
   We prove density first prove for a few special cases of f.
   Take $f = f_1 f_2$, which acts diagonally. Then, by induction, $[M_{ij}[k], f_1 f_2] = [M_{ij}[k], Q_{12}] = M_{ij}[k+1]$ and $[M_i[k], f] = M_i[k+1]$ and similarly for $M_j[k]$. So we have density for this f.
   Now, let $f = t_1 t_{12'}$. Then for $(x,y) \in V_1 \perp V_2$, $(x,y)(1-f) = (x-y, y+x)$. Since $(1-f)^2 = -2f$, $L_d(1-f)$ contains $2L_d$ and the diagonal $M_{ij}[1-r]$, which generate $[L_d, R]$ (see the first paragraph). So, we have density for this f.
   Finally, let f be an arbitrary fourvolution in R. Then $|L : L(f-1)|$ has order $|L : 2L|^{\frac{1}{2}}$ so $L(f-1) \leq [L, Q]$ implies that $L(f-1) = [L, Q]$. □

Corollary 7.6. *For $i = 1, 2$ and for all integers $j \geq 0$, $M_i[1-r] \cap L[j] = M_i[1-r+j]$.*

Proof. Fix i. We choose f_i for the twisting since it preserves the $M_i[k]$. The equalities are valid for j even since $L[2k] = 2^k L$, for all $k \geq 0$ and $M_i[1-r]$ is a direct summand of L. Now, $M_i[1-r] \cap L[1] \geq M_i[2-r]$. Applying one more twist, which is a scaled isometry, we get $M_i[2-r] \cap L[2] \geq M_i[3-r]$. Since $M_i[2-r] \cap L[2] = M_i[2-r] \cap 2L = M_i[2-r] \cap 2M_i[1-r] = 2M_i[1-r] = M_i[3-r]$, whence all our containments are equalities. □

   We give a fairly complete account of minimal vectors.

Lemma 7.7. (i) *A minimal vector of L_d has norm $2^{\lfloor \frac{d}{2} \rfloor}$ and is in $M_1[1-r]$ or $M_2[1-r]$ or has the form $x_1 + x_2$, where each x_i projects to a minimal vector of $M_i[1-r]$, for $i = 1, 2$. Its norm is $\mu(M_i[1-r]) = 2^{1-r}\mu(M) = 2^{\lfloor \frac{d}{2} \rfloor}$.*

(ii) *The minimal vectors span L_d.*

Proof. (i) Suppose that the minimal vector x is not in $M_1[1-r]$ or $M_2[1-r]$. Write $x = x_1 + x_2$, where x_i is the projection of x to V_i, $i = 1, 2$. Since $x_i \in M_i[-r]$, x_i has norm at least $\frac{1}{2}\mu(M_i[1-r])$, whence $(x, x) \geq \mu(M_i[1-r])$. It follows that these inequalities are equalities. The last statement follows from induction and 7.4.

Easily, (i) implies (ii) since L_d is the sum of three sublattices spanned by minimal vectors. □

Corollary 7.8. *L_d is a lattice of BW-type.*

Definition 7.9. A minimal vector in L_d has *type* 1, 2 *or* 3, respectively, as it is in $M_1[1-r], M_2[1-r]$ or in neither. These three types partition $MinVec(L_d)$.

Lemma 7.10. *For $d \geq 2$, L_d is an indecomposable lattice.*

Proof. As in the proof of 5.1, we see that the minimal vectors of L are partitioned into equivalence classes by membership in the L_i. However, it is clear from 7.7 and induction that there is just one equivalence class in the sense of 5.1. □

The following terminology will be useful. It applies to lattices used in 7.2 and later.

Definition 7.11. A lattice M is a *scaled BW-lattice*, abbreviated *sBW lattice*, if there is an integer $s > 0$ so that $M \cong \sqrt{s}BW_{2^e}$, for some $e > 0$. A sublattice M of a BW-lattice L is called a *suitably scaled Barnes-Wall sublattice (relative to L)*, abbreviated *ssBW sublattice*, if M is a sBW lattice and $\mu(M) = \mu(L)$.

We use the notation $BW_{2^p,q}$, $p \leq q$, for a scaled copy of BW_{2^p} whose isometry type is suitable as a sublattice of BW_{2^q}, i.e. a sBW lattice with minimum norm $2^{\lfloor \frac{q}{2} \rfloor}$.

8 The groups R_{2^d}, G_{2^d} and invariant lattices

Notation 8.1. In this section, $L = L_d$ and $d \geq 2$ have the meaning of 7.1 and 7.2.

Definition 8.2. We define $R := R_{2^d} := \langle Q_{12}, t_i, t_{ij} \rangle \cong 2_+^{1+2d}$, where Q_{12} and the involutions are as in 7.2. We define $G_{2^d} := N_{Aut(L)}(R_{2^d})$.

Definition 8.3. Elements and subsets of G_{2^d} are called *lower* if in R and are otherwise called *upper*. In particular, a fourvolution 6.1 may be called upper or lower.

Theorem 8.4. *For $d \geq 2$, $G_{2^d} \cong BRW^0(2^d, +) \cong 2_+^{1+2d}\Omega^+(2d, 2)$.*

Proof. The cases $d \leq 3$ have been discussed earlier (and the case $d = 4$ was treated explicitly in [19]). We may assume that $d \geq 4$. Since G_{2^d} is finite, containing R as a normal subgroup, G_{2^d} is contained in $\widetilde{G} = BRW^0(2^d, +)$, the natural $2^{1+2d}_+\Omega^+(2d,2)$ subgroup of $GL(2^d, \mathbb{C})$ containing R; see Appendix A2.)

Let D be the dihedral group of order 8 described in 7.2. Then $D \leq G := Aut(L)$. Let $t \in D$ be a noncentral involution. We claim that $C_{G_{2^d}}(t)R/R$ corresponds to a maximal parabolic in $\widetilde{G}/R$. For standard theory about parabolic subgroups, see [8].

Suppose $t = t_1$ or t_2. By induction, $Aut(M_i[1-r])$ contains a copy of $G_{2^{d-1}}$ as $N_{Aut(M_i[1-r])}(Q_i)$, and $S := Stab_G(M_1[1-r] \perp M_2[1-r])$ contains a group T of the form $[2^{1+2(d-1)}_+ \times 2^{1+2(d-1)}_+].[\Omega^+(2(d-1),2) \times 2]$. Also, since t_{ij} interchanges $M_1[1-r]$ and $M_2[1-r]$, it normalizes this group. Its image in $\widetilde{G}/R$ is a maximal parabolic, the stabilizer of a singular vector.

Suppose that $t = t_{ij}$ or $t_{ij'}$. Then the above argument goes through with $M_i[1-r]$, t_i replaced by $M_{ij}[-r]$, t_{ij}, and gives a distinct subgroup of the form $[2^{1+2(d-1)}_+ \times 2^{1+2(d-1)}_+].[\Omega^+(2(d-1),2) \times 2]$ containing R. (Proof of distinctness: in both cases, the center of the respective stabilizer is $\{\pm 1, \pm t\}$.)

Therefore, G/R contains two different maximal parabolics of $\widetilde{G}/R$, whence $G = \widetilde{G}$, [8], so we are done. $\square$

Remark 8.5. Note that the 8.4 uses only a basic result about orthogonal groups (maximality of certain stabilizers) but nothing very explicit about their interior structure, nor about particular elements. This is possible since we have a suitable uniqueness statement.

Lemma 8.6. *The subgroup of $Aut(L)$ which is trivial on $L/L[1]$ is just R if $d \geq 3$. If $d = 2$, it contains R with index 6 and modulo it, $Aut(L)$ maps onto Sym_3. In the notation 7.2, $L[1] = M_1[-r] + M_2[-r] + M_{12}[1-r]$.*

Proof. Let T be the subgroup trivial on $L/L[1]$. Note that $[L, R] = L[1]$, 7.4. Therefore, $T \geq R$.

Assuming $T > R$, we have a normal nontrivial 2-group T/R in G_{2^d}/R_{2^d}. If $d = 2$, the latter quotient is simple, the shape of G_{2^d} given in 8.4 shows that this is impossible. Assume that $d = 2$. Simplicity does not hold. The statement is an exercise (note that the isometry group is W_{F_4}, order $2^7 3^2$, which acts on $L/L[1]$ as $GL(2,2)$). $\square$

Notation 8.7. For $x \in MinVec(L)$, let $SF(x) := x^R$. Call this the *sultry frame containing x*. (See 9.1). From 8.6 and the structure of $R \cong 2^{1+2d}_+$, $SF(x)$ is a double orthogonal basis, of cardinality 2^{d+1}.

Proposition 8.8. *Let $x, y \in MinVec(L)$. Equivalent are (i) $y \in SF(x)$; (ii) $x - y \in L[1]$.*

Proof. Trivially, (i) implies (ii). For (ii), we use a familiar argument. Let $z \in SF(x)$. First we note that $z \pm y \in L[1]$ is 0 or has norm at least $2\mu(L)$. Assuming $y \neq \pm z$, we have $(z \pm y, z \pm y) = 2\mu(L) \pm 2(z,y) \geq 2\mu(L)$, whence $(z,y) = 0$. This is not the case for every $z \in SF(x)$. $\square$

Proposition 8.9. *The number of minimal vectors is* $(2^d + 2)(2^{d-1} + 2)\cdots(2^2 + 2)(2 + 2)$. *The values for small d are:*

d	$|MinVec(L)|$	*Prime Factorization*
0	2	2
1	4	2^2
2	24	$2^3 3$
3	240	$2^4 3.5$
4	4320	$2^5 3^3 5$
5	146880	$2^6 3^3 5.17$
6	9694080	$2^7 3^4 5.11.17$
7	1260230400	$2^8 3^4 5^2 11.13.17$
8	325139443200	$2^9 3^5 5^2.11.13.17.43$
9	167121673804800	$2^{10} 3^5 5^2 11.13.17.43.257$
10	171466837323724800	$2^{11} 3^8 5^2 11.13.17.19.43.257$
11	351507016513635840000	$2^{12} 3^8 5^4 11.13.17.19.41.43.257$
12	144047575367287967232000	$2^{13} 3^9 5^4 11.13.17.19.41.43.257.683$

Proof. Use 7.7, 8.7, 8.8 and induction. □

Corollary 8.10. *If* $x \in MinVec(L)$, $Stab_{G_{2^d}}(x+L[1])/R_{2^d}$ *is a maximal parabolic subgroup of* G_{2^d}/R_{2^d} *of the shape* $2^{\binom{d}{2}}{:}GL(d,2)$.

Proof. The pairs $\{\pm x\}$ of minimal vectors in this coset is an orbit of R_{2^d} for which a point stabilizer E is elementary abelian of order 2^{1+d}; see 8.8. These pairs of vectors are exactly the minimal vectors of the total eigenlattice of E, so as a set are stable under $N_{G_{2^d}}(E)$, which has the indicated properties. □

Corollary 8.11. *When d is even,* $L \cap 2L^* = L[1]$, *whence the lower group is normal in* $Aut(L)$ *and* $G_{2^d} = Aut(L)$.

Proof. When d is even, the duality level is 1, whence $L \cap 2L^* = L \cap 2L[-1] = L \cap L[1] = L[1]$ is invariant by the entire automorphism group. Now use 8.6. □

When d is even, this result essentially solves the problem of determining the automorphism group. The case d arbitrary is harder (it is finally proved in 11.9, which does not use 8.11).

Proposition 8.12. *For an odd prime, p,* L/pL *is an absolutely irreducible module for* G_{2^d}.

Proof. This is trivial, since R acts absolutely irreducibly. □

Lemma 8.13. *For all k,* $L[k]/L[k+1]$ *is an absolutely irreducible* $\mathbb{F}_2$*-module for* G_{2^d}.

Proof. This is easy to check for $d \leq 4$, so we assume that $d \geq 5$ and use induction.

We may assume that $k = 0$. Let $D \cong Dih_8$ be as in 7.2. For a noncentral involution t of D, we get by induction that $C_{G_{2^d}}(t)$ acts irreducibly on each $L^{\pm}(t)/[L^{\pm}(t), C_R(t)]$.

Since L is a sum of fixed point sublattices for the noncentral involutions of D, see 6.33, it follows that $L/L[1]$ has two absolutely irreducible composition factors for $C_{G_{2^d}}(t)$, each of dimension 2^{d-2}.

The group $C_{G_{2^d}}(t)$, of shape $[2_+^{1+2(d-1)} \times 2_+^{1+2(d-1)}].\Omega^+(2(d-1), 2)$ (discussed in the proof of 8.4), acts on $L/L[1]$ and has exactly one irreducible submodule, namely $Tel(L,t)/L[1]$, and two composition factors (this follows by induction on d, since on any eigenspace for t, we know the irreducible quotients for any lattice invariant under $C_{G_{2^d}}(t)$). These irreducible submodules distinct as t ranges over a set of generators for D (e.g. t_1, t_{12} for the group of 7.2).

It follows that $L/L[1]$ is irreducible for the action of G_{2^d}. We now prove absolute irreducibility. If K is an extension field of $\mathbb{F}_2$ and $K \otimes L/L[1]$ decomposes, then its restriction to $C_{G_{2^d}}(t)$ would have over 4 composition factors (since $O_2(C_{G_{2^d}}(t))$ acts nontrivially), which is impossible since, by induction, the composition factors for $C_{G_{2^d}}(t)$ are absolutely irreducible of dimension 2^{d-2}. $\square$

Proposition 8.14. *Let M be a lattice in $\mathbb{Q} \otimes L$ which is invariant under G_{2^d}. Then there is a rational number r so that $rM = L$ or $L[1]$.*

Proof. We may assume that $M \leq L$. By 8.12, we may assume that L/M is a power of 2. For some positive integer n, $2^n L \leq M$. Now use 8.13 and the fact that $[L[k], R_{2^d}] = L[k+1]$. $\square$

Later, in 11.9, we prove that G_{2^d} is all of $Aut(L_d)$.

9 Sultriness

When f is a fourvolution on a lattice L, $1 - f$ (actually, any of $\pm 1 \pm f$) is an endomorphism of L which is also an isometry scaled by $\sqrt{2}$. Next, we see that a sultry transformation is naturally interpreted as a *scaled lift of a transvection*, a point which suggested the term "sultry".

Theorem 9.1. *The function $\gamma_f : Aut(L) \to Aut(L)$, $x \mapsto (1-f)^{-1} x (1-f)$, normalizes $R = R_{2^d}$ and G_{2^d}. Furthermore, γ_f is the identity on $C_R(f)$ and if $x \in R \setminus C_R(f)$, γ_f takes x to $fx \in R \setminus C_R(f)$, hence normalizes and induces an outer automorphism on the dihedral group $\langle f, x \rangle$. Hence, on $R/Z(R)$, γ_f acts as the transvection associated to the nonsingular point $Z(R)f$ of $R/Z(R)$.*

Proof. Since $\frac{1}{\sqrt{2}}(\pm 1 \pm f)$ is orthogonal, the image of $\gamma := \gamma_f$ is a subgroup of the orthogonal group. Since any $\pm 1 \pm f$ carries each $L[k]$ onto $L[k+1]$, the image of γ stabilizes L. We conclude that γ takes $Aut(L)$ onto itself.

We calculate that $x(1-f) = x - xf = x - f^{-1}x = x + fx = (-f + 1)(fx)$, which proves the remaining statement. $\square$

10 Proof of uniqueness

Notation 10.1. Given $d \geq 3$ and L_1, L_2, we let $\mathfrak{X} := \mathfrak{X}(L_1, L_2)$ be the set of all X-quadruples of the form (L, L_1, L_2, t); see 2.3.

Theorem 10.2. *We use the notation in 2.3, 6.1, 6.2, 7.2 and 9.1. Suppose that $d \geq 3$ and (L_1, L_2) is an orthogonal pair of lattices, so that each L_i is BW-type of rank 2^{d-1}.*

(i) $\mathfrak{X}$ *is an orbit under the natural action of $F_1 \times F_2$, where $F_i := Stab_{Aut(L_i)}$ $(L_i[1-r])$ (see 6.2; by induction, $F_i \cong G_{2^{d-1}}$). Define $Q_i := C_{F_i}(L_i/L_i[1])$.*

The elements of $\mathfrak{X}$ are in correspondence with each of the following sets.

(a) F_1/Q_1;

(b) F_2/Q_2;

(c) *Pairs of involutions $\{s, -s\}$ in the orthogonal group on V which interchange L_1 and L_2;*

(d) *Dihedral groups of order 8 which are generated by the SSD involutions associated to L_1, L_2 and involutions as in (c).*

(ii) (a) *The subgroup G_L^0 of $F_1 \times F_2$ which stabilizes L has structure $Q_1 \times Q_2 \leq G_L^0$ and $G_L^0/(Q_1 \times Q_2)$ is the diagonal subgroup of $F_1/Q_1 \times F_2/Q_2$ with respect to the isomorphism induced by s, an involution as in (i.c).*

(b) *The subgroup G_L of $Aut(L_1 \perp L_2) \cong Aut(L_i) \wr 2$ which stabilizes L is $G_L^0 \langle s \rangle$. We have $G_L \cong [2_+^{1+2(d-1)} \times 2_+^{1+2(d-1)}] . [\Omega^+(2(d-1), 2) \times 2]$.*

(c) *The subgroup of G_L which acts trivially on $L/L[1]$ is $R := \langle Q_{12}, s, t_i \rangle$, where t_i is the SSD involution associated to L_i and $Q_{12} := \{xx^s | x \in Q_1\}$. The quotient $G_L/R \cong 2^{2d-2} {:} \Omega^+(2d-2, 2)$ is a maximal parabolic subgroup of $Out^0(2_+^{1+2d}) \cong \Omega^+(2d, 2)$. (See Appendix A2).*

Remark 10.3. *The extension in (c) is split, despite G_{2^e} being nonsplit over R_{2^e} for $e \geq 4$. See Appendix A2.*

Proof. (i) We prove the classification by induction. For $d = 2$, $Aut(L_{D_4}) \cong 2_+^{1+4}[Sym_3 \wr 2]$ and for $d = 3$, $Aut(L_{E_8}) \cong W_{E_8}$. When $d = 4$, the main theorem follows from the arguments of [19].

For the rest of the proof, we assume that $d \geq 4$. By induction, a lattice satisfying the $X(2^{d-1})$ condition is uniquely determined up to isometry. This applies to the lattices L_1, L_2.

Let $(L, L_1, L_2, t) \in \mathfrak{X}$. Then $\det(L)$ and $|L : L_1 \perp L_2|$ are determined. Define $G := Aut(L)$.

Let p_i be the orthogonal projection of L to $V_i := \mathbb{Q} \otimes L_i$, for $i = 1, 2$. Since L is integral, L^{p_i}, the projection of L to V_i is contained in L_i^*. When d is odd, L^{p_i} must be $L_i^* = L_i[-1]$, by determinant considerations.

Assume now that d is even. There is a subgroup H of $C_G(t)$, $H \cong BRW^0(2^{d-1}, +)$, so that H acts faithfully on both V_i and stabilizes L, L_1 and L_2. By 7.14 and $X(2^d)(f)$, the projection is $L_i[-1]$, where the twist is with respect to a lower fourvolution in H. Since the groups Q_i act trivially on $L_i[-1]/L_i$, for $i = 1, 2$, the group $\langle Q_1, Q_2, t \rangle$ stabilizes L and in fact acts trivially on $L/L_1 \perp L_2$. We define a group $R := \langle Z(Q_1), Z(Q_2), Q_{12}, t \rangle \cong 2_+^{1+2d}$, where $Q_{12} := \{xx^t | x \in Q_1\}$.

Let D be the dihedral group which is generated by t and either t_i, the involution generating $Z(Q_i)$ (or what is the same, the RSSD involution associated to L_i). It follows that L is determined by L_1, L_2, D in the sense that L lies between $L_1 \perp L_2$ and $L_1[-1] \perp L_2[-1]$ and $L/[L_1 \perp L_2]$ is the fixed point submodule for the action of D (equivalently, of t) on $[L_1[-1] \perp L_2[-1]]/[L_1 \perp L_2]$. Recall from earlier in the proof that D does determine the $L_i[-1]$ by use of X(f).

Now, to what extent do L and $L_1 \perp L_2$ determine D? The answer is: up to conjugacy in $Aut(L_1 \perp L_2) \cong Aut(L_1) \wr 2$ (note that $d \geq 4$ here). Our group D is generated by the center of the natural index 2 subgroup of $Aut(L_1 \perp L_2)$ and a wreathing involution. In general, wreathing involutions in a wreath product of groups $K \wr 2$ form an orbit under the action of either direct factor isomorphic to K in the base group of the wreath product. This proves correspondence of $\mathfrak{X}$ with (c) and (d). The stabilizer subgroup is diagonal in the base group $K \times K$, and either direct factor represents all cosets of the stabilizer (whence the equivalence of $\mathfrak{X}$ with (a) and (b))

It follows that, up to isometry preserving $L_1 \perp L_2$, D, hence L, is determined by the pair of indecomposable lattices L_1 and L_2.

Proof of statement (ii) is easy. The statement about parabolic subgroups is proven with a standard result from the theory of Chevalley groups, e.g. [8]. Independently of that theory, the maximality could be proved directly by showing that there is no system of imprimitivity on the set of isotropic points. This is an exercise with Witt's theorem. $\square$

11 Minimal vectors, the zoop2 property and $Aut(BW_{2^d})$

We continue to use the notations of 8.1 and 8.2.

Remark 11.1. For $L = BW_{2^d}$ a Barnes-Wall lattice and $k \in \mathbb{Z}$, we have $MinVec(L[k]) = MinVec(L)[k]$ (see 6.5).

Theorem 11.2. *We use notation of 7.2. The group G_{2^d} acts transitively on the set of minimal vectors.*

Proof. We use notation of 7.9. It is clear that the minimal vectors of types 1 and 2 are in a single G_{2^d}-orbit, say $\mathcal{O}$. Consider a minimal vector $u + v$ of type 3. We assume that L corresponds to the involution $s = t_{ij'}$ in the sense of 10.2 (i). Then v and $u^{t_{i'j}}$ differ by an element of $M_2[1-r]$, so by induction, these are in the same orbit under Q_2, equivalently under Q_{12}. Therefore, u and v are in the same R-orbit.

By induction, we have transitivity on the minimal vectors of type 3 by the group RF_{12}, where the second factor is the natural diagonal subgroup of $F_1 \times F_2$. Call $\mathcal{O}'$ the orbit containing the type 3 minimal vectors.

Suppose that G_{2^d} is not transitive. *First Contradiction.* Then $MinVec(L)$ is the disjoint union of two orbits $\mathcal{O}$ and $\mathcal{O}'$ and so G_{2^d} preserves the $\mathbb{Z}$-span of $\mathcal{O}$, which is just $M_1[1-r] \perp M_2[1-r]$, an orthogonal sum of two orthogonally indecomposable lattices. Thus $G'_{2^d} \geq R$ leaves both summands invariant, which is impossible since R is irreducible on $\mathbb{C} \otimes L$. *Second Contradiction.* The lower involutions form a conjugacy class in G_{2^d}, so there is $g \in G_{2^d}$ which conjugates t_1 to t_{12}. Then g takes the set of minimal vectors fixed by t_1 (those of type 2) to those fixed by t_{12}, which are contained in those of type 3. Therefore $\mathcal{O}$ and $\mathcal{O}'$ are not distinct orbits. Transitivity follows. $\square$

Now we give a few results about stablilzers in G_{2^d}. These will be strengthened later.

Lemma 11.3. (i) *If F is a sultry frame and $x \in F$, then $\{g \in R | x^g = \pm x\} = \{g \in R | y^g = \pm y \text{ for all } y \in F\}$ is a maximal elementary abelian subgroup of R. Call it R_F. The quotient R/R_F operates regularly on the eigenlattices.*

(ii) *Define $C_F := C_{G_{2^d}}(F/\{\pm 1\}) := \{g \in G_{2^d} | y^g = \pm y \text{ for all } y \in F\}$. This is elementary abelian and has shape $2^{1+d+\binom{d}{2}}$.*

(iii) *Its normalizer $N_F := N_{G_{2^d}}(C_F) = \text{Stab}_{G_{2^d}}(F)$ in G_{2^d} satisfies $N_F/C_F \cong AGL(d, 2)$. We have $R_F \leq C_F$.*

Proof. (i) Set $P := \{g \in R | x^g = \pm x\}, Q := \{g \in R | y^g = \pm y \text{ for all } y \in F\}$. Observe that $Q \leq P$ and Q is elementary abelian. Transitivity of R on F and normality of P in R implies that $P = Q$ has order 2^{d+1}.

(ii) This follows from (i) and the order of the unipotent radical for the stabilizer of a maximal totally isotropic subspace for $\Omega^+(2d, 2)$.

(iii) This follows from the actions of R on R_F together with the structure of the stabilizer of a maximal totally isotropic subspace for $\Omega^+(2d, 2)$. $\square$

Definition 11.4. Suppose that F is a frame. A subset $S \subseteq V$ has the *zop2 property (with respect to F)* if $|(x, y)|$ is 0 or a power of 2, for all $x \in F$ and $y \in S$. We say that S has the *zoop2 property* if is has the zop2 property and just one power of 2 occurs among the scalars $|(x, y)|$, $x \in F, y \in S$.

Lemma 11.5. *For all integers p, q, any minimal vector of $L[q]$ has the zoop2 property with respect to any sultry frame in $L[p]$.*

Proof. We may assume that $p = 0$ and $q \in \{0, 1\}$. The property is easy to check for $d \leq 3$. We assume $d \geq 4$. Say $x \in MinVec(L)$, $y \in SF(x)$ and $z \in MinVec(L[q])$ so that $(x, z) \neq 0 \neq (y, z)$. Take a lower involution t so that t fixes x and y. Then $x, y \in L^+(t)$, a ssBW lattice, and z projects to a minimal vector in $L^+(t)[q]$, so we are done by induction. $\square$

Definition 11.6. Given a sultry frame F of 2^{d+1} elements, there are 2^d subsets which form a basis. Suppose X is such a set. If $v \in V$, we write $v = \sum_{x \in X} a_x x$ and define the *support of v* to be the set $\{x \in X | a_x \neq 0\}$. This depends on the double basis F, not on the choice $X \subset F$.

Lemma 11.7. *Suppose that $d \geq 2$. Let $x \in MinVec(L)$ and $A(x) := \{y \in MinVec(L) | (x, y) = \frac{1}{2}(x, x)\}$. Then $A(x) \cup \{x\}$ spans a lattice isometric to the Hamming code lattice described in 5.6. In particular, there is a labeling of $SF(x)/\{\pm 1\}$ with $\mathbb{F}_2^d$ so that the elements of $A(x)$ have support which is an affine 2-space.*

Proof. Define J_0 to be the square lattice spanned by $SF(x)$ and J the lattice spanned by $A(x)$ and $SF(x)$. In a natural way, J/J_0 corresponds to a nonzero code in $\frac{1}{2}J_0/J_0 \cong \mathbb{F}_2^{2^d}$. Since $\mu(L) = (x, x)$, this code has minimum weight at least

4. For $y \in A(x)$, $supp(y)$ is a 4-set with respect to the double basis $SF(x)$, 11.5. Therefore the minimum weight of C is 4.

Note that we have an action of $AGL(d, 2)$ on $\frac{1}{2}J_0/J_0$ by coordinate permutations. This follows from 11.3. This action is triply transitive. Since C has minimum weight 4, its weight 4 codewords forms a Steiner system with parameters $(3, 4, 2^d)$ which is stable under this action of $AGL(d, 2)$. Such a system is unique since in $AGL(d, 2)$, the stabilizer of three points fixes a unique fourth point. Therefore, C is the code $\mathcal{H}_d^e$, up to equivalence.

Finally, we must show that $A(x) \cup \{x\}$ spans J. If $d = 2$, $L \cong L_{D_4}$ and the result is easy to check directly. We assume $d \geq 3$. Let $y \in SF(x), y \neq \pm x$. Since $d \geq 3$, we may choose a lower involution t which fixes both x and y (in the notation of 11.3, $t \in R_F$). Let L^+ be the sublattice of points of L fixed by t, a sBW lattice. Then, induction implies that the sublattice of L^+ spanned by $A(x) \cap L^+$ contains y. We conclude that $SF(x) \subset span(A(x) \cup \{x\})$, and we are done. $\square$

Proposition 11.8. *Suppose that $d \geq 4$. For $x \in MinVec(L)$,*

$$Stab_{Aut(L)}(x) \leq Stab_{Aut(L)}(SF(x)).$$

Proof. Define $A(x) := \{y \in MinVec(L) | (x, y) = \frac{1}{2}(x, x)\}$. By 11.7 lattice $J := span(A(x) \cup \{x\})$ contains $SF(x)$ and is a copy of the lattice in 5.6. Since $d \geq 4$, given a weight 4 codeword, there exists another weight 4 codeword which meets it in a 1-set (this is not so for $d = 3$). Therefore, $SF(x) = \{z \in MinVec(J) | (z, J) \leq \frac{1}{2}(z, z)\mathbb{Z}\}$, we are done (see the proof of 5.6). It follows from 5.1 that $Stab_{Aut(L)}(x) \leq Stab_{Aut(L)}(J)$. $\square$

Corollary 11.9. *For $d \geq 4$, $Aut(L_d) = G_{2^d}$.*

Proof. This follows since G_{2^d} is transitive on minimal vectors and the stabilizer of some minimal vector in G is contained in G_{2^d}. $\square$

Definition 11.10. Let $x \in MinVec(L)$ and $SF(x)$ its sultry frame. Let $q \in \mathbb{Z}$ and $k \in \mathbb{Z}$. Define $A(L, x, q, k) := \{z \in MinVec(L[q]) | (z, y) \in \{0, \pm 2^k\}$ for all $y \in SF(x)\}$. This is the *level k layer in $MinVec(L[q])$ with respect to x or $SF(x)$.*

Lemma 11.11. *Suppose that the group G_0 factorizes as $G_0 := GZ$, where G, Z are subgroups so that $[G, Z] = 1$. Suppose that G_0 acts on the set X and that G stabilizes and acts transitively on a set of Z-orbit representatives. Let S be the set of all G-invariant sets of orbit representatives. Then Z acts transitively on S.*

Proof. A member of S is determined by any element of X which it contains. Therefore the members of S partition X.

Suppose that $X_1, X_2 \in S$. Let A be a Z-orbit and let a_i be the unique element of $A \cap X_i$, for $i = 1, 2$. Take $z \in Z$ so that $a_1^z = a_2$. Then X_2 and X_1^z are both G-invariant sets of orbit representatives and contain a_2, hence are equal. $\square$

Note that the next result deals with minimal vectors in all sultry twists of L.

Notation 11.12. For integers $d \geq 2$ and $p, q \in \mathbb{Z}$, define $I(d,p,q)$ to be the set of integers listed below. Here, $r \in \{0,1\}$ is the remainder of d modulo 2, $m := \lfloor \frac{d}{2} \rfloor$ and $s \in \{0,1\}$ is the remainder of $p - q$ modulo 2. We define

$$I(d,p,q) := \left\lfloor \frac{p+1}{2} \right\rfloor + \left\lfloor \frac{q+1}{2} \right\rfloor + \{-rs, 0, 1, \ldots, m\}.$$

As usual $a + \{b, c, \ldots\}$, means $\{a + b, a + c, \ldots\}$. We call $I(d,p,q)$ the *interval of exponents for dot products of minimal vectors*. (See 11.14 for an explanation of this term.)

Example 11.13. Some examples:

$$I(3,p,p) = \begin{cases} \{p, p+1\} & p \text{ even}; \\ \{p+1, p+2\} & p \text{ odd}. \end{cases}; I(3,p,p+1) = \{p, p+1, p+2\};$$

$$I(4,p,p) = \begin{cases} \{p, p+1, p+2\} & p \text{ even}; \\ \{p+1, p+2, p+3\} & p \text{ odd}. \end{cases}; I(4,p,p+1) = \{p+1, p+2, p+3\}.$$

Lemma 11.14. *The set of integers*

$$\{(x,y) | x \in MinVec(L[p]), y \in MinVec(L[q])\}$$

is $\{0, \pm 2^k | k \in I(d,p,q)\}$; *see* 11.12.

Proof. Let $m := \lfloor \frac{d}{2} \rfloor$ and $r := d - 2m \in \{0, 1\}$.

First we take $p = q = 0$. Then for $x \in MinVec(L)$, with respect to a basis $\Omega \subseteq SF(x)$ an element $y \in MinVec(L[q])$ has the form $y = 2^{-t} \sum_{u \in A} u\varepsilon_B$, where $A \subseteq \Omega$ is an affine subspace of Ω of dimension $a \leq d$, and a satisfies $m = m - 2t + a$, or $a = 2t$. Then $(x, y) = 0$ or $\pm 2^{m-t}$. Thus, the values of a, t, $m - t$ which occur are $\{0, 2, \ldots, 2m\}$, $\{0, 1, \ldots, m\}$, $\{0, 1, \ldots, m\}$, respectively.

Next, if $p = 0$ and $q = -1$, a similar discussion applies, but here $\mu(L[-1]) = \frac{1}{2}\mu(L)$, so we get the condition $m - 1 = m - 2t + a$, or $a = 2t - 1$, whence odd parity for a. Therefore, the values of a, t, $m - t$ which occur are $\{1, 3, \ldots, d + r - 1\}$, $\{1, \ldots, d - m\}$, $\{2m - d, 2m - d + 1, \ldots, m - 1\}$, respectively.

Suppose that $p = -1$ and $q = 0$. We get the condition $m = m - 1 - 2t + a$, or $a = 2t + 1$ is odd. Therefore the values of $a, t, m - t$ which occur are $\{1, 3, \ldots, d + r - 1\}$, $\{0, 1, \ldots, d - m - 1\}$, $\{2m - d + 1, 2m - d + 2, \ldots, m\}$, respectively.

For the general case, just observe that $I(d, p, q+2) = 1 + I(d, p, q)$, $I(d, p+2, q) = 1 + I(d, p, q)$, and $I(d, p+1, q+1) = 1 + I(d, p, q)$. □

Lemma 11.15. *Let $G = AGL(d, 2)$ act naturally on the permutation module $A := \mathbb{F}_2^\Omega$, where $\Omega := \mathbb{F}_2^d$ with the natural G-action. Let B be the submodule generated by the affine subspaces of codimension 1. For $d \geq 3$, $H^1(G, B) = 0$.*

Proof. We have an exact sequence $0 \to B \to A \to A/B \to 0$. From this, the long exact cohomology sequence gives the exact sequence $H^0(G, A/B) \to H^1(G, B) \to H^1(G, A)$. The right term is, by the Eckmann-Shapiro lemma, isomorphic to

$H^1(G_0, \mathbb{F}_2)$, where $G_0 \cong GL(d, 2)$ is the stabilizer of 0 in G. This is isomorphic to $Hom(G_0, \mathbb{F}_2)$, which is trivial for $d \geq 3$. The module A/B is indecomposable for $AGL(d, 2)$, with a faithful module of dimension d as the socle and quotient the trivial 1-dimensional module. Since the fixed points are 0, $H^0(G, A/B) = 0$. Exactness implies that $H^1(G, B) = 0$. □

Theorem 11.16. *Let $d \geq 2$ and let $F := SF(x)$, for a minimal vector $x \in L[p]$ (see 8.7). Also, define $H := Stab_{G_{2^d}}(F)$ (denoted N_F in 11.3 (ii)).*

(i) *There exists a basis X contained in F and labeling of X by $\mathbb{F}_2^d$ so that with respect to X, H is the monomial group $\mathcal{E}_{\mathcal{C}_X}{:}AGL(d, 2)$ (see 5.5), where $\mathcal{C}_X$ is the code generated by affine subspaces of codimension 2 in X and where $AGL(d, 2)$ is the natural subgroup of permutation matrices. The code $\mathcal{C}_X$ has parameters $[2^d, 1 + d + \binom{d}{2}, 2^{d-2}]$.*

(ii) *Any two labelings as in (i) are conjugate by the action of H.*

(iii) *For fixed q, k, the sets $A(L, x, q, k)$ (see 11.10) are the elements of $L[q]$ which are all linear combinations of X of the form $2^{-t} \sum_{x \in A} x \varepsilon_B$, where A is an affine subspace of X of dimension a, $p + a - 2t = q$ and $k = p + \lfloor \frac{d}{2} \rfloor - 2t = \lfloor \frac{d}{2} \rfloor + q - a$ and ε_B effects sign changes exactly at indices in $B \subseteq A$; here B is in the code $\mathcal{C}_A$, which is spanned by all $A \cap S$, where S is an affine subspace of codimension 2 in X.*

(iv) *For a fixed integer a, the sets $A(L, x, q, k)$ are nonempty exactly for the indices $k \in I(d, p, q)$ and they are the orbits of H on $MinVec(L[q])$.*

Proof. (i): We use notation of 7.2. We may and do assume that $d \geq 4$. There is by induction a basis X_1 of V_1 contained in F and labeling of X_1 by $\Omega_1 := \mathbb{F}_2^{d-1}$ so that we get an identification of the stabilizer of $SF(x) \cap V_1$ with $\mathcal{E}_{\mathcal{C}_{\Omega_1}}{:}AGL(d-1, 2)$, in analogous notation.

The frame is a double basis for the total eigenspace of E_1, a maximal elementary abelian subgroup of a lower group R_1 on M_1. Using our standard diagonal notation 5.2, 10.2, take involution $s = t_{12'}$ in dihedral group D and the corresponding subgroup E_{12} of R_{12}. Then s interchanges M_1 and M_2. Let $t \in D$ be the SSD involution associated to M_1. Then $E := \langle E_1, t \rangle$ is a maximal elementary abelian group in R and its total eigenlattice has the frame F as a double basis. Identify Ω_1 with a codimension 1 affine subspace of $\Omega := \mathbb{F}_2^d$. We define Ω_2 to be the complement in Ω of Ω_1. Choose any vector $v_0 \in \Omega_2$. Let $v_1 \in X_1$ be a frame vector labeled by 0 and let $v_2 := v_1^s \in X_2 := SF(x) \cap V_2$. Since the action of s is an isomorphism of the transitive $C_H(s)$-sets X_1 and X_2, the labeling on X_1 transfers uniquely to X_2 and we translate this labeling to X_2 via vector addition by v_0 to make a labeling of X_2 by Ω_2. The resulting labeling of X is uniquely determined (depending on v_0, s, X_1).

From A.3, we see that in G_{2^d}, a frame stabilizer contains a subgroup J isomorphic to $AGL(d, 2)$ in the normalizer of E which permutes a basis of the eigenlattice. Its intersection, K, with a natural $G_{2^{d-1}}$ subgroup is an analogous $AGL(d-1, 2)$ subgroup. Let Z be the group generated by $\{\pm 1_V\}$.

There are just two J-invariant sets of Z-orbit representatives in F. When one of them is restricted to K, we get two orbits. If X_1 is one of these, the other is X_1^s or $-X_1^s$. We replace s by $-s$ if necessary to arrange for the other to be X_1^s.

Then $s \in J$. The labeling on X_1 now extends to all of X, which is an H-invariant set.

(ii): Let ℓ, ℓ' be two labelings for which H is the indicated monomial group. We shall transform one to the other by action of H. Call the *domain* of a labeling to be the points of $SF(x)$ which get a label.

The stabilizer H_ℓ in H of the labeling ℓ (equivalently, of its domain) is a complement to the normal subgroup of sign changes. Such a subgroup is isomorphic to $AGL(d, 2)$. We first note that any two complements are conjugate. This follows from a cohomology argument, 11.15. From this, we may and do arrange for the two labelings to have the same domain, which we call D. Since H acts 3-transitively and leaves invariant a unique Steiner system with parameters $[3, 4, 2^d]$, addition of labels of vectors is determined by H once an origin is chosen. Given an origin, a partial labeling of D by a basis of $\mathbb{F}_2^d$ determines the labeling. Any two such choices lie in one orbit under the action of H.

(iii) and (iv): It is clear from induction and the form of the types 1, 2 and 3 minimal vectors that a minimal vector has the zoop2 property 11.4 with respect to a given sultry frame. So, the nonempty sets $A(L, x, q, k)$, for $k \in I(d, p, q)$, partition $MinVec(L[q])$. It remains to show that they are orbits for the frame stabilizer.

The action of $AGL(d, 2)$ is transitive on affine subspaces of given dimension. Write $v = v_1 + v_2$, where v_i is the projection to V_i, $i = 1, 2$. Either $v = v_1, v = v_2$ or $v_1 \neq 0 \neq v_2$ and there exist integers t_i and affine subspaces A_i of X_i and $B_i \in \mathcal{C}_{A_i}$ so that $v_i = 2^{-t_i} \sum_{y \in A_i} y \varepsilon_{B_i}$. The zoop2 property implies that $t_1 = t_2$ and $\dim(A_1) = \dim(A_2)$. Call these common values t, a, respectively. We assume that $v_1 \neq 0 \neq v_2$.

If there exists an affine hyperplane X' of X so that $U := supp(v) \subseteq X'$, we use induction since the v is a minimal vector in the sBW sublattice of rank $d - 1$ supported by U. Suppose that no such X' exists. Then we are in the third case $v_1 \neq 0 \neq v_2$ and we use notation $v_1 \neq 0 \neq v_2$ as above. Let X' be any affine hyperplane. We claim that $|U \cap X'| = \frac{1}{2}|U|$. Suppose otherwise. Then, replacing X' by its complement, we may assume that $|U \cap X'| < \frac{1}{2}|U|$. Then the sublattice S of L supported by X' has a vector in S^* of norm less than $\frac{1}{2}\mu(L)$, a contradiction. The claim follows. We get a final contradiction by using 4.6. □

Remark 11.17. The results 11.16 (iii), (iv), were proved in [7]; see Théorème I.5, Théorème II.2.

12 Orbits on norm 4 frames in L_{E_8}.

We give an application of our theory by giving a short proof that the Weyl group of E_8 has just four orbits on plain frames 6.15 of norm 4 vectors in L_{E_8}, equivalently, of D_1^8-sublattices. This result can be deduced from a classification of $\mathbb{Z}_4$ codes [9].

Definition 12.1. Let L be any lattice. If M is a sublattice, $2L \leq M \leq L$, the *d-invariant of the frame F (relative to M)* is the dimension of the span of $F + M/M$. Also, we say two plain frames E, F are *congruent* if and only if $E + M = F + M$.

The *d-invariant* of a plain frame F is the dimension of the subspace of $L/2L$ spanned by $F + 2L$, i.e., the relative d-invariant for $M = 2L$.

Remark 12.2. Now suppose that $L = BW_{2^3}$. The d-invariant of a frame is a number between 1 and 4 since the image is not trivial and spans a totally singular subspace.

Remark 12.3. It is easy to see that the Weyl group of E_8 is transtive on frames of roots. This follows from Witt's theorem since the Weyl group induces the full orthogonal group on L_{E_8} modulo 2 and any frame of roots spans an index 16 sublattice with all even inner products, hence corresponds mod 2 to a totally isotropic subspaces with nonsingular vectors. The next result refers to action of the proper subgroup G_{2^3} on frames of roots and norm 4 vectors.

Proposition 12.4. (i) *In the action of G_{2^3} on frames of norm 2 vectors, there are four orbits. They are distinguished by their d-invariants relative to the sultry twist $L[1]$.*

(ii) *In the action of W_{E_8} on frames of norm 4 vectors, there are four orbits. They are distinguished by their d-invariants.*

Proof. (i) It is easy to determine the orbits of G_{2^3} on frames of roots. They are represented by the following vectors with respect to $x_1, \ldots, x_8$, a standard orthogonal basis of roots (see 5.4):

$F_1 : \pm x_1, \ldots, \pm x_8$.

$F_2 : \pm x_i, i \notin A; \frac{1}{2}\sum_{j \in A} \pm x_j$, where A is a 4-set of indices representing a Hamming codeword, and evenly many signs over A are minus.

$F_3 := \pm x_i, i \in B; \frac{1}{2}(00aaaa00), \frac{1}{2}(0000pqrs), \frac{1}{2}(00tu00cc)$, where B is a 2-set of indices (which we take to be $\{1, 2\}$) and the indicated partition of the eight indices into 2-sets has the property that the union of any two of them is a Hamming codeword. Also, $a, b, c, p, q, r, s, t, u \in \{\pm 1\}$ and where $p = -q, r = -s, t = -u$.

$F_4 := \pm x_1$ and $\pm \frac{1}{2}(01111000), \pm \frac{1}{2}(0001, -1, 110), \pm \frac{1}{2}(0, -1, 0, 0, 1, 0, 1, 1)$.

The proof is an easy exercise with the action of the monomial group $H \cong 2^7{:}AGL(3,2)$, a subgroup of G_{2^3}, where the group of sign changes at evenly many indices is indicated by 2^7. Since G_{2^3} is transitive on roots, an orbit of such a frame has a member containing x_1. We now restrict ourselves to transformations by elements of $H \leq G_{2^3}$. If the remaining members of the frame are the x_i, we are in case F_1. If not, one can arrange for the next member of the frame to be something of the form mentioned in case F_2, supported by a 4-set, A. If all remaining members of the frame are some $\pm x_j$ or supported by the same 4-set, we are in the orbit of F_2. If not, similar reasoning brings us to case F_3 or F_4.

One must show that these frames represent different orbits, and that is accomplished by showing that their images in $L/L[1]$ span subspaces of dimensions 1, 2, 3 and 4, respectively. (This is verified by Smith canonical forms, easy to do by hand or with a software package like Maple): in our notation, $L[1]$ is the $\mathbb{Z}$-span of the $x_i \pm x_j$ and $\frac{1}{2}(x_1 + \cdots + x_8)$. These dimensions are the d-invariants of the original orbits.

(ii) Let $\mathcal{O}_i$, for $i = 1, \ldots, r$ be the orbits. Since W_{E_8} induces the full orthogonal group on $L/2L$, any orbit has a representative contained in $L[1]$ since

$L[1]/2L$ is a maximal totally singular subspace. Now consider the subgroup G_{2^3}, which is normalized by (the nonorthogonal transformation) $1 - f$, where f is a fourvolution. The action of $(1 - f)$ takes the set of 240 roots bijectively to the union of the nonempty sets $\mathcal{O}_i \cap L[1]$, and this correspondence preserves orbits of G_{2^3}. We are done by (i). $\square$

13 Clean pictures, dirty pictures and transitivity

We next prove transitivity results for certain kinds of sublattices. In particular, we can classify certain scaled embeddings of BW_{2^k} in BW_{2^d}, for certain $k \leq d$. See A.4 for the clean and dirty terminology.

Theorem 13.1. *Let $L = BW_{2^d}$, for $d \geq 4$. There is a G_{2^d}-invariant bijection between sublattices of L which are ssBW of rank 2^{d-1} and noncentral lower involutions, via the SSD correspondence.*

Proof. Let M be such a sublattice and $t = t_M$ the associated SSD involution. Since t normalizes R and has trace 0, it is dirty (see Appendix A2), whence there is an element $g \in R$ so that $[t, g] = -1$. We may arrange for g to be an involution. Then g interchanges M and $N := L \cap M^{\perp}$, whence N is a $ssBW_{2^{d-1}}$. By 6.33, the condition $\det(L^{\pm}(g)) = \det(M)$ implies that L is part of an X-quadruple $(L, L^+(g), L^-(g), t)$, whence the classification 10.2 implies that t is lower. $\square$

Remark 13.2. There are cases of sublattices X of BW_{2^d} of rank 2^{d-1} which satisfy $L/[X \perp X^{\perp}]$ elementary abelian, but X is not isometric to a scaled $BW_{2^{d-1}}$. For $d = 3$, one can take X to be the sublattice spanned by a root system of type A_1^4 which is not contained in a D_4 subsystem. Such a sublattice is SSD and corresponds to a SSD involution of trace 0 which is upper with respect to any conjugate of G_{2^3} which contains it. The noncentral involutions of R_3 have trace 0 and fixed point sublattice isometric to L_{D_4}.

Theorem 13.3. *Suppose that $L = BW_{2^d}$ and that M, M' are sublattices which are the fixed point lattices for clean isometries of order 2. If $\operatorname{rank}(M) = \operatorname{rank}(M')$, then there is an isometry g of L so that $M' = M^g$.*

Proof. Such sublattices correspond to SSD involutions with nonzero traces. Now use A.8. $\square$

The following is an application of 13.3.

Corollary 13.4. *Suppose that $d \geq 5$ is odd. Then in BW_{2^d} any two ssBW sublattices of rank 2^{d-2} are in the same orbit under G_{2^d}.*

Proof. Such sublattices must be SSD. $\square$

Definition 13.5. Let $L = BW_{2^d}$. A *first generation sublattice* of L is a sublattice L_1 so that there exists a sublattice L_2 and an involution t so that $(L, L_1, L_2, t) \in \mathfrak{X}$.

A chain of lattices $L = L(0) \geq L(1) \geq \cdots \geq L(d)$ is a *generational chain* if there exists an elementary abelian group $E \leq R$ and a chain of subspaces

$E = E(d) > E(1) > \cdots > E(0) = \langle -1 \rangle$ so that for each k, $|E(k)| = 2^{k+1}$ and $L(k)$ is the total eigenlattice of $E(k)$; see 6.8.

In each $L(k)$, each orthogonally indecomposable summand is a ssBW sublattice, all of common rank 2^{d-k} if $k \leq d-2$, and $L(d-1)$ is a direct sum of isometric rank 1 lattices. Call $L(k)$ a kth *generation sublattice* and $E(k)$ *its defining lower group*. A sublattice is *ancestral* if it is a kth generation sublattice, for some k.

Theorem 13.6. *Let $d \geq 4$. If $L = BW_{2^d}$ and Z is a kth-generation sublattice, $k \leq d-2$, then the stabilizer of Z in $Aut(L)$, is just $N_{Aut(L)}(E)$, where E is its defining lower group, as in 13.5. It contains R_{2^d} and its image in G_{2^d}/R_{2^d} is a maximal parabolic which modulo the unipotent radical has shape $GL(k,2) \times \Omega^+(2(d-k),2)$. The kth generation sublattices are in G_{2^d}-equivariant bijection with the elementary abelian subgroups of R_{2^d} which contain $Z(R_{2^d})$.*

Proof. The direct summands of Z realize all the linear characters of E which do not have -1 in their kernel. Thus, Z determines E. By definition of ancestral sublattices, E determines Z. $\square$

Definition 13.7. A sublattice of $L = BW_{2^d}$ is an *k-generation ancestor lookalike* if it is an orthogonal direct sum of 2^k copies of ssBW lattices, all of rank 2^{d-k}.

The transitivity situation for lookalikes is unclear. Here is a simple result.

Proposition 13.8. *For $L = BW_{2^3}$, there is just one orbit of the automorphism group on third generation ancestral lookalike sublattices and there are four orbits for G_{2^3}. For BW_{2^4}, there are at least 4 orbits of the automorphism group on third generation ancestral lookalike sublattices.*

Proof. For the case $L = BW_{2^3} \cong L_{E_8}$, this was covered in 12.4.

Now take the case $L = BW_{2^4}$. Let F be such a frame. Then $F + L[1]$ spans a totally singular subspace of $L/L[1]$. Since $Aut(L)$ induces on $L/L[1]$ its simple orthogonal group, we may assume that F lies in the ancestor sublattice $L_1 + L_2 \cong \sqrt{2}L_{E_8} \perp \sqrt{2}L_{E_8}$.

Since norm 4 elements in $L_1 + L_2$ are indecomposable, we have $F = F_1 \cup F_2$ where $F_i := F \cap L_i$. By using the ideas in the proof of 12.4, we find that the dimension of the span of $F_2 + L_1[1]$ in $L_1/L_1[1]$ can be 1, 2, 3 or 4. We conclude that the image of F in $L/L[1]$ spans a space of dimension at most 8 and dimensions 1,2,3 and 4 actually do occur. This gives a lower bound of 4 on the number of orbits. $\square$

14 The Ypsilanti lattices

We now set up a procedure for creating many isometry types of lattices in sufficiently large dimensions divisible by 8. Here is a rough idea. We take several isometric "good" lattices (indecomposable, high minimum norm, elementary abelian discriminant group) and study overlattices L of their orthogonal direct sum $L_1 \perp \cdots \perp L_s$. We consider conditions like X (3.3) but without (e). A suitable concept of avoidance allows us to build many lattices L with enough but not too many minimal vectors. We gain enough control over the automorphism groups to get a fairly high lower bound on the number of isometry types.

We start with a generalization of the maps $f - 1$ where f is a fourvolution.

14.1 Michigan lattices and Washtenawizations

Definition 14.1. A *2-special endomorphism* on a lattice L is an endomorphism p so that
  (i) $(xp, yp) = 2(x, y)$ for all $x, y \in L$;
  (ii) $Lp^2 = 2L$ (thus, $\frac{1}{2}p^2 \in Aut(L)$);
  (iii) there is an integer r so that $L^* = Lp^{-r}$ (r is called the *duality level*).

If L has a 2-special endomorphism, call L a *2-special lattice*. Call L *normalized* if the duality level is 0 or 1.

Remark 14.2. A 2-special lattice is scale-isometric by a power of a 2-special endomorphism to a normalized lattice.

Notation 14.3. We adapt notations used earlier and set $L[k] := Lp^k$, for $k \in \mathbb{Z}$. When $C_{Aut(L)}(L[k]/L[k+1])$ is independent of $k \in \mathbb{Z}$, we define $Lower(L) := C_{Aut(L)}(L/L[1])$ and $Upper(L) := Stab_{Aut(L)}(L[-1])/Lower(L)$.

Notation 14.4. The sublattice of the lattice L spanned by the minimal vectors is denoted $SMV(L)$. When L has a 2-special endomorphism, define $SMV(L, L[1]) := SMV(L) + L[1]/L[1]$ and define $mvd(L, L[1])$ to be the dimension of $SMV(L, L[1])$. This number is called *the mv-dimension* and is positive if $L \neq 0$. In case p or $L[1]$ is understood, we write $mvd(L)$ for $mvd(L, L[1])$ and note that this invariant could depend on choice of 2-special endomorphism.

Define the *Washtenaw number* or *Washtenaw ratio* of $L \neq 0$ to be the ratio

$$Washtenaw(L) := 2\operatorname{mvdim}(L)/\operatorname{rank}(L) = \operatorname{mvdim}(L)/\dim(L/L[1]) \in (0, 1].$$

Definition 14.5. A *Michigan lattice* is a lattice M
  (i) with a 2-special endomorphism, p;
  (ii) $SMV(M)$ has finite index in M;
  (iii) $Aut(M)$ fixes each Mp^k, $k \in \mathbb{Z}$;
  (iv) $g \in Aut(M)$ is trivial on Mp^k/Mp^{k+1} if and only if g is trivial on $Mp^\ell/Mp^{\ell+1}$, for all $k, \ell \in \mathbb{Z}$.

Note that a Michigan lattice L is indecomposable if $SMV(L)$ is indecomposable.

Definition 14.6. We are given a normalized Michigan lattice M such that $SMV(M)$ is indecomposable. Let $t \geq 3$ be an integer.

Let $M_1, \ldots, M_{2^t}$ denote pairwise orthogonal copies of M, identified by isometries $\psi_i : M \to M_i$, with 2-special endomorphism p_i corresponding to p by ψ_i. The direct sum has a 2-special endomorphism, q, which is the direct sum of the p_i.

A *degree t Washtenawization* of M is a lattice W contained in $\mathbb{Q} \otimes (M_1 \perp \cdots \perp M_{2^t})$ so that

(i) W contains $(M_1 \perp \cdots \perp M_{2^t})[1-r]$ and is a sublattice of $(M_1 \perp \cdots \perp M_{2^t})[-r]$; ($r$ is the duality level of M) and the quotient $M/(M_1 \perp \cdots \perp M_{2^t})[1-r]$ is elementary abelian of dimension $2^{t-2} \operatorname{rank}(M)$;

(ii) For all i, $W \cap (\mathbb{Q} \otimes M_i) = M_i[1-r]$;

(iii) $\mu(W) = 2^{1-r}\mu(M)$;

(iv) $SMV(W) = \sum_{i=1}^{2^t} SMV(M_i)$ and $Washtenaw(W) = \frac{1}{2} Washtenaw(M)$;

(v) $Aut(W)$ has the form $[\prod_{i=1}^{2^t} Lower(M_i)].[Upper(M) \times Aut(\mathcal{C})]$, where $\mathcal{C}$ is an indecomposable (4.8) self orthogonal doubly even binary code of length 2^t; furthermore, $Aut(M)$ embeds in $Aut(W)$ by diagonal action.

A *minimal Washtenawization* is a degree 3 Washtenawization, using the extended Hamming code (which is essentially the only choice here). It is unique up to isometry.

Remark 14.7. By 5.1, Washtenawizations are indecomposable, since the code is indecomposable. In the notation of 14.6, the duality level of W is $1 - r$ and $|Upper(W)|$ divides $|Upper(M)|(2^t!)$. Also, $Aut(W)$ permutes the set $\{M_1, \ldots, M_{2^t}\}$.

Proposition 14.8. *For all $t \geq 3$, degree t Washtenawizations exist.*

Proof. Let M be a normalized Michigan lattice. Take the lattice W between $(M_1 \perp \cdots \perp M_{2^t})[1-r]$ and $(M_1 \perp \cdots \perp M_{2^t})[-r]$ which corresponds to some indecomposable doubly even self orthogonal code, $\mathcal{C}$ (for example, see 4.9). Since nonzero code words have weight at least 4, the minimal vectors of W lie in $SMV((M_1 \perp \cdots \perp M_{2^t})[1-r])$ (use 5.8).

Since q acts diagonally as p on $(M_1 \perp \cdots \perp M_{2^t})[1-r]$, the definition of W implies that the image of $SMV((M_1 \perp \cdots \perp M_{2^t})[1-r])$ in W/Wp has dimension $2^{t-1}\operatorname{mvdim}(M)$. This implies that $Washtenaw(W) = \frac{1}{2} Washtenaw(M)$.

Since $Aut(W)$ permutes the minimal vectors, it permutes the indecomposable direct summands of the lattice they generate, which are just the 2^t $SMV(M_i)$, which in turn define the M_i as the summands of W (as abelian groups) which contain the $SMV(M_i)$. It follows that $Aut(W)$ is contained in a natural wreath product $Aut(M_i) \wr Sym_{2^t}$ which permutes $\{M_1, \ldots, M_{2^t}\}$. Obviously, $Aut(W)$ contains a group G_0 of the form indicated in 14.6(v). Now, use 5.8(ii) and the fact that $Aut(M)$ leaves each twist $M[k]$ invariant. $\square$

14.2 Overlattices of direct sums of 2-special lattices

Notation 14.9. Throughout this section, M is a normalized 2-special lattice (14.1) and M_1, M_2 are pairwise orthogonal lattices isometric to M with duality level $r \in \{0, 1\}$. Let t be an isometry of order 2 which interchanges them.

Definition 14.10. The ith *admissible component group* K_i is the full general linear group on $M_i[-r]/M_i[1-r]$ when the duality level of M is 0 and when the duality level of M is 1, it is the full orthogonal group on the nonsingular quadratic space $M_i[-r]/M_i[1-r]$, $x + M[1-r] \mapsto 2^{r-1}(x,x) \pmod{2}$.

Notation 14.11. Let $d \geq 5$ be an integer and let M_1, M_2 be isometric normalized 2-special lattices of ranks 2^{d-1} and duality level 1. Set $V_i := \mathbb{Q} \otimes M_i$. Let $\mathfrak{Y} := \mathfrak{Y}(M_1[1-r], M_2[1-r])$ denote the set of even integral lattices M which contain $M_1[1-r] \perp M_2[1-r]$ and satisfy $M \cap V_i = M_i[1-r]$ for $i = 1, 2$ and whose projection to V_i is $M_i[-r]$. This is a set of rank 2^d unimodular lattices. (Note differences with 10.1, which results in unimodular lattices for ranks 2^d, d odd only.)

Remark 14.12. A member L of $\mathfrak{Y}$ is determined by an isomorphism of vector spaces $\zeta : M_1[-r]/M_1[1-r] \to M_2[-r]/M_2[1-r]$, namely $L/(M_1[1-r] + M_2[1-r])$ is just the diagonal in the identification of the two $M_i[-r]/M_i[1-r]$ based on ζ. We may write $L/(M_1[1-r] + M_2[1-r]) = \{(x + M_1[1-r], (x + M_1[1-r])^\zeta) | x \in M_1[1-r]\}$.

Conversely, given a linear isomorphism ζ, we get an $L \in \mathfrak{Y}$ by taking the diagonal as above provided (a) when $d-1$ is odd, no condition; (b) when $d-1$ is even, ζ is an isometry of nonsingular quadratic spaces $M_1[-r]/M_1[1-r] \to M_2[-r]/M_2[1-r]$.

The reason for the isometry condition in (b) is that the nonsingular cosets (respectively, the singular cosets) of the two $M_i[-r]/M_i[1-r]$ must be matched to create a diagonal which gives an even lattice L. In (a), since the two $M_i[-r]$ are even integral lattices, any matching by a linear isomorphism results in an element of $\mathfrak{Y}$, whence no conditions are demanded. The requirement in (b) of taking $M_1[-r]/M_1[1-r]$ to $M_2[-r]/M_2[1-r]$ comes from the definition of $\mathfrak{Y}$, 14.11.

Notation 14.13. We use the notations $L \mapsto \zeta(L), \zeta \mapsto L(\zeta)$ to express the bijection between $\mathfrak{Y}$ and such isomorphisms.

Such ζ are in bijection with K_1 and with K_2 (see 14.10) by $\zeta \mapsto \zeta_i \in K_i$, where the latter are defined by the formulas $\zeta : x + M_1[1-r] \mapsto (x^t + M_2[1-r])^{\zeta_2} = y^t + M_2$, where $y + M_2 = (x + M_1)^{\zeta_1 t}$, where t is as in 14.9. Call ζ_i the K_i-component of ζ, or of $L = L(\zeta)$.

14.3 Avoidance

Definition 14.14. We say that two subspaces of a vector space *avoid* each other if their intersection is 0. If $g : V \to V'$ is an invertible linear transformation, $W \leq V$ and $W' \leq V'$, we say that g is a (W, W')-*avoiding map* if $W^g \cap W' = 0$. Let $A(W_1, W_2)$ be the set of (W, W')-avoiding maps in the set of linear isomorphisms from V onto V'.

We need some terminology for discussing asymptotic behavior.

Notation 14.15. Suppose that $f(x)$ is a real-valued function on $(0, \infty)$. The *dominant term* in $f(x)$ (abbreviated $DT(f(x))$) is the expression of the form $a_0 \log_2(x)^{a_1} 2^{a_2 x} x^{a_3}$ which is asymptotic to $f(x)$ (the a_i are constants). We may indicate dependence on the variable x by DT_x. (This definition applies to a limited family of real-valued functions, but suffices for our purposes.)

Similarly, if f is as above, we define the *dominant term of the logarithm* (DTL or DTL_x) of $2^{f(x)}$ to be $DT(f(x))$. For example,

$$DTL(2^{(0.43)\log_2(2x-3)2^{3x-4} + 2^{2x} - \log_2(x+1)^5 x^3 - \log_2(x)^7(x^2+1)}) = \frac{0.43}{16} \log_2(x) 2^{3x-4}.$$

Proposition 14.16. *Suppose that $a \le b$ are positive integers. Suppose that $V := \mathbb{F}_2^{2b}$ has a maximal Witt index nonsingular quadratic form and that W_1 and W_2 are two a-dimensional totally singular subspaces. We set $q := \frac{a}{b}$ and think of q as a constant and a as a function of b.*

(i) *Let H be the stabilizer in $O(V)$ of W_1. Then,*

$$DTL_b(|H|) = DTb(\tfrac{1}{2}a(3a-1) + 2(b-a)b) = b^2(2 - 2q + \tfrac{3}{2}q^2).$$

(ii) *For an integer k, let $A(W_1, W_2; k)$ be the set of avoiding maps as in 14.14 so that $\dim(W_1^g \cap W_2^\perp) = k$. Then $A(W_1, W_2; k)$ is nonempty precisely for $k = 0, 1, \ldots, \min\{a, b-a\}$ and for each such k, $A(W_1, W_2; k)$ is a regular orbit for the action of H.*

Proof. (i): We have $DTL_k(|\Omega^+(2k,2)|) = 2k^2 - k$. We may assume $W_1 = W_2$. Let H be the subgroup of the orthogonal group which fixes W_1 globally. It follows from A.1 that $DTL(|H|)$ is the DTL of $\tfrac{1}{2}a(3a-1) + 2(b-a)b - (b-a)$.

(ii): By Witt's theorem, two nonavoiding maps g, g' are in the same H-orbit if the dimensions of the images of W_1 under g, g' intersect $W_1^\perp$ in spaces of the same dimensions. All H-orbits are regular. For a nonempty $A(W_1, W_2; k)$, we have $k \le \dim(W_1) = a$ and since the image of an avoiding W_1 in $V/W_2^\perp$ has dimension at most $a = \dim(V/W_2^\perp)$, we have $k + a = k + \dim(W_2) \le b$, the dimension of any maximal totally isotropic subspace. The value $k = \min\{a, b-a\}$ can be achieved. □

Corollary 14.17. *We use the notations of 14.16 and assume that $q \le \tfrac{1}{2}$. Then*

$$DTL_b(|A(W_1, W_2)|) = v(q)\log_2(b)b^2, \text{ where } v(q) := (2 - 2q + \tfrac{3}{2}q^2).$$

Proof. Note that $q \le \tfrac{1}{2}$ means $a = \min\{a, b-a\}$ in 14.16. □

14.4 Down Washtenaw avenue to Ypsilanti

We next create large families of lattices in dimensions $2^d \gg 0$.

Definition 14.18. Let W be a normalized Michigan lattice which has duality level $r = 1$ and Washtenaw ratio $q \le \tfrac{1}{2}$.

Take orthogonal copies M_1, M_2 of W and consider the set $\mathfrak{Y} := \mathfrak{Y}(M_1, M_2)$ as in 14.11. Consider the associated maps $\zeta(L), L \in \mathfrak{Y}$ (see 14.13) which are avoiding maps 14.14 for the subspaces $SMV(M_1[-1], M_1), SMV(M_2[-1], M_2)$ of $M_1[-1]/M_1, M_2[-1]/M_2$, respectively. The corresponding lattices form a subset $\mathfrak{Y}_{av}(M_1, M_2)$ of $\mathfrak{Y}(M_1, M_2)$ in the notation of 14.11. Their ranks are $2\,\mathrm{rank}(W)$. They are called *Ypsilanti lattices*. Let $IsomTypes(M_1, M_2)$ be the set of isometry types of lattices in $\mathfrak{Y}_{av}(M_1, M_2)$.

When W is a Washtenawization of a BW lattice, the Ypsilanti lattices of rank $2^d = 2\,\mathrm{rank}(W)$ are called the *Ypsilanti cousins* of BW_{2^d}.

Lemma 14.19. *We use the notations of 14.18.*
  (i) *If $N \in \mathfrak{Y}_{av}(M_1, M_2)$, $SMV(N) = SMV(M_1) \perp SMV(M_2)$.*
  (ii) *$N \in \mathfrak{Y}_{av}(M_1, M_2)$ is indecomposable.*

Proof. (i) Obviously, $\mu(N) \geq \mu(M_i), i = 1, 2$. Consider a vector $x = x_1 + x_2 \in N \setminus (M_1 \perp M_2)$. Then the x_i have norms at least $\mu(M_i[-1]) = \frac{1}{2}\mu(M_i)$. For (x, x) to equal $\mu(M_i)$, we need x_i to be a minimal vector of $M_i[-1]$ for $i = 1, 2$. This is not the case since N was defined with an avoiding map.

(ii) Use 5.9. □

Lemma 14.20. *Suppose that we are given $q = 2^{-j}$ for some $j > 0$. For all $k \geq 5 + 3j$, there exists a Michigan lattice $W(k)$ so that* $\text{rank}(W(k)) = 2^k$, $Washtenaw(W(k)) = q$ *and the duality level of $W(k)$ is 1. We may also arrange for* $\mu(W(k)) = 2^{1-r+\lfloor \frac{j}{2} \rfloor + \lfloor \frac{e}{2} \rfloor}$, *where $e = k - 3j$ if k is even and $e = k - 3j - 1$ if k is odd.*

Proof. If we start with BW_{2^e} and perform the minimal Washtenawization procedure s times, we get a lattice $W(e, s)$ of rank 2^{e+3s}. We may take a degree 4 Washtenawization to $W(e, s)$ and get a lattice $W'(e, s + 1)$ of rank 2^{e+3s+4}.

Each Washtenawization changes duality level. We define $W(k)$ according to the following cases. When k is even, we require $k - 3j \geq 4$, which means k is at least 8. When k is odd, we require $k - 3j - 1 \geq 4$, which means that k is at least 9.

$$W(k) := \begin{cases} W(k - 3j, j) & k \text{ even}; \\ W'(k - 3j - 1, j) & k \text{ odd}. \end{cases}$$ □

Definition 14.21. We call a sequence of lattices as in 14.20 *the j-Washtenaw series, for the fixed ratio $q = 2^{-j}$*. It starts at rank 2^{5+3j}. The isometry types of certain members of the series depend on choice of indecomposable doubly even code of length 16. Ypsilanti cousins associated to such series are called *Ypsilanti j-cousins*. The set of such isometry types is denoted $Ypsi(2^d, j)$.

Lemma 14.22. *We use the notations of 14.18, 14.20 and let $W(k)$ be the Washtenaw series.*

(i) *If N and N' are two cousins of rank 2^d, $Isom(N, N')$ is contained in the group $G_0(e, h)$ of orthogonal transformations which stabilize $L_1 \perp \cdots \perp L_{2^h}$, the indecomposable direct summands of $SMV(N) = SMV(N')$ (in fact, the L_i are the pairwise isometric scaled Barnes-Wall lattices, of rank 2^e, on which the Washtenawizations M_1, M_2 were based; the notation means $d = k + 1 = e + h$, with $h = 3j$ or $3j + 1$).*

(ii) $DTL_d(|G_0(e, h)|)$ *is bounded above by a constant times d^2.*

Proof. (i) Given $N, N' \in \mathfrak{Y}_{av}(M_1, M_2)$, an isometry of N to N' takes $SMV(N)$ to $SMV(N')$. Both of these equal $SMV(M_1 \perp M_2)$.

(ii) This follows from $DTL_f(|\Omega^+(2f, 2)|) = 2f^2$ and boundedness of h. □

Notation 14.23. When $W(k)$ runs through the j-Washtenaw series 14.21, we let $\Upsilon(2^d, j) := |Ypsi(2^d, j)|$.

Lemma 14.24. $\Upsilon(2^d, j) \geq |\mathfrak{Y}_{av}(M_1, M_2)|/|G_0(e, h)|$, *whence* $DTL_d(\Upsilon(2^d, j)) \geq \frac{1}{16}\upsilon(2^{-j})d\, 2^{2d}$, *as in 14.17.*

Proof. In 14.17, take $b = 2^{d-2}$, because the admissible component group 14.10 is $O^+(2^{d-1}, 2)$ since the duality level has been arranged to be 1. Then use 14.22(ii) and 14.23. □

Remark 14.25. For a fixed large value of d, we can make the families $Ypsi(2^d, j)$, $q = 2^{-j}$, for all $1 \leq j \leq \lfloor \frac{d-5}{3} \rfloor$. This would make roughly $d/3$ times as many as one of the $Ypsi(2^d, j)$, so would not increase the DTL.

We summarize our counting in dimensions 2^d.

Theorem 14.26. *For any $j > 0$, the number of isometry types of Ypsilanti lattices in dimension 2^d has DTL at least $\frac{1}{16} v(2^{-j}) d\, 2^{2d}$. In particular, the number of indecomposable even unimodular lattices in dimensions 2^d has DTL at least $c\, d\, 2^{2d}$, for any $c \in (0, \frac{1}{8})$.*

Remark 14.27. With a bit more work, we could define lattices like Ypsilanti cousins for $d < 9$, though we would not expect them to represent more than a fraction of $mass(2^d)$ isometry types. In dimension 32, the mass formula gives value about 10^7 and the number of isometry types (still not known) has been bounded below by about 10^{10} (see [24]).

14.5 From dimensions 2^d to arbitrary dimensions

Notation 14.28. For an integer $n > 0$ divisible by 8, let 2^d be the largest power of 2 less than or equal to n. Fix some q^{-j}, $j > 0$. Let $Ypsi(n, j)$ be the set of isometry types of even integral unimodular lattices which contain a Ypsilanti j-cousin of rank 2^d as an orthogonal direct summand. Clearly, $\Upsilon(n, j) := |Ypsi(n, j)| \geq \Upsilon(2^d, j)$.

Corollary 14.29. *We use the notation of 14.28. For any constant $c \in [0, \frac{1}{32})$, we take $j > 0$ so that $q = 2^{-j}$ satisfies $2 - q + \frac{3}{2}q^2 > 64c$.*
Then $\log_2(\Upsilon(n, j)) \geq c \log_2(n)\, n^2$.

Proof. Take the integer d which satisfies $2^d \leq n < 2^{d+1}$. Then $d < \log_2(n) \leq d+1$ and $2^d > \frac{n}{2}$. We have $\Upsilon(n, j) \geq \Upsilon(2^d, j)$ and $DTL(\Upsilon(n, j)) \geq DTL(\Upsilon(2^d, j)) > 4c\, d\, 2^{2d} \geq 4c(\log_2(n) - 1)(\frac{n}{2})^2$ whose DT is at least $c \log_2(n) n^2$. □

14.6 Number of Ypsilanti cousins compared with the mass formula

Notation 14.30. We follow the notations of [31], pp. 54, 90, except we write $mass(n)$ instead of "M_n". Stirling's formula $(n! \sim n^{n+\frac{1}{2}} e^{-n}(2\pi)^{\frac{1}{2}})$ implies that $DTL(n!) = \log_2(n) n$. Let B_j be the jth Bernoulli number. Let $n \in 8\mathbb{Z}$, $k := \frac{n}{8}$.

Proposition 14.31. $DTL_n(mass(n)) = \frac{1}{4} \log_2(n)\, n^2$.

Proof. We have $mass(n) = \frac{B_{2k}}{8k} \prod_{j=1}^{4k-1} \frac{B_j}{4j}$. Because $\zeta(2j) \in (1, 2)$ for all $j \geq 1$, the formula $B_j = 2\zeta(2j).(2j)!/(2\pi)^{2j}$ shows that $DTL_j(B_j) = DT(\log_2(2j)2j)$.

We have

$$\log_2(mass(n)) = \log_2(B_{2k}) + \sum_{j=1}^{4k-1} \log_2(B_j) - (8k+1) - \log_2((4k-1)!) - \log_2(k).$$

Since $DTL_k(B_{2k}) = DTL_k((4k)!)$, $DTL_n(mass(n)) = DT_n(\sum_{j=1}^{4k-1}\log_2(2j)2j)$. The latter summation can be thought of as Riemann sums, which can be estimated with integrals (think of $\int 2x \ln(2x)\,dx = \int 2x \ln(x)\,dx + \ln(2)\int 2x\,dx = x^2 \ln(x) - \frac{1}{2}x^2 + \ln(2)x^2 + c$, which has dominant term $x^2 \ln(x)$). We conclude that $DTL_n(mass(n)) = DT_n(\log_2(8k)(4k)^2) = \frac{1}{4}\log_2(n)n^2$. □

Proposition 14.32. *For positive integers n,q, define $A(n,q) := \sum_{i\geq 0}\lfloor\frac{n}{q^i(q-1)}\rfloor$ and let $\mathcal{P}$ be the set of prime numbers at most $n+1$. Set $f(n) := \prod_{q\in\mathcal{P}} q^{A(n,q)}$. Then a finite subgroup of $GL(n,\mathbb{Q})$ has order dividing $f(n)$.*

Proof. This is a result of Minkowski [28]. See the discussions in exercises for Section 7 of [6]. □

Lemma 14.33. $DTL_n(f(n)) = n\log_2(n)$.

Proof. Well known? A proof may be deduced from [29], Th. 8.8(b), p. 369. □

Remark 14.34. The DTL of $n\log_2(n)$ is small compared to $DTL(mass(n))$. It follows that the DTL of the number of isometry types of rank n even unimodular lattices is the same as that of $DTL(mass(n))$.

We summarize:

Corollary 14.35. *For any $a \in (0,\frac{1}{8})$, there is an integer j so that $DTL_n(\Upsilon(n,j)) \geq a \cdot DTL(mass(n))$. Furthermore, when n is a power of 2, and $b \in (0,\frac{1}{2})$, there is an integer j so that $DTL_n(\Upsilon(n,j)) \geq b \cdot DTL(mass(n))$.*

Remark 14.36. We conclude with some numerical comparisions.

Asymptotics for $\Upsilon(2^d,j)$, $\Upsilon(n,j)|$ and $mass(n)$.

j	$q=2^{-j}$	$v(q)$ = constant coefficient of $16\ DTL(\Upsilon(2^d,j))$ (see 14.24)	Lower bound for $DTL(\Upsilon(2^d,j))/DTL(mass(2^d))$
1	.5000000000	1.375000000	.3437500000
2	.2500000000	1.593750000	.3984375000
3	.1250000000	1.773437500	.4433593750
4	.06250000000	1.880859375	.4702148438
5	.03125000000	1.938964844	.4847412109
6	.01562500000	1.969116211	.4922790527
7	.007812500000	1.984466553	.4961166382
8	.003906250000	1.992210388	.4980525970
9	.001953125000	1.996099472	.4990248680
10	.0009765625000	1.998048306	.4995120764

Appendices

A1 Group orders

Proposition A.1. (i) *For q a power of 2, the order of $\Omega^+(2n,q)$ is $q^{n(n-1)}(q^n - 1)\prod_{i=1}^{n-1}(q^{2i} - 1)$.*

(ii) *The stabilizer in $\Omega^+(2n,q)$ of an isotropic point has shape $q^{2(n-1)}{:}[\Omega^+(2(n-1),q) \times q - 1]$.*

(iii) *The stabilizer in $\Omega^+(2n,q)$ of a maximal totally singular subspace has shape $q^{\binom{n}{2}}{:}GL(n,q)$, and this is a maximal subgroup.*

(iv) *The stabilizer in $\Omega^+(2n,q)$ of a totally singular subspace of dimension $m < n$ has the form RL, where the unipotent radical has order $2^{\binom{m}{2}+2m(n-m)}$ and $L \cong \Omega^+(2(n-m),2) \times GL(m,2)$. These are maximal subgroups.*

Proof. These are well-known properties of the orthogonal groups. Proofs may be obtained from [8, 17]. □

A2 $Aut^0(2_\varepsilon^{1+2d})$, $Out^0(2_\varepsilon^{1+2d})$ and $BRW^0(2^d, \varepsilon)$.

Basic theory of extraspecial groups extended upwards by their outer automorphism group has been developed in several places. We shall use [3, 4, 5, 13, 14, 15, 16].

Notation A.2. Let $R \cong 2_\varepsilon^{1+2d}$ be an extraspecial group which is a subgroup of $G := GL(2^d, \mathbb{F})$, for a field $\mathbb{F}$ of characteristic 0. Let $N := N_G(R) \cong \mathbb{F}^\times . 2^{2d} O^\varepsilon(2d, 2)$. The *Bolt-Room-Wall group* is a subgroup of this of the form $2_\varepsilon^{1+2d}.\Omega^\varepsilon(2d, 2)$. If $d \geq 3$ or $d = 2, \varepsilon = -$, N' has this property. For the excluded parameters, we take a suitable subgroup of such a group for larger d. We denote this group by $BRW^0(2^d, +)$ or $\mathcal{D}(d)$. It is uniquely determined up to conjugacy in G by its isomorphism type if $d \geq 3$ or $d = 2, \varepsilon = -$. It is conjugate to a subgroup of $GL(2^d, \mathbb{Q})$ if $\varepsilon = +$. Let $R = R_{2^d}$ denote $O_2(G_{2^d})$. We call R_{2^d} the *lower group* of $BRW^0(2^d, +)$ and call G_{2^d}/R_{2^d} the *upper group* of $BRW^0(2^d, +)$.

For $g \in N$, define $C_{R \text{ mod } R'}(g) := \{x \in R | [x, g] \in R'\}$, $B(g) := Z(C_{R \text{ mod } R'}(g))$ and let $A(g)$ be some subgroup of $C_{R \text{ mod } R'}(g)$ which contains R' and complements $B(g)$ modulo R', i.e., $C_{R \text{ mod } R'}(g) = A(g)B(g)$ and $A(g) \cap B(g) = R'$. Thus, $A(g)$ is extraspecial or cyclic of order 2. Define $c(d) := \dim(C_{R/R'}(g))$, $a(g) := \frac{1}{2}|A(g)/R'|$, $b(g) := \frac{1}{2}|B(g)/R'|$. Then $c(d) = 2a(d) + 2b(d)$.

Corollary A.3. *Let L be any $\mathbb{Z}$-lattice invariant under $H := BRW^0(2^d, +)$. Then H contains a subgroup $K \cong AGL(d, 2)$ and L has a linearly independent set of vectors $\{x_i | i \in \Omega\}$ so that there exists an identification of Ω with $\mathbb{F}_2^d$ which makes the $\mathbb{Z}$-span of $\{x_i | i \in \Omega\}$ a permutation module for $AGL(d, 2)$ on Ω.*

Proof. In H, let E, F be maximal elementary abelian subgroups such that $R = EF$ and let K be their common normalizer. It satisfies $K/R \cong GL(d, 2)$.

Now, let z generate $Z(R)$ and let E_1 complement $\langle z \rangle$ in E and F_1 complement $\langle z \rangle$ in F. The action of K on the hyperplanes of E which complement $Z(R)$ satisfies $N_K(E_1)F = K, N_F(E_1) = Z(R)$. Now consider the action of $N_K(E_1)$ on the hyperplanes of F which complement $Z(R)$. We have that $K_1 := N_K(E_1) \cap N_K(F_1)$ covers $N_K(E_1)/E$. Therefore, $K_1/Z(R) \cong GL(d, 2)$. Let K_0 be the subgroup of index 2 which acts trivially on the fixed points on L of E_1, a rank 1 lattice. So, $K_0 \cong GL(d, 2)$. Let x be a basis element of this fixed point lattice. Then the semidirect product $F_1{:}K_0$ is isomorphic to $AGL(d, 2)$ and $\{x^g | g \in F_1\}$ is a permutation basis of its $\mathbb{Z}$-span (and the x^g are eigenvectors for all the characters of E which are nontrivial on $Z(R)$). $\square$

Definition A.4. We use the notation of A.2. An element $x \in N$ is *dirty* if there exists g so that $[x, g] = xz$, where z is an element of order 2 in the center. If g can be chosen to be of order 2, call x *really dirty* or *extra dirty*. If x is not dirty, call x *clean*.

Lemma A.5. *Let $\mathbb{F}_2^{2d}$ be equipped with a nondegenerate quadratic form with maximal Witt index. The set of maximal totally singular subspaces has two orbits under $\Omega^+(2d, 2)$ and these are interchanged by the elements of $O^+(2d, 2)$ outside $\Omega^+(2d, 2)$.*

Proof. This is surely well known. For a proof, see [17]. $\square$

Theorem A.6. *We use the notation of A.2, A.4. Let $g \in N$. Then $\operatorname{Tr}(g) = 0$ if and only if g is dirty. Assume now that g is clean and has finite order. Then $\operatorname{Tr}(g) = \pm 2^{a(g)+b(g)} \eta$, where η is a root of unity. If $g \in BRW(d, +)$, we may take $\eta = 1$. Furthermore, every coset of R in $BRW(d, \varepsilon)$ contains a clean element and if g is clean, the set of clean elements in Rg is just $g^R \cup -g^R$.*

Proof. [15]. $\square$

Lemma A.7. *Suppose that t, u are involutions in $\Omega^+(2d, 2)$, for $d \geq 2$. Suppose that their commutators on the natural module $W := \mathbb{F}_2^{2d}$ are totally singular subspaces of the same dimension, e. Suppose that $e < d$ or that $e = d$ and that $[W, t]$ and $[W, u]$ are in the same orbit under $\Omega^+(2d, 2)$. Then t and u are conjugate. If $e = d$, t and u are conjugate if and only if $[W, t]$ and $[W, u]$ are in the same orbit under $\Omega^+(2d, 2)$, and there are two such orbits.*

Proof. Well know. Use induction on d. $\square$

Corollary A.8. *Suppose that t, u are clean involutions in H with $\operatorname{Tr}(t) = \operatorname{Tr}(u) \neq 0$. Then t and u are conjugate in G_{2^d} unless possibly d is even and $\operatorname{Tr}(t) = \operatorname{Tr}(u) = \pm 2^{\lfloor \frac{d}{2} \rfloor}$. If the latter holds, t, u lie in one of two classes of involutions.*

Proof. We may assume that t, u are noncentral. These involutions are not lower and have the same dimension of fixed points on $R/R' \cong \mathbb{F}_2^{2d}$. Let $T, U \leq R$ be their respective centralizers in R. Since both t, u are clean, $[R, t]$ and $[R, u]$ are elementary abelian subgroups of T, U, respectively. From A.7, we deduce that Rt and Ru are conjugate in G_{2^d} unless possibly $\operatorname{Tr}(t) = \operatorname{Tr}(u) = \pm 2^{\lfloor \frac{d}{2} \rfloor}$.

We may assume that $Rt = Ru$. Now use A.6 to deduce that t is R-conjugate to u or $-u$. The trace condition implies that t is conjugate to u.

The exceptional case $\operatorname{Tr}(t) = \operatorname{Tr}(u) = \pm 2^{\lfloor \frac{d}{2} \rfloor}$ leads to two classes; see A.7. $\square$

Remark A.9. The extension $1 \to R_{2^d} \to G_{2^d} \to \Omega^+(2d, 2) \to 1$ is nonsplit for $d \geq 4$. This was proved first in [4], then later in [7] and in [13] (for both kinds of extraspecial groups, though with an error for $d = 3$; see [14] for a correction). The article [13] gives a sufficient condition for a subextension $1 \to R_{2^d} \to H \to H/R_{2^d} \to 1$ to be split, and there are interesting applications, e.g. to the centralizer of a 2-central involution in the Monster. A general discussion of exceptional cohomology in simple group theory is in [16].

A3 Indecomposable integral representations for a group of order 2

Proposition A.10. Let G be a cyclic group of order 2 and M a finitely generated $\mathbb{Z}$-free G-module. Then M is a direct sum of modules isomorphic to $\mathbb{Z}[G]$, the group algebra; the $\mathbb{Z}$-rank 1 trivial module; the $\mathbb{Z}$-rank 1 nontrivial G-module.

Proof. [10], Section 74. The case where G has order any prime number is treated. □

References

[1] E. Bannai, Positive definite unimodular lattices with trivial automorphism groups, Mem. Amer. Math. Soc. 85 (1990), no. 429. MR1004770 (90j:11030)

[2] E. S. Barnes and G. E. Wall, Some extreme forms defined in terms of abelian groups, J. Aust. Math. Soc. 1 (1959), 47-63.

[3] B. Bolt, T. G. Room and G. E. Wall, On the Clifford collineations, transform and similarity groups, I. J. Aust. Math. Soc. 2 (1961/1962), 60-79.

[4] B. Bolt, T. G. Room and G. E. Wall, On the Clifford collineations, transform and similarity groups, II. J. Aust. Math. Soc. 2 (1961/1962), 80-96.

[5] B. Bolt, On the Clifford collineation, transform and similarity groups. III; Generators and involutions, J. Austral. Math. Soc. 2 (1961/1962), 334-344.

[6] N. Bourbaki, Élements de Mathématique, Groupes et algèbres de Lie, Chapitres 2 et 3, Diffusion C.C. L. S., Paris, 1972.

[7] M. Broué and M. Enguehard, Une famille infinie de formes quadratiques entière; leurs groupes d'automorphismes, Ann. scient. Éc. Norm. Sup., 4eme série, t. 6 (1973), 17-52.

[8] R. Carter, Simple Groups of Lie Type, Wiley-Interscience, London ,1972.

[9] J. Conway and N. Sloane, Self-dual codes over the integers modulo 4, J. Combin. Theory Ser. A 62 (1993), 30-45 .

[10] C. Curtis and I. Reiner, Representation Theory of Groups and Associative Algebras, Interscience, 1962.

[11] C. Dong, R. Griess. Jr. and G. Hoehn, Framed vertex operator algebras, codes and the moonshine module, Comm. Math. Phys. 193 (1998), 407-448.

[12] D. Gorenstein, Finite Groups, Harper and Row, New York, 1968.

[13] R. L. Griess, Jr., Automorphisms of extra special groups and nonvanishing degree 2 cohomology, Pacific J. Math. 48 (1973), 403-422.

[14] R. L. Griess, Jr., On a subgroup of order $2^{15}|GL(5,2)|$ in $E_8(C)$, the Dempwolff group and $Aut(D_8 \circ D_8 \circ D_8)$, J. Algebra 40 (1976), 271-279.

[15] R. L. Griess, Jr., The monster and its nonassociative algebra, in Proceedings of the Montreal Conference on Finite Groups, Contemp. Math. 45 (1985),121-157.
[16] R. L. Griess, Jr., Sporadic groups, code loops and nonvanishing cohomology, J. Pure Appl. Algebra 44 (1987), 191-214.
[17] R. L. Griess, Jr., Elementary abelian subgroups of algebraic groups, Geometria Dedicata, 39 (1991), 253-305.
[18] R. L. Griess, Jr., Twelve Sporadic Groups, Springer, 1998.
[19] R. L. Griess, Jr, Pieces of Eight, Advances in Mathematics 148 (1999), 75-104.
[20] R. L. Griess, Jr., Positive definite lattices of rank at most 8, J. Number Theory 103 (2003), 77-84.
[21] B. Gross, Group representations and lattices, J. Amer. Math. Soc. 3 (1990), no. 4, 929–960.
[22] R. W. Hamming, Error detecting and error correcting codes, Bell Syst. Tech. J. 29 (1950), 147-160. MR 12:35.
[23] B. Huppert, Endliche Gruppen I, Springer, Berlin, 1968.
[24] O. King, A mass formula for unimodular lattices with no roots. (English. English summary) Math. Comp. 72 (2003), no. 242, 839–863 (electronic). 11H55 (11E41) MR1954971 (Review)
[25] M. Kneser, Theorie der Kristalgitter, Math. Ann. 127 (1954), 105-106.
[26] J. MacWilliams and N. Sloane, The Theory of Error Correcting Codes, North-Holland, 1977.
[27] J. Milnor and D. Husemoller, Symmetric Bilinear Forms, Ergebnisse der Mathematick und Ihrer Grenzgebiete, Band 73, Springer, New York, 1973.
[28] H. Minkowski, Zur Theorie der positiven quadratischen Formen, Jour. für die reine und angew. Math. 101 (1887), 196-202.
[29] I. Niven, H. S. Zuckerman, Hugh L. Montgomery, An Introduction to the Theory of Numbers, Fifth Edition, Wiley, New York, 1991.
[30] B. L. Rothschild and J. H. van Lint, Characterizing finite subspaces, J. Combin. Theory Ser. A 16 (1974), 97-110.
[31] J.-P. Serre, A Course in Arithmetic, Springer Verlag, Graduate Texts in Mathematics 7, 1973.

B.3 Involutions on the Barnes-Wall Lattices and Their Fixed Point Sublattices, I.

Robert L. Griess Jr.*

Abstract

We study the sublattices of the rank 2^d Barnes-Wall lattices BW_{2^d} which occur as fixed points of involutions. They have ranks 2^{d-1} (for dirty involutions) or $2^{d-1} \pm 2^{k-1}$ (for clean involutions), where k, the defect, is an integer at most $\frac{d}{2}$. We discuss the involutions on BW_{2^d} and determine the isometry groups of the fixed point sublattices for all involutions of defect 1. Transitivity results for the Bolt-Room-Wall group on isometry types of sublattices extend those in [GrPO2^d]. Along the way, we classify the orbits of $AGL(d, 2)$ on the Reed-Muller codes $RM(2, d)$ and describe *cubi sequences* for short codewords, which give them as Boolean sums of codimension 2 affine subspaces.

Keywords and Phrases: Barnes-Wall lattices, involutions, Bolt-Room-Wall group, fixed point lattice.

1 Introduction

We continue to study the Barnes-Wall lattices BW_{2^d} and their isometry groups, which are the Bolt-Room-Wall groups $BRW^+(2^d) \cong 2^{1+2d}_+ \Omega^+(2d, 2)$ for $d \geq 2, d \neq 3$ and W_{E_8} for $d = 3$. In particular, we classify involutions in $BRW^+(2^d)$ and determine properties of their fixed point sublattices, including automorphism groups. For background, we analyze words of the Reed-Muller code $RM(d, 2)$ in some detail and in particular determine the orbits of $AGL(d, 2)$.

We shall be using the Barnes-Wall-Ypsilanti uniqueness theory as developed in [GrPO2^d]. We recommend this article for background and terminology. *Notational warning:* $O(L)$ means orthogonal group on a quadratic space L but $O(G)$ means $O_{2'}(G)$ for a finite group G.

The main results of this article are described below. See 3.18, 3.19.

Theorem 1.1. *The orbits for the action of $AGL(d, 2)$ on the Reed-Muller code $RM(2, d)$ are as follows (for each category, there is one orbit for each allowed value of k):*

---
*Department of Mathematics University of Michigan, Ann Arbor, MI 48109, USA. rlg@umich.edu

Short sets of defect $k = 0, \ldots, \lfloor \frac{d}{2} \rfloor$, *which are of the form* $S_1 + \cdots + S_k$, *where the* S_i *are affine codimension 2 spaces which are linearly coindependent with respect to an origin in their common intersection; such a set has cardinality (or Hamming weight)* $2^{d-1} - 2^{d-k-1}$.

Long sets, which are complements of short sets.

Midsets, of cardinality 2^{d-1}, *which are either affine hyperplanes (defect 0) or nonaffine midsets of the form* $S + H$, *where* H *is an affine hyperplane and* S *is a short set of weight* $2^{d-1} - 2^{d-k-1}$, *for a unique* $k \in \{1, \ldots, \lfloor \frac{d-1}{2} \rfloor\}$. *(Note:* $k \neq \frac{d}{2}$ *here.)*

Some background in the structure of BRW groups is required to state our main results. We refer the reader to the Appendix for a summary and notations. For definitions of clean and dirty, see A.3 and for defect, see A.5.

Theorem 1.2. (i) *When d is odd, the conjugacy classes for involutions in the BRW group $BRW^+(2^d)$ are represented by the transformations:*

(Split Case) ε_X, *where X is a codeword as listed in 1.1, one for each value of the defect,* $k \leq \frac{d-1}{2}$.

(Nonsplit Case) $\eta_{d,2k,\varepsilon}$, *for* $k = 1, \ldots, \frac{d-1}{2}$, $\varepsilon = \pm$.

(ii) *When d is even, the conjugacy classes for involutions in the BRW group $BRW^+(2^d)$ are represented by the transformations:*

(Split Case) ε_X, *where X ranges over the codewords listed in 1.1, but one for each value of the defect, k, together with the single clean involution ε_Y^τ, where Y is a short codeword with defect $k = \frac{d}{2}$ and τ is an outer automorphism of $BRW^+(2^d)$.*

(Nonsplit Case) $\eta_{d,2k,\varepsilon}$, *for* $k = 1, \ldots, \frac{d}{2}$, *where* $\varepsilon = \pm$ *except for* $k = \frac{d}{2}$ *when* $\varepsilon = +$ *only.*

The next result extends transitivity results in [GrPO2^d] to a wider class of sublattices.

Procedure 1.3. (Conjugacy for involution fixed point sublattices and recognition criteria for such) Two RSSD sublattices M_1, M_2 of BW_{2^d} are in the same orbit of G_{2^d} if and only if their associated involutions are conjugate. We may use 1.2 as a guide to orbits of $BRW^+(2^d)$ on RSSD sublatttices. In particular, whether two given RSSD sublattices are in the same orbit of $BRW^+(2^d)$ may be decided *within the lattice* by surveying a family of RSSD sublattices of BW_{2^d}. It is unnecessary to examine the explicit representation of the group $BRW^+(2^d)$. See 4.1.

Definition 1.4. In general, if X is a subobject of Y, the *inherited group* means the image in $Sym(X)$ of $Stab_{Aut(Y)}(X)$.

In the next result, this applies to the containment $L^\varepsilon(t) \leq L := BW_{2^d}$.

Theorem 1.5. *Consider a clean involution t of defect 1 on $L := BRW^+(2^d)$.*

When the trace of t is positive, the rank of $L^+(t)$ is $2^{d-2}3$. The automorphism group is inherited when $d \geq 2, d \neq 3$ and for $d = 3$ it is W_{B_6}.

When the trace of t is negative, the rank of $L^+(t)$ is 2^{d-2} and the fixed point sublattice is a scaled version of $BW_{2^{d-2}}$, whose automorphism group is $BRW^+(2^{d-2})$ if $d \neq 5$ and is W_{E_8} if $d = 5$.

Theorem 1.6. *The automorphism groups of the involution fixed point sublattices is inherited when the involution is dirty, split, of defect 1 and when $d \geq 5$ is odd.*

Theorem 1.7. *The automorphism groups of the involution fixed point sublattices is not inherited when the involution is nonsplit of defect is 1 and $d \geq 5$. The fixed point sublattices are isometric to $ssBW_{2^{d-2}} \perp ssBW_{2^{d-2}}$.*

The author thanks Alex Ryba for many useful discussions. The author has been supported by NSA grant USDOD-MDA904-03-1-0098.

2 Notation and terminology

We mention some special terminology, definitions and notation; see [GrPO2^d].

$BRW^0(2^d, \pm)$	Bolt, Room and Wall group, [GrPO2^d]
BW_{2^d}, the Barnes-Wall lattice in dimension 2^d	[GrPO2^d]
clean	an element of $BRW^0(2^d, \pm)$ not conjugate to its negative
D, a lower dihedral group	a dihedral group of order 8 in the lower group R
defect of an involution	A.5
density, commutator density	[GrPO2^d]
determinant of a lattice, L	$\|\mathcal{D}L\|$
diagonal	3.14
dirty	an element of $BRW^0(2^d, \pm)$ conjugate to its negative
$\mathcal{D}L$, discriminant group of an integral lattice L	$\mathcal{D}L = L^*/L$
ε_S	3.14
fourvolution	a linear transformation whose square is -1
$G = G_{2^d}$	$BRW^+(2^d)$
inherited	1.4
L^*, the dual of the lattice L	$\{x \in \mathbb{Q} \otimes L \mid (x, L) \leq \mathbb{Z}\}$
lower	in R
$R = R_{2^d}$	$O_2(BRW^+(2^d))$
sBW, sBW_{2^k}	scaled copy of some BW_{2^k} $\cong \sqrt{s}BW_{2^k}$ for some integer $s > 0$
ssBW, $ssBW_{2^d}$ (for a sublattice of BW_{2^d})	suitably scaled copy of BW_{2^k} = a scaled BW_{2^k} with scale 2^h, $h = \frac{d-k}{2}$ for $d-k$ even; $h = \frac{d-k-1}{2}$ for $d-k$ odd, d even; $h = \frac{d-k-1}{2} + 1$ for $d-k$ odd, d odd
SSD, semiselfdual, RSSD, relatively semiselfdual	applies to certain sublattices

total eigenlattice, $Tel(E), Tel(L,E)$	of an integral lattice; there are associated involutions the sum of the eigenlattices of an elementary abelian 2-group or involution E on the lattice L
upper	in G but not in R

Conventions. Our groups and most endomorphisms act on the right, often with exponential notation. Group theory notation is mostly consistent with [Gor, Hup, Gr12]. The commutator of x and y means $[x,y] = x^{-1}y^{-1}xy$ and the conjugate of of x by y means $x^y := y^{-1}xy = x[x,y]$. These notations extend to actions of a group on an additive group.

Here are some fairly standard notations used for particular extensions of groups: p^k means an elementary abelian p-group; $A.B$ means a group extension with normal subgroup A and quotient B; $p^{a+b+\cdots}$ means an iterated group extension, with factors $p^a, p^b, \ldots$ (listed in upward sense); $A{:}B, A{\cdot}B$ mean, respectively, a split extension, nonsplit extension.

3 Preliminaries

3.1 Groups

Definition 3.1. The *Dickson invariant* is a natural homomphism $O^+(2d,2) \to \mathbb{Z}_2$ which has the property that it is nontrivial on orthogonal transvections. (For an exact definition, see [Dieud]). The kernel is the subgroup $\Omega^+(2d,2)$. Elements of the latter group are called *even* and elements of $O^+(2d,2)$ which are not even are called *odd*.

This notion extends to the full holomorph $2^{1+2d}.O^+(2d,2)$ in $GL(2^d,\mathbb{C})$, so that the BRW group $BRW^+(2^d)$ is considered its even subgroup [GrMont].

Notation 3.2. From now on, $d \geq 2$, $G_{2^d} := BRW^+(2^d)$, $R_{2^d} := O_2(G_{2^d})$. Reference to d will typically be suppressed and we use G for G_{2^d} and R for R_{2^d}.

Lemma 3.3. *Let t be an isometry of V, a vector space in characteristic 2 with an alternating bilinear form. Then $[V,t] = \mathrm{Im}(t-1)$ is totally isotropic.*

Proof. Let $x, y \in V$. Then $(x(t-1), y(t-1)) = (x,y) - (x, yt) - (xt, y) + (xt, yt)$. Since we are in characteristic 2 and t is an isometry, the first and last terms cancel. Since $t^2 = 1$, the middle two terms cancel. □

Remark 3.4. When t leaves invariant a quadratic form associated to the alternating bilinear form, the totally isotropic space of 3.3 may be totally singular or not.

Notation 3.5. Let R be an extraspecial group and H a subgroup of R which contains $Z(R)$. Then H has a central product decomposition, $H = AB$, where $A = Z(H)$ and $B = Z(R)$ or B is extraspecial. Clearly, $A \cap B = Z(R)$. The group B is not unique if $A > Z(R)$, but the set of such B forms an orbit under $Stab_{Aut(R)}(H)$ if A is elementary abelian. We call such a decomposition of H a *CMZ-decomposition* (for complement modulo the center) and such a B is called a *CMZ-subgroup*.

Lemma 3.6. *An involution t which acts on an extraspecial group $R \cong 2_+^{1+2d}$ as an even automorphism fixes a noncentral involution if $d \geq 2$.*

Proof. If t is inner, this is obvious. Suppose that t acts nontrivially on the Frattini factor of R. Since $[R,t]$ is not contained in $Z(R)$ and is normal in R, $Z(R) \leq [R,t]$. Also, $[R,t]$ is abelian (by 3.3). Since t inverts a set of generators for $[R,t]$, it inverts $[R,t]$, so centralizes $\Omega_1([R,t])$. Also, $[R,t]$ is noncyclic since for even orthogonal transformations, the space of fixed points is even dimensional (see A.5). This completes the proof. □

Lemma 3.7. *Let t be an upper involution in the automorphism group of an extraspecial 2-group of plus type. Then t centralizes a maximal elementary abelian subgroup if and only if its image in the outer automorphism group is even and $[R,t]$ is elementary abelian.*

Proof. The necessity follows from the well-known facts that $\Omega^+(2d,2)$ has two orbits on maximal totally singular subspaces and that they are fused by $O^+(2d,2)$ [GrElAb].

We now prove sufficiency. We may assume that the order of the extraspecial group R is 2^{1+2d}, for $d \geq 2$ (there are no even upper involutions for $d = 1$). Let t be an upper involution.

The action of t fixes a noncentral involution $u \in R$, by 3.6. So, t acts on $C_R(u)/\langle u \rangle \cong 2_+^{1+2(d-1)}$. If $u \notin [R,t]$, then t acts evenly on this extraspecial group and we finish by induction. Therefore, we are done if t fixes an involution outside $[R,t]$, so suppose that none exist. Then since R has plus type, $[R,t]$ has order 2^{d+1}. Since t inverts $[R,t]$, we are done since $[R,t]$ is elementary abelian by hypothesis, so is maximal elementary abelian. □

Proposition 3.8. *We are given $V = \mathbb{F}^{2d}$ with quadratic form q and associated bilinear form $(\cdot,\cdot)$ so that $V = I \oplus J$ is a decompostion into maximal totally singular d-dimensional subspaces. Define $Inv(V, I)$ to be the set of involutions t in G, the orthogonal group for q, so that t is trivial on I and V/I and $[V,t] = I$. Then*

(0) $Inv(V, I) \neq \emptyset$ if and only if d is even.

(1) Assume that d is even. Then $Inv(V, I)$ is in bijection with these two sets:

(1.a) the set of $2d \times 2d$ matrices of the form $I_{2d} + N$, where N has rank d and is supported in the upper right $d \times d$ submatrix, which is alternating.

(1.b) The set of all sequences $v_1, w_1, \ldots, v_d, w_d$ with each $v_j \in J, w_j \in I$ so that $[v_i, t] = w_i$ for all i and $(v_i, w_j) = 0$ except for $\{i,j\}$ of the form $\{2k-1, 2k\}$ for $k = 1, \ldots, \frac{d}{2}$ in which case $(v_i, w_j) = 1$.

Proof. For (0), use A.5. The proof of (1) is formal. □

Definition 3.9. A *natural BRW subgroup* of G is a subgroup of the form $C_G(S)$, where S is a plus type extraspecial subgroup of R. Natural BRW subgroups occur in pairs, each member being the centralizer in G of the other.

We need to discuss normalizers of lower elementary abelian subgroups in G and centralizers of clean upper involutions.

Proposition 3.10. *Let E be a lower elementary abelian group of order 2^{a+b}, where $2^a = |Z(R) \cap E|$. Let $N := N_G(E)$ and $C := C_G(E)$. Suppose that $b \geq 1$. Then N and C have the following structure.*

There are subgroups $S, T \leq R$ and $P \leq G$ so that

(i) T and S are extraspecial of respective orders $2^{1+2(d-b)}, 2^{1+2b}$ (though $T = 1$ if $b = d$), $[T, S] = 1$ and $R = TS$;

(ii) $EZ(R)$ is maximal elementary abelian in S; it follows that $TEZ(R) = C_R(E)$.

(iii) the group $P := C_N(C_R(E)/EZ(R)) \cap N_N(E_0)$, where E_0 complements $Z(R) \cap E$ in E, satisfies $P \cap S = EZ(R)$ and $P/T \cong 2^{\binom{b}{2}+b(2d-2b)}{:}GL(2b, 2)$;

(iv) $C_C(S) = C_G(S)$ is the natural BRW-subgroup containing T;

(v) $C_P(T)S/S$ has the form $2^{\binom{b}{2}}{:}GL(2b, 2)$.

(vi) $C = O_2(P)C_G(S)$;

(vii) if $a = 0$, $N = CP$ and if $a = 1$, $N = CSP$.

Definition 3.11. *Given an involution t in an orthogonal group over a field of characteristic 2, an MNS-subspace for t (minimal nonsingular) is a nontrivial, nonsingular subspace which is t-invariant, and no proper subspace of it has these properties.*

Lemma 3.12. *Let t be an involution in the orthogonal group $\Omega^\varepsilon(2e, 2)$ and S a MNS-subspace for t. Suppose that t acts nontrivially on S.*

Either S has dimension 2 and a basis u, v so that $u^t = v$ and $(u, v) = 1$, so that u and v are both singular or both nonsingular; or S has dimension 4 and a basis u_1, u_2, v_1, v_2 of singular vectors so that $v_1^t = v_2, u_1^t = u_2$ and the Gram matrix for this basis is

$$\begin{pmatrix} 0 & 0 & 1 & 0 \\ 0 & 0 & 0 & 1 \\ 1 & 0 & 0 & 0 \\ 0 & 1 & 0 & 0 \end{pmatrix}.$$

Furthermore both of the above spaces are MNS-subspaces.

Proof. We may suppose that $\dim(S) \geq 4$ and that for every singular vector $v \in S$, $(v, v^t) = 0$, then try to get the last conclusion. We note that S is spanned by its singular vectors.

Take a singular vector v_1 not fixed by t and define $v_2 := v_1^t$. Choose a singular vector $u_1 \in S$ so that $(v_1, u_1) = 1$ and $(v_2, u_1) = 0$. Using t-invariance, we find that the sequence $v_1, v_2, u_1, u_2 := u_1^t$ has Gram matrix

$$\begin{pmatrix} 0 & 0 & 1 & 0 \\ 0 & 0 & 0 & 1 \\ 1 & 0 & 0 & b \\ 0 & 1 & b & 0 \end{pmatrix}.$$

This matrix is nonsingular, whence S has dimension just 4. Now, if $b \ne 0$, $\text{span}\{v_1 + u_2, u_1 + v_2\}$ is a 2-dimensional MNS-subspace. Therefore, $b = 0$. Since $S(t-1)$ is totally singular, S is minimal. $\square$

Lemma 3.13. *Let u be an involution in $\Omega^+(2e, 2)$ of defect e. There exists a maximal totally singular subspace F so that $F \cap F^u = 0$.*

Proof. Take an MNS-subspace for S. Then t acts nontrivially on S since the defect is e. Also, t leaves invariant the summands of the decomposition $S \perp S^\perp$. We are therefore done by induction if we check it for the cases of 3.11. This is trivial for the 2-dimensional case and for the 4-dimensional case, take the span of the second and third basis elements. $\square$

Notation 3.14. On the rational vector space spanned by a Barnes-Wall lattice, we take a sultry frame F containing a basis labeled by affine space $\mathbb{F}_2^d$ [GrPO2^d]. For a subset S of the index set, define the orthogonal involution ε_S to be the map which is -1 at frame elements labeled by a member of S and 1 on the other frame elements. The set of such linear maps, for $S \in RM(2, d)$, forms the *diagonal group*, denoted $\mathcal{E}$ or $\mathcal{E}_d$. It is a subgroup of $BRW^+(2^d)$. The *defect of the codeword* c is the defect of the involution ε_c.

Definition 3.15. Recall that an involution in the BRW group $BRW^+(2^d)$ is *dirty* if it is conjugate to its negative and otherwise, it is *clean*; A.3. These properties are equivalent to having nonzero, zero trace, respectively, on the natural 2^d-dimensional module. Furthermore, if the trace is nonzero, it has the form $\pm 2^{d-k}$, where k is the defect A.5 of the involution. We call such an involution a (d, k)-*involution*. Any involution in the lower coset of such is also called a (d, k)-*involution*. (This terminology applies to dirty involutions in such a coset.)

The dimension of the space of commutators of a defect k diagonal involution with the translation group of $AGL(d, 2)$ is $2k$ since the translation group can be interpreted as a complement in R_{2^d} to the diagonal subgroup corresponding to $RM(1, d)$. The terms clean and dirty apply to codewords, according to whether the corresponding involutions are clean or dirty.

The term *absolute clean trace* or *positive clean trace* applies to any element of $BRW^+(2^d)$ and means, the absolute value of the trace of any clean element in its lower coset. So, the absolute clean trace is a power of 2 even if the element is dirty. We let $\mathcal{D}$ and $\mathcal{C}$, respectively, denote the set of dirty and clean codewords in $RM(2, d)$.

Proposition 3.16. *Let $u \in G$ be a clean (d, k)-involution, $k > 0$. Then*

(i) $C_G(u)$ has the following form: it is a subgroup of $N_G(E)$, where $E = [R, u]$ is a rank $2k + 1$ elementary abelian group as in 3.10; $C_G(u)$ corresponds to the natural $Sp(2k, 2)$ subgroup of $N_G(E)/C_G(E) \cong GL(2k, 2)$ associated to the identification of $R/C_R(E)$ with $E/Z(R)$ derived from commutation with t;

(ii) The involution $uR \in G/R$ has centralizer $C_G(u)R/R$.

Proof. (i) It is clear from 3.10 that $C_G(u)$ has this form, except possibly for the replacement of $GL(2k, 2)$ by $Sp(2k, 2)$. It is clear that commutation by u gives a linear isomorphism of S/E onto $E/Z(R)$ which makes these two spaces into dual modules for $C_G(u)$. The action of $C_G(u)$ is therefore symplectic on both. It suffices to show that there is a subgroup of $C_G(u)$ which acts on both as the full group $Sp(2k, 2)$.

We take an elementary abelian subgroup F of S so that $FZ(R) = F \times Z(R)$ is maximal elementary abelian and so that $F \cap F^u = 1$ (see 3.13). Then u acts on $H := C_{C_G(u)}(T) \cap N_G(F) \cap N_G(F^u)$, which has shape $2 \times GL(2(d-k), 2)$ (the shape is clearly of the form $2.GL(2k, 2)$ but is actually a direct product; see [GrPO2^d] or the Appendix). Clearly, $C_H(u)$ has shape $2 \times Sp(2k, 2)$.

(ii) This follows from noticing that the set of clean elements in uR is just the union of the R-conjugacy class of u with the R-conjugacy class of $-u$. □

Remark 3.17. The exact structure of centralizers for dirty involutions is not needed in this article, but we give a sketch.

There are three main kinds of dirty involutions: lower involutions (defect 0); upper split (positive defect, with elementary abelian commutator subgroup on R); (upper) nonsplit (positive defect, with exponent 4 commutator subgroup on R).

The centralizer of a lower involution has shape $[2 \times 2^{1+2(d-1)}]2^{2(d-1)}.\Omega^+(2(d-1), 2)$.

Let t be a dirty split upper involution. Then $t = ru$, where u is an upper involution and r is a lower involution from $R \setminus [R, u]$. The structure of $C_G(u)$ is discussed in 3.16. We have $C_G(t) \leq C_G(u)$, $C_R(t)$ has index 2 in $C_R(u)$ and $C_G(t)R/R$ is a natural subgroup of $C_G(u)R/R$ of shape $2^{2(d-2k)}{:}\Omega^+(2(d-2k), 2)$.

Let t be a nonsplit involution. Let S be a maximal extraspecial subgroup of $C_R(t)$. Then $C_R(S) \geq [R, t] = [C_R(S), t]$. Also, $C_R(t) = S \times E$, where E is elementary abelian and a complement in $\Omega_1([R, t])$ to $Z(R)$. We say t has *plus type* or *minus type* according to the type of the extraspecial group S. Now, $N_G([R, t]) \geq R$ and $N_G([R, t])/R$ modulo its unipotent radical has the form $\Omega^+(2(d-2k), 2) \times GL(2k-1, 2)$. The image of $C_G(t)$ in the latter quotient has the form $\Omega^+(2(d-2k), 2) \times O(2k-1, 2)$.

3.2 The codes $RM(2, d)$ and the diagonal group

Our vector spaces are finite dimensional. We shall mix styles at times, so that a codeword may be written in lower case (when we think of it as a vector) or upper case (if we think of it as a geometric structure, like an affine subspace).

Notation 3.18. The *Reed-Muller code* $RM(k, d)$ is the binary code indexed by affine space $\mathbb{F}_2^d$ and spanned by all affine subspaces of codimension k. Its dimension is $\sum_{i=0}^{k} \binom{d}{i}$.

Definition 3.19. A *midset* is a codeword in $RM(2, d)$ of size 2^{d-1}. A midset is *nonaffine* if it is not a codimension 1 affine subspace. A codeword is *short* if its weight is less than 2^{d-1}. A codeword is *long* or *tall* if its weight is more than 2^{d-1}.

Lemma 3.20. *Let $t \in G$ be an involution so that $[R, t]$ is elementary abelian and $\mathcal{E}$ a given diagonal group. Then there is a conjugate of t in $\mathcal{E}$, unless possibly*

d is even and t has defect $\frac{d}{2}$, in which case there exists another diagonal group containing t.

Proof. Use 3.7 and the fact that $C_R(t)$ is nonabelian if and only if $C_R(t)$ contains representatives of both G-conjugacy classes of maximal elementary abelian subgroups of R. $\square$

Notation 3.21. We will study the action of $AGL(d, 2)$ on $\mathbb{F}_2^d$ and various codes. Let $T := T(d, 2)$ denote the translation subgroup and $GL(d, 2)$ the stabilizer of some origin (understood from context).

Definition 3.22. Linear subspaces U_i of a vector space are *independent* if their sum is their direct sum. Linear subspaces U_i of a vector space are *coindependent* if their annihilators in the dual space are independent.

This definition extends to a collection of affine subspaces U_i of a vector space, provided their common intersection is nonempty. One then chooses any origin in $\bigcap_i U_i$ and uses the above definition (which is independent of choice of origin).

Lemma 3.23. *Suppose that we have $k \geq 1$ linearly coindependent codimension 2 affine subspaces $S_1, \ldots, S_k$ in $\mathbb{F}_2^d$ with nonempty common intersection. Then $|S_1 + \cdots + S_k| = 2^{d-1} - 2^{d-k-1}$. (Note: $k \leq \frac{d}{2}$ here.)*

Proof. Let $a(d, k)$ be $2^{d-1} - 2^{d-k-1}$. We use induction on k. The result is trivial for $k = 1, 2$. We may assume that the spaces contain a common origin, so are linear.

Assume that $k \geq 3$ and that the formula holds by induction for $k - 1$. We have $S_k \cap (S_1 + \cdots + S_{k-1}) = S_1 \cap S_k + \cdots + S_{k-1} \cap S_k$, which, by induction on d and coindependence in $S_k \cong \mathbb{F}_2^{d-2}$, has cardinality $a(d-2, k-1)$. It follows that $|S_1 + \cdots + S_k| = 2^{d-2} + a(d, k-1) - 2a(d-2, k-1) = a(d, k)$. $\square$

Definition 3.24. A set of codimension 2 subspaces as in 3.23 is called a *cubi sequence of codimension 2 spaces*. Their Boolean sum is called a *a cubi sum*. [1]

Notation 3.25. Let c be a clean codeword of defect k. Let

$$Cubi(c) := \left\{ (S_1, \ldots, S_k) \mid \bigcap_{i=1}^{k} S_i \neq \emptyset, S_1, \ldots, S_k \text{ are coindependent affine} \right.$$

$$\left. \text{codimension 2 subspaces, and } \sum_{i=1}^{k} S_i = c \right\},$$

the set of *cubi expressions* of c, i.e. the set of ordered cubi sequences as above whose sum is c.

Corollary 3.26. *Given any integer $j \in [0, \frac{d}{2}]$, there is an involution of defect j in the diagonal group.*

---

[1] We chose the term *cubi* because our theory suggested the remarkable cubi sculpture series by David Smith. See also the footnote at 3.34.

Proof. If $j = 0$, take a lower involution. Suppose $j > 0$. Then take $\varepsilon_{S_1+\cdots+S_j}$, in the notation of 3.23. □

Next, we show explicitly how to realize a dirty class associated to the clean class within the diagonal group.

Lemma 3.27. *Given $d \geq 3$ and $k \geq 1$ and a length k cubi sequence in $\mathbb{F}_2^d$, there exist hyperplanes whose sum with the cubi sum has cardinality 2^{d-1}. In fact, any hyperplane which neither contains nor avoids the cubi intersection meets this condition.*

Proof. Let $S_1, \ldots, S_k$ be our cubi sequence and let $U := \bigcap_{i=1}^{k} S_i$. Let $\mathcal{N}$ be the set of hyperplanes which neither contain U nor avoid U. Then $|\mathcal{N}| = 2^{d+1} - 2^{2k+1}$. This is positive for $d \geq 3$ and $k \geq 1$.

Let $H \in \mathcal{N}$. Then the spaces $S_i \cap H$ have codimension 2 in H. They are coindependent with respect to H since $H \cap U$ has codimension 1 in U. Therefore, 3.23 gives $|H \cap (S_1 + \cdots + S_k)| = |(S_1 \cap H) + \cdots + (S_k \cap H)| = 2^{d-2} - 2^{d-k-2}$. Consequently, $|H + S_1 + \cdots + S_k| = 2^{d-1} + 2^{d-1} - 2^{d-k-1} - 2(2^{d-2} - 2^{d-k-2}) = 2^{d-1}$. □

Remark 3.28. The codeword of weight 2^{d-1} constructed in the proof of 3.27 is not a hyperplane, since the Boolean sum of two distinct nondisjoint hyperplanes is a hyperplane and $|S_1 + \cdots + S_k| < 2^{d-1}$.

We next need to work from a nonaffine midset to the class of clean codewords that it comes from.

Definition 3.29. Let $d \geq 3$. Given a nonaffine midset a, a hyperplane h so that $a + h$ is clean is called a *cleansing hyperplane for h*. It follows that if a has defect k, and h is cleansing, then $|a \cap h| = 2^{d-2} \mp 2^{d-k-2}$. (Note that $d - k \geq 2$ for $d \geq 3$.)

Lemma 3.30. *Every coset of $RM(1,d)$ in $RM(2,d)$ contains a clean codeword.*

Proof. Take a nontrivial coset, say $u + RM(1,d)$ and take a complement S in $RM(1,d)$ to the 1-space spanned by the universe. The subgroup of the diagonal group corresponding to S has 1-dimensional fixed point sublattice, so the sum of the traces of its elements is 2^d. Assume that the lemma is false. Then every element of $\langle u, S \rangle \setminus S$ gives a diagonal map of trace 0. Therefore the sum of the traces for the subgroup of the diagonal group corresponding to $\langle u, S \rangle$ is 2^d, which is impossible since this number must be divisible by 2^{1+d}. □

Lemma 3.31. *If $c \in RM(2,d)$ is clean, the number of its conjugates by R is 2^{2k}, where c has defect k.*

Proof. This is just the correspondence of the R-orbit of c under the action of conjugation on $RM(2,d)$ with the cosets of $C_R(c)$ in R, together with the definitions of defect and cleanliness. □

Proposition 3.32. *In a given coset $c + RM(1,d)$, where c is clean and has defect k, the number of clean codewords is 2^{2k+1} and the number of dirty codewords is $2^{d+1} - 2^{2k+1}$.*

Proof. If $c \in RM(2,d)$, the number of its transforms by R is 2^{2k}, by 3.31. The coset $c + RM(1,d)$ also contains the same number of transforms of the complement $c + \mathbb{F}_2^d$, which is also clean.

We use the irreducible module for G, which is a 2^d-dimensional complex vector space, and the trace function Tr on it. The previous paragraph implies that the sum $s(c) := \sum_{v \in c + RM(1,d)} \text{Tr}(v)^2$ is at least $2 \cdot 2^{2k+2(d-k)} = 2^{2d+1}$.

Since the group $RM(1,d)$ acts on the 2^d-dimensional complex vector space so as to afford all linear characters nontrivial on the center, each with multiplicity 1, it follows from orthogonality relations for the group generated by R and c that each $s(c) = 2^{2d+1}$. The coset therefore has 2^{2k+1} clean elements and $2^{d+1} - 2^{2k+1}$ dirty elements. □

Corollary 3.33. *The number of cleansing hyperplanes for a dirty codeword* $s \in RM(2,d)$ *is* 2^{2k+1}, *where k is the defect of any clean involution in the coset* $s + RM(1,d)$. *Thus the set $\mathcal{N}$ of 3.27 is the full set of noncleansing hyperplanes.*

Example 3.34. Let $d = 4, k = 1$ and let S be a defect 1 (nonaffine) midset. There are 8 cleansing hyperplanes. Write $S = A + H$, where A is short and H a cleansing hyperplane of S (this involves half the cleansing hyperplanes). Then A is a 4-set (hence an affine hyperplane) and $S \cap H$ is a 2-set. This set is stable by translation with elements of the core. Therefore, S is a union of four cosets of $S \cap H$. The assignment $H \mapsto S \cap H$ is one-to-one from the set of cleansing hyperplanes such that $S + H$ is short. By counting, this is a bijection. The union of any two sets $S \cap H$, as H varies, is an affine 2-space. Therefore, S is the disjoint union of a pair of disjoint, nonparallel affine 2-spaces, in three different ways. [2]

Corollary 3.35. *Given cleansing hyperplanes H_1, H_2 for the dirty codeword S, if $H_1 \cap S = H_2 \cap S$, then $H_1 = H_2$, i.e., for cleansing hyperplanes, H, the map $H \mapsto H \cap S$ is monic.*

Proof. If H_1 and H_2 are distinct, then, since they meet, their sum is a hyperplane. Since $H_1 + H_2$ is contained in the complement of S, it equals the complement of S. This is a contradiction since S is not affine. □

Procedure 3.36. *We now have a procedure to determine the orbit of a dirty codeword.* It depends only on examining the code, not the action of the group $AGL(d, 2)$. Call such a codeword v. Add to v all of the $2^{d+1} - 2$ affine hyperplanes. A nonempty set of these will be cleansing and the corresponding sums will have weight of the form $2^{d-1} \pm 2^{d-k-1}$, which will give the defect k. This procedure is exponential in d.

Lemma 3.37. *Two short (resp. long) clean codewords of the same defect are in the same orbit under $AGL(d, 2)$. A short clean codeword is a cubi sum.*

Proof. We interpret these codewords by their actions on the commutator quotient of R. The result follows from transitivity of the natural action of $GL(d, 2)$ on alternating matrices of the same rank. □

---

[2] These configurations also suggest the David Smith cubi theme; see 3.24.

Lemma 3.38. *Suppose that we are given* $(S_1, \ldots, S_k) \in Cubi(c)$ *as in 3.25. The subspace* $\bigcap_{i=1}^{k} S_i$ *has dimension* $d - 2k$ *and is the subgroup of the group of translations which fixes* c. *This subspace depends on* c *only, not on a choice from* $Cubi(c)$.

Proof. Clearly, the above intersection is a linear subspace and translations by it fix each S_i, hence also fix c. Since the space of commutators of the translation group with c has dimension $2k$, no translations outside this subspace fixes c. Therefore, this intersection depends on c only. □

Lemma 3.39. *The stabilizer in* $AGL(d, 2)$ *of the clean codeword* c *of defect* k *is transitive on* $Cubi(c)$, *and the the stabilizer of a member of* $Cubi(c)$ *has shape* $2^{d-2k}.2^{2k(d-2k)}[(\prod_{i=1}^{k} GL(2, 2)) \times GL(d - 2k, 2)]$.

Proof. The initial 2^{d-2k} refers to the group of translations which stabilize $\bigcap_{i=1}^{k} S_i$. The result follows from transitivity of $GL(d, 2)$ on ordered direct sums of k 2-spaces in the dual. □

Definition 3.40. The *core* of a clean codeword is $\bigcap_{i=1}^{k} S_i$, where $(S_1, \ldots, S_k) \in Cubi(c)$. The definition is independent of choice from $Cubi(c)$, by 3.38.

Theorem 3.41. *The stabilizer of a clean codeword of defect k in $AGL(d, 2)$ is a group of the form* $[2^{(1+2k)(d-2k)}]{:}[Sp(2k, 2) \times GL(d - 2k, 2)]$. *It has two orbits on* $\mathbb{F}_2^d$, *namely the core and its complement.*

Proof. The second statement follows from the structure of the stabilizer, which we now discuss.

We may think of our clean codeword c as a cubi sum for cubi sequence $(S_1, \ldots, S_k)$. Choose an origin in the core 3.40, i.e., the $(d-2k)$-space $U := S_1 \cap \cdots \cap S_k$.

Let H be the stabilizer of c in $AGL(d, 2)$. Then $H_t := H \cap T$ is transitive on U. The last paragraph implies that $H = H_t H_0$ where H_0 is the stabilizer of the origin. So, H_t corresponds to U and H_0 lies in the stabilizer in $GL(d, 2)$ of the subspace U, a parabolic subgroup P of the form $2^{2k(d-2k)}{:}[GL(2k, 2) \times GL(d-2k, 2)]$. Note that $O_2(P)$ is a tensor product of irreducibles for the two factors, so is irreducible.

We next argue that H_0 is a natural $2^{2k(d-2k)}{:}[Sp(2k, 2) \times GL(d-2k, 2)]$-subgroup of P.

Consider $C_G(t)$, where t is the diagonal matrix ε_c. Then we have the CMZ decomposition 3.5 for $C_R(t)$ and a related one for R: $R = R_1 R_0$, where $[R_0, R_1] = 1$, $C_R(t) = C_1 R_0$, where R_0 is extraspecial, and $C_1 \leq R_1$ and C_1 is elementary abelian and contains $Z(R)$. There is a corresponding product $J_0 J_1$ of commuting natural BRW subgroups, with $R_i = O_2(J_i), i = 1, 2$. We have $|C_1| = 2^{2k+1}$ and $C_1 = [R, t] = [R_1, t]$. The action of t preserves R_1 and the maximal elementary abelian subgroup C_1. Also, t acts on $N_{J_1}(C_1) \cong 2^{1+4k} 2^{\binom{2k}{2}} GL(2k, 2)$. There is a pair of maximal elementary abelian subgroups B_1, B_2 so that $R_1 = B_1 B_2, B_1 \cap B_2 = Z(R)$ and t interchanges B_1 and B_2 (see 3.13).

Choose $D_i \leq B_i$ so that $B_i = D_1 \times Z(R)$ and t interchanges D_1 and D_1. The common stabilizer of D_1 and D_2 in $Aut(R_1)$ has the form $2 \times GL(2k, 2)$.

The action of t has fixed point subgroup of the form $2 \times Sp(2k,2)$ because D_1 and D_2 are in t-invariant duality. Therefore, the image of H in the left factor of $P/O_2(P) \cong GL(2k,2) \times GL(d-2k,2)$ contains a copy of $Sp(2k,2)$. Since the image of H in the left factor stabilizes a nondegenerate form, the image is exactly $Sp(2k,2)$.

We claim that the stabilizer of c in $AGL(d,2)$ contains the natural $GL(d-2k,2)$ subgroup which commutes with the above copy of $Sp(2k,2)$. This follows since the stabilizer of a member of $Cubi(c)$ involves a copy of $GL(d-2k,2)$ which acts faithfully on the core and commutes with the action of the above $Sp(2k,2)$, which acts trivially on the core and faithfully on a complement to the core (meaning, on a linear complement, assuming the origin is chosen from the core).

The claim implies that H maps onto the right factor of $P/O_2(P) \cong GL(2k,2) \times GL(d-2k,2)$. It follows that $O_2(P)$ is an irreducible module for H (a tensor product of irreducibles for the factors $Sp(2k,2)$ and $GL(d-2k,2)$), whence $H \cap O_2(P)$ is either 1 or $O_2(P)$. The latter group preserves all cosets of U in $\mathbb{F}_2^d$ and each S_i is a union of such cosets, whence $O_2(P) \leq H$. $\square$

Lemma 3.42. *Two dirty codewords of the same defect are in the same orbit under $AGL(d,2)$.*

Proof. This is obvious from 3.27 and how the stabilizer of the core in $AGL(d,2)$ acts on $\mathbb{F}_2^d$. $\square$

Remark 3.43. The main theorems 1.1 and 1.2 follow from 3.37, 3.42, A.14, A.13. Note that, as a corollary, we get the well-known result that the minimum weight codewords in $RM(2,d)$ are the affine codimension 2 subspaces.

Proposition 3.44. *Let c be a clean codeword of defect k.*

(i) The stabilizer in $AGL(d,2)$ of the coset $c + RM(1,d)$ is $T(d,2)S$, where $T(d,2)$ is the full translation group and S is the stabilizer of c in $AGL(d,2)$ (see 3.41).

(ii) Let $s \in c + RM(1,d)$ be a dirty codeword. The commutator space $[T(d,2),s]$ has dimension $2k$ The stabilizer of s in $AGL(d,2)$ is a subgroup of S of index $2^{d+1} - 2^{2k+1}$ of shape $[2^{(1+2k)(d-2k-1)}]{:}[Sp(2k,2) \times AGL(d-2k-1,2)]$. It is $Stab_S(h)$, where $h = s + c$ is an affine codimension 1 subspace which meets the core of c in a codimension 1 subspace of it. The initial $2^{1 \cdot (d-2k-1)}$ corresponds to translations by the intersection of the core of c with a cleansing hyperplane.

Proof. (i) This is clear since the set of clean elements in $c + RM(1,d)$ is just the set of 2^{2k} $T(d,2)$-transforms of c.

(ii) Since s is dirty, $d - 2k > 0$.

Consider the set $\mathcal{P}$ of all pairs $(s,r) \in c + RM(1,d)$ so that s is dirty, r is short and clean (whence $s + r$ is a hyperplane, so is a cleansing hyperplane; 3.29). We refer to 3.41. Let H be the stabilizer of this coset in $AGL(d,2)$. Then H acts transitively on $\mathcal{P}$, which has cardinality $(2^{d+1} - 2^{2k+1})2^{2k}$, so $Stab_H((s,r))$ has index $2^{d+1} - 2^{2k+1}$ in $Stab_H(r)$, which has form $[2^{(1+2k)(d-2k)}]{:}[Sp(2k,2) \times GL(d-2k,2)]$.

Now, consider a hyperplane h in $\mathbb{F}_2^d$ which meets U in a codimension 1 subspace of U. By 3.27, $r + h$ is a midset, so $(r + h, r) \in \mathcal{P}$. Since $H_{(r+h,h)}$ stabilizes

h, it follows that $H_{(r+h,r)}$, hence every $H_{(s,r)}$, has the form $[2^{(d-2k-1)+2k(d-2k)}]$: $[Sp(2k,2) \times AGL(d-2k-1,2)]$. □

4 The conjugacy classes of involutions in G_{2^d} and orbits on RSSD sublattices

We continue to let $G := G_{2^d}$, $R := R_{2^d}$ and let $t \in G$ be an involution. We summarize the conjugacy classes of involutions.

Suppose that t centralizes a maximal elementary abelian subgroup (so is in a diagonal group). For each maximal elementary abelian subgroup E of $C_R(t)$, we have representatives of $\lfloor \frac{d}{2} \rfloor$ clean classes of upper involutions in a diagonal group $C_G(E)$. Upper involutions of the same defect and trace are conjugate in G except for the case where d is even and the involutions have full defect $\frac{d}{2}$. Two such involutions are clean and are conjugate if and only if their traces are equal and maximal elementary abelian subgroups in their lower centralizers are in the same orbit under the even orthogonal group.

Suppose that t does not centralize a maximal elementary abelian subgroup. Then $[R,t]$ is abelian of exponent 4 and has order 2^{1+2k} for some $k \geq 1$. It is now clear from A.14, A.13, that t is conjugate to some $\eta_{2k,\pm}$; see A.7.

Procedure 4.1. In [GrPO2^d], we showed that two RSSD sublattices in BW_{2^d} which had the same rank, but unequal to 2^{d-1} (the clean case), are in the same orbit under $BRW^+(2^d)$ with the exception of two orbits for maximal defect $\frac{d}{2}$. Also, [GrPO2^d] treats the case of rank 2^{d-1} sublattices which are fixed points of lower involutions. We now give a procedure for determining when two RSSD sublattices are in the same orbit of $BRW^+(2^d)$ *which depends only on examining a restricted set of sublattices, not the whole group* $BRW^+(2^d)$. Besides the two given RSSD sublattices, we need to examine only the ones associated to lower involutions, which may be constructed directly, by induction.

Recall that for $d > 3$, the lower involutions in $BRW^+(2^3)$ are those RSSD involutions associated to $ssBW_{2^{d-1}}$ sublattices [GrPO2^d].

Here we deal with the general dirty case, i.e., rank 2^{d-1}, which represents many orbits. Their associated RSSD involutions are dirty, so if diagonalizable are conjugate to elements of the diagonal group supported by a midsize codeword. We assume that $d > 3$.

We are given a dirty RSSD sublattice. Multiply this involution by all lower involutions.

Suppose that a nonempty set of such products are clean involutions with common defect $k \in [0, \frac{d}{2}]$. Since the defect k is less than $\frac{d}{2}$, k determines the orbit of the sublattice, by 3.44. If $k = \frac{d}{2}$, there are two orbits, depending on which maximal elementary abelian lower group corresponds to the RSDD involution.

Suppose that no such product is clean. Then the involution is some $\eta_{2k,\pm}$. The subgroups $C_R(t)$ and $[R,t]$ determine k and the sign $\pm$ and so the orbit of the sublattice.

For completeness, we treat the case $d = 3$.

Proposition 4.2. *In $BW_{2^3} \cong L_{E_8}$, the orbits of W_{E_8} on RSSD sublattices are* (i) *those of $BRW^+(2^3)$ on RSSD sublattices of even rank, i.e., one for rank 2, three for rank 4 and one for rank 6; and* (ii) *four orbits, of respective ranks $1, 3, 5, 7$, which are sublattices generated by a root, a set of three orthogonal roots, and the annihilators of such sublattices.*

Proof. Note that the determinant 1 subgroup of W_{E_8} contains a natural $BRW^+(2^3)$ subgroup of odd index. For rank 2 and 6 sublattices, we are in the clean cases in $BRW^+(2^3)$. For rank 4, we are in the dirty cases, of which there are just three, associated to a nonsplit involution (see A.15(ii)), to a lower involution and an upper dirty involution.

There are two orbits of W_{E_8} on 4-sets of mutually orthogonal pairs consisting of roots and their negatives. One of these 4-sets spans a sublattice of BW_{2^3} which is a direct summand and the other spans a sublattice properly contained in a D_4-sublattice. These cases correspond in the above sense to the nonsplit and lower cases. The third case gives rank 4 sublattices not spanned by roots (A.15).

Now consider the case of odd rank fixed point sublattice, M. It suffices to do the ranks 1 and 3 cases. We use a lemma that if g is in a Weyl group and V is the natural module, then g is a product of reflections for roots which lie in $[V, r]$ [Car]. At once, this implies that the rank 1 lattice here is spanned by a root. Suppose now that $\text{rank}(M) = 3$. Let Φ be the set of roots in M. If there is a pair of nonorthogonal linearly independent roots, then Φ has type A_3 or $A_2 A_1$. Since $\mathcal{D}M$ is an elementary abelian 2-group, neither of these is possible. We conclude that Φ has type $A_1 A_1 A_1$. Since M is even, it must equal the sublattice spanned by Φ. We are done since W_{E_8} has a single orbit on subsets of three orthogonal roots in a root system of type E_8. $\square$

Remark 4.3. For simplicity, discuss the main theorems for ranks at most 3 so that we may later use the assumption $d \geq 4$, as needed.

When $d = 1$, the fixed point sublattice of any involution is 0 or a rank 1 lattice.

Assume $d = 2$. The dirty involutions in BW_{2^2} and their fixed point sublattices are analyzed in A.15. If $t \in BRW^+(2^2)$ is clean, its fixed point sublattice has rank 1 or 3. In these respective cases, the sublattice is spanned by a vector of norm 2 or 4 or is the orthogonal of such a rank 1 sublattice, so is a root lattice of type B_3 or C_3. See the proof of 4.2.

When $d = 3$, all fixed point sublattices are accounted for in the proof of 4.2. They are all orthogonal direct sums of indecomposable root lattices.

4.1 Containments in $RM(2, d)$

Lemma 4.4. *Let $A, B \in RM(2, d)$ and suppose that $0 \neq A < B \neq \mathbb{F}_2^d$. Let X^c denote the complement of the subset X of $\mathbb{F}_2^d$. Then one of the following holds:*

(i) *A is a codimension 2 subspace and B is a midset; or B^c is a codimension 2 subspace and A^c is a midset.*

Furthermore, (i) *happens for affine hyperplanes B for any $d \geq 3$, and for nonaffine midsets B exactly when B has defect 1 and $d \geq 3$, respectively.*

(ii) A is short and B is long, of respective cardinalities $2^{d-1} - 2^{d-k-1}$, $2^{d-1} + 2^{d-r-1}$, where $(k,r) = (1,1), (1,2), (2,1)$ or $(2,2)$. We summarize:

(k,r)	$\lvert A \rvert$	$\lvert B \rvert$	$\lvert A+B \rvert$
$(1,1)$	$2^{d-1} - 2^{d-2} = 2^{d-2}$	$2^{d-1} + 2^{d-2} = 2^{d-2}3$	2^{d-1}
$(2,1)$	$2^{d-1} - 2^{d-3} = 2^{d-3}3$	$2^{d-1} + 2^{d-2} = 2^{d-2}3$	$2^{d-3}3$
$(1,2)$	$2^{d-1} - 2^{d-2} = 2^{d-2}$	$2^{d-1} + 2^{d-3} = 2^{d-3}5$	$2^{d-3}3$
$(2,2)$	$2^{d-1} - 2^{d-3} = 2^{d-3}3$	$2^{d-1} + 2^{d-3} = 2^{d-3}5$	2^{d-2}

Note that cases $(1,2)$ and $(2,2)$ are dual in the sense that A and $A+B$ may be interchanged. Note that the case $(1,1)$ corresponds to (i) for the midset $A+B$ containing B^c. Note also that A in case $(1,2)$ and $A+B$ in case $(2,2)$ are codimension 2 affine spaces.

Proof. If B is a midset, and A is not a codimension 2 affine subspace, then $A < B$ implies that A has cardinality $2^{d-1} - 2^{d-k-1}$ for an integer k and A is a cubi sum in the sense of 3.37. Since $A + B = A \setminus B$ is also a codeword, it has cardinality $2^{d-1} - 2^{d-r-1}$ for an integer $r \geq 1$. It follows that $k = r = 1$. Then A and A^c are affine codimension 2 subspaces. Therefore, if B is a midset, (i) holds.

It is obvious that (i) happens in an essentially unique way when B is an affine hyperplane. Assume B is a midset but not affine. The codimension 2 affine subspaces A and A' whose union is B are not translates of each other. Let A'' be a translate of A' which meets A nontrivially. The intersection has codimension 1 or 2 in each of A or A'' and it is an exercise to show that for codimension 1, this situation does happen in an essentially unique way, and that it does not happen for $k = 2$ (reason: such subspaces are affinely coindependent and so an associated linear system expressing their intersection has a solution).

Assume that neither A nor B is a midset. In case both are long, we may replace with complements to assume both are short. In any case, we may assume that A is short, of cardinality $2^{d-1} - 2^{d-k-1}$, for some integer k, $0 < k \leq \frac{d}{2}$.

First assume that B is short, say of cardinality $2^{d-1} - 2^{d-r-1}$, for $r > k$. Then $A + B$ has cardinality $2^{d-k-1} - 2^{d-r-1} = 2^{d-r-1}(2^{r-k} - 1)$. Since $A + B$ is short, there exists an integer $s \leq \frac{d}{2}$ so that $2^{d-r-1}(2^{r-k} - 1) = 2^{d-1} - 2^{d-s-1} = 2^{d-s-1}(2^s - 1)$. If both sides are powers of 2, then $r = k+1$, $s = 1$ and $d - r - 1 = d - s - 1$ implies that $r = s = 1$ and $k = 0$, a contradiction. Therefore both sides are not powers of 2 and so $r = s$ and $s = r - k$ and so $s = r$ and $k = 0$, a final contradiction.

Therefore B is long, of cardinality $2^{d-1} + 2^{d-r-1}$, for $r > 0$. Then $A + B$ has cardinality $2^{d-r-1} + 2^{d-k-1}$. Since $r \geq 1, k \geq 1$, this number is at most 2^{d-1} and is less than 2^{d-1} if $(r, k) \neq (1, 1)$.

Suppose that $r = k$. Then $2^{d-r-1} + 2^{d-k-1} = 2^{d-r}$ is 2^{d-1} or 2^{d-2}, implying $r = k = 1, r = k = 2$, respectively.

Suppose that $r < k$. Then $A + B$ is short and there exists an integer $s \leq \frac{d}{2}$ so that $2^{d-r-1} + 2^{d-k-1} = 2^{d-1} - 2^{d-s-1}$, and $2^{d-k-1}(2^{k-r} + 1) = 2^{d-s-1}(2^s - 1)$. Now, $2^{k-r}+1$ is odd, so it follows that $s = k$, $k - r = 1$ and $s = 2$. So, $k = 2, r = 1$.

Suppose that $r > k$. Then $A + B$ is short and there exists an integer $s \leq \frac{d}{2}$ so that $2^{d-r-1} + 2^{d-k-1} = 2^{d-1} - 2^{d-s-1}$, and $2^{d-r-1}(2^{r-k} + 1) = 2^{d-s-1}(2^s - 1)$. It follows that $s = r$, $r - k = 1$ and $s = 2$. So, $k = 1, r = 2$. $\square$

4.2 About defect 1 midsets

Lemma 4.5. *Let $d \geq 3$. Suppose that B is a midset of defect 1. Then B contains affine hyperplanes of codimension 2. Suppose that A is an affine codimension 2 space contained in B. There exists a unique hyperplane H so that $B \cap H = A$. (The other two hyperplanes which contain A are cleansing hyperplanes for B 3.29.)*

Proof. Let A and A' be any pair of disjoint codimension 2 subspaces. Then $A + A'$ is a midset and it has defect 0, 1 or 2 if A' has a translate which meets A in codimension 0, 1 or 2, respectively. The first statement follows from 3.32 and transitivity of $AGL(d, 2)$ on midsets of a given defect 3.42.

For the second, consider the three hyperplanes H_1, H_2, H_2 which contain A. Suppose that $H_1 \cap B > A$. Then $|H_1+B| = |H_1|+|B|-2|H_1 \cap B| = 2^d - 2|H_1 \cap B| < 2^{d-1}$, whence H_1 is a cleansing hyperplane, and so $|H_1+B| = 2^{d-1}-2^{d-1-1} = 2^{d-2}$ and $|H_1 \cap B| = \frac{1}{2}(2^{d-1} + 2^{d-1} - 2^{d-2}) = 2^{d-3}3$. This means that at most two of the H_i meet B in a set larger than A. Therefore, since $H_i \setminus A$ for $i = 1, 2, 3$, partition $\mathbb{F}_2^d \setminus A$, exactly two of the H_i meet B in a set larger than A and so there exists an H which meets B in A, and by above counting, it is unique. $\square$

5 More group theory for BRW groups

We list some assumed results from group theory.

Lemma 5.1. (i) *A faithful module for $\prod_1^k Sym_3$ in characteristic 2 has dimension at least $2k$.*

(ii) *A faithful module for $Sp(2k, 2)$ in characteristic 2 has dimension at least $2k$.*

Proof. Let $K_1 \times \cdots \times K_k$ be the natural direct product of $K_i \cong Sp(2, 2) \cong Sym_3$ in $Sp(2k, 2)$. Clearly, (i) implies (ii). We prove (i).

We may assume that the field F is algebraically closed and that $k \geq 2$. Let M be a module of minimal dimension. Consider the decomposition $M = M' \oplus M''$, where $M' = [M, O_3(K_1)]$ and $M'' = C_M(O_3(K_1))$.

Clearly, $\dim(M')$ is a positive even integer. Suppose $M'' \neq 0$. Then by induction applied to the action of $K_2 \times \cdots \times K_k$ on M'', we have $\dim(M'') \geq 2(k-1)$ and we are finished. Suppose $M'' = 0$. Then we may decompose $M'' = P \oplus Q$ where P and Q represent the two distinct linear characters of $O_3(K_1)$. The actions of $K_2 \times \cdots \times K_k$ on P and Q are faithful and equivalent since P and Q are interchanged by elements of K_1. We now finish by induction. $\square$

Lemma 5.2. *Let $\mathbb{F}_2^{2m}$ have a nonsingular quadratic form of type $\nu = \pm$ and let $sv(m, \nu)$, $av(m, \nu)$ denote the number of singular and nonsingular vectors in the case of type $\nu = \pm$. Then $sv(m, \nu) = (2^m - \nu 1)(2^{m-1} + \nu 1)$ and $av(m, \nu) = (2^m - \nu 1)2^{m-1}$.*

Proof. Well-known. Note that $sv(m, \nu) + av(m, \nu) + 1 = 2^{2m}$. $\square$

Lemma 5.3. *Let $k \geq 2$. Let U be the essentially unique $2k+1$ dimensional $\mathbb{F}_2$-module for $Sp(2k,2)$ with socle of dimension 1 and quotient the natural $2k$-dimensional module. Then (i) U is the natural module for $O(2k+1,2)$; (ii) The orbits of $Sp(2k,2)$ on U consist of the two 1-point orbits lying in the radical, and the singular points and the nonsingular points. Each of the latter orbits form coset representatives for the nontrival cosets of the radical.*

Proof. This is mainly the 1-cohomology result [Poll], plus a standard interpretation of Ext^1. □

5.1 For clean involutions

We use the following notation throughout this subsection.

Notation 5.4. We have the clean upper involution t of defect $k \geq 1$. Take a CMZ decompostion $C_R(t) = PZ$. Denote by q_t the quadratic form on $Z = Z_t$ described in 5.3. The subscript indicates dependence on the involution, t. Call $z \in Z$ *singular* or *nonsingular*, according to the value of $q_t(z)$.

Lemma 5.5. *Use the notation of 5.4. For all $k \geq 1$, the set map $x \mapsto [x,t]$ takes $R \setminus C_R(t)$ to the set of nonsingular vectors in Z with respect to the invariant quadratic form.*

For $k \geq 2$, the action of $C_G(t)$ as $Sp(2k,2)$ on Z is indecomposable; the upper Löwey series has factors of dimensions $1, 2k$.

Proof. Let f be the commutator map $R \to Z$ defined by $f(x) := [x,t]$. Every coset of $Z(R)$ in Z contains an element of $\text{Im}(f)$. If $f(x) = f(y)$, we have $1 = f(x)f(y) = [x,t][y,t]$, which is congruent to $f(xy)$ modulo $\langle -1 \rangle$. If $f(xy) \in \langle -1 \rangle$, then $xy \in C_R(t)$. Therefore f maps $R/C_R(t)$ isomorphically onto $Z/\langle -1 \rangle$. Also, the image of f is a set of cardinality 2^{2k} which contains 1 and is invariant under $C_R(t)$, which acts on Z as $Sp(2k,2)$, i.e., $\text{Im}(f)$ is a $C_R(t)$-invariant transversal to $Z(R)$ in R.

We compute that $(*)$ $f(xy) = [xy,t] = [x,t]^y[y,t] = f(x)^y f(y)$.

We claim that Z is an indecomposable module for $Sp(2k,2)$. Suppose it is decomposable. Then $\text{Im}(f)$ must be either a subspace of Z complementing $Z(R)$ or essentially a coset of some $C_R(t)$-invariant subspace, say Z_0, namely it is the set Y which is the notrivial coset with -1 replaced by 1. Then there exists a homomorphism $h : R \to Z_0$ with the property that $f(x) = -h(x)$ if $x \notin C_R(t)$ and $h(x) = 1$ if $x \in C_R(t)$.

It can not be a subspace since $[R,t]$ is normal in R. So, the second alternative applies to $\text{Im}(f)$. Now, we shall get a contradiction, using $(*)$.

Note that we have an alternating bilinear form g on Z with values in $Z(R)$, defined by $g(a,b) := [a',t,b']$ where priming on $a \in Z$ means an element $a' \in R$ so that $f(a') = a$. It helps to think of the Hall commutator identity $[x,y^{-1},z]^y [y,z^{-1},x]^z [z,x^{-1},y]^x = 1$.

There is a g-totally singular subspace of dimension $k+1$ in Z, say W. Assuming that $\text{Im}(f) = Y$, we take any elements a,b,c in R so that $abc = 1$ and none of a,b,c is in $C_R(t)$. Then $f(a)f(b)f(c) = (-1)^3 h(a)h(b)h(c) = -1$. From $(*)$, we get $f(c) = f(ab) = f(a)^b f(c)$. Now choose a, b, c so that $f(a), f(b), f(c) \in W$ (this is possible since $k \geq 2$). Then $g(f(a), f(b)) = 1$ implies that $f(a)^b = f(a)$, which implies that $f(c) = f(a)f(b)$, in contradiction with $f(a)f(b)f(c) = -1$. This proves that Z is indecomposable.

At this point, we know that $\text{Im}(f)$ is one of two orbits for $Sp(2k, 2)$ in Z, the singular one and the nonsingular one. We claim that it is the singular one. Suppose otherwise. Take W and a, b, c as above. Then $(*)$ implies that (in additive notation) the sum of two orthogonal nonsingular vectors is nonsingular, a contradiction. □

Definition 5.6. Let ϕ be a linear character of Z which is nontrivial on $Z(R)$. Then $Ker(\phi)$ is a nonsingular quadratic space by restriction of q_t; see 5.4. Its *type* is plus or minus, according to the Witt index of the restriction of q_t.

Lemma 5.7. *Consider $X := \{(\varphi, z) | \varphi \in Hom(Z, \mathbb{F}_2), z \in Z\}$ and let $Y_{\varepsilon, \zeta, \eta} := \{(\varphi, z) \in X | \varphi(Z(R)) \neq 1, z \neq 1, q_t(z) = \zeta, type(\varphi) = \varepsilon, \varphi(z) = \eta\}$, for $\zeta, \eta \in \mathbb{F}_2$. Then $C(t)$ is transitive on $Y_{\varepsilon, \zeta, \eta}$, for $\zeta, \eta = 0, 1$.*

The orbit lengths are
$|Y_{\varepsilon,0,0}| = (2^{2k-1} + \varepsilon 2^{k-1})sv(k, \varepsilon);$
$|Y_{\varepsilon,0,1}| = (2^{2k-1} + \varepsilon 2^{k-1})av(k, \varepsilon);$
$|Y_{\varepsilon,1,0}| = (2^{2k-1} + \varepsilon 2^{k-1})av(k, \varepsilon);$
$|Y_{\varepsilon,1,1}| = (2^{2k-1} + \varepsilon 2^{k-1})sv(k, \varepsilon).$

Note that rows 2 and 3 are equal and rows 1 and 4 are equal.

Proof. It is well-known that $C(t)/O_2(C(t)) \cong Sp(2k, 2)$ acts with two orbits on characters of Z which take nontrivial value on $Z(R)$. These orbits have respective stabilizers the natural subgroups $O^\varepsilon(2k, 2)$ and respective lengths $2^{2k-1} + \varepsilon 2^{k-1}$. The rest follows from 5.2. □

Notation 5.8. Let $t \in G$ be an involution. Then $C_G(t)$ acts on each eigenlattice $L^\varepsilon(t)$. Its image in $O(L^\varepsilon(t))$ is denoted G_ε.

Lemma 5.9. *The action of $C_G(t)$ on $L^\pm(t)$ is irreducible. The center of G_ε is just $\{\pm 1\}$.*

Proof. The second statement follows from orthogonality of the representation plus absolute irreducibility, which we now prove. We prove irreducibility for a natural subgroup of $C_G(t)$ of the form AB, where $[A, B] = 1$, $A \cong 2^{1+2(d-k)}$, $Z \leq B, B/Z \cong Sp(2k, 2)$; see 3.5, 5.5. Every faithful irreducible of A has dimension 2^{d-2k}. The central involution of R is in Z and so every irreducible of B on $\mathbb{Q} \otimes L$ involves an orbit of characters of Z of cardinality $2^{2k-1} \pm 2^{k-1}$, and both orbit lengths occur with multiplicity 2^{d-2k}. Therefore, just two irreducibles for AB occur in $\mathbb{Q} \otimes L$, and they have respective dimensions $2^{d-2k}(2^{2k-1} \pm 2^{k-1}) = 2^{d-1} \pm 2^{d-k-1}$. The conclusion follows. □

Lemma 5.10. *Assume t is clean with positive trace. Let $z \in Z$, $z \neq \pm 1$. The trace of z on $L^\pm(t)$ is $\pm 2^{d-k-1}$ if z is q_t-singular and is $\mp 2^{d-k-1}$ if z is q_t-nonsingular.*

Proof. We use the subgroup denoted AB in the proof of 5.9. For AB, the module $L^\varepsilon(t)$ decomposes as a tensor product of irreducibles. It suffices to prove that the trace of z on the tensor factor irreducible for B is $\pm 2^{k-1}, \mp 2^{k-1}$, respectively.

Note that $2^{2k} - 1 - sv(k, \varepsilon) = (2^k - \varepsilon)(2^k - \varepsilon - (2^{k-1} + \varepsilon)) = (2^k - \varepsilon)2^{k-1}$.

We use 5.2 to deduce that

$$|Y_{\varepsilon,0,0}| = 2^{k-1}(2^k + \varepsilon)sv(k,\varepsilon) = 2^{k-1}(2^k + \varepsilon)(2^k - \varepsilon)(2^{k-1} + \varepsilon)$$

and

$$|Y_{\varepsilon,0,1}| = 2^{k-1}(2^k + \varepsilon)(2^{2k} - 1 - sv(k,\varepsilon)) = 2^{k-1}(2^k + \varepsilon)(2^k - \varepsilon)2^{k-1}.$$

Let $\Phi_\varepsilon := \{\varphi | \varphi(Z(R)) \neq \{1\}\}$. A given singular $z \in Z$ is in the kernel of $2^{k-1}(2^{k-1} + \varepsilon) = 2^{2k-2} + \varepsilon 2^{k-1}$ characters in Φ_ε and outside the kernel of 2^{2k-2} characters in Φ_ε. It follows that the trace of z on $L^\varepsilon(t)$ is $\varepsilon 2^{k-1}$. Singular and nonsingular elements of $Z \setminus Z(R)$ are paired by congruence modulo $Z(R)$. Therefore, nonsingular elements have trace $-\varepsilon 2^{k-1}$. $\square$

5.2 For dirty involutions

We assume the following notation throughout this subsection.

Notation 5.11. Let t be a dirty split upper involution of defect k. A *UL factorization of t* is an expression $t = u\ell$, where u is a clean involution and ℓ is a lower involution (note that all of t, u, ℓ commute). Write $\mathcal{UL}(t)$ for all pairs (u, ℓ) as above. Let $\mathcal{U}(t)$ be the set of u and let $\mathcal{L}(t)$ be the set of ℓ which arise this way. We have $|\{\mathcal{UL}(t)\}| = 2^{1+2(d-2k)+2k} - 2^{1+2k}$.

We get a result for traces of u and ℓ on $L^\varepsilon(t)$ which is similar to 5.10.

Lemma 5.12. *On $L^\varepsilon(t)$, the trace of $z \in Z \setminus Z(R)$ is 0 and the trace of ℓ is $\pm 2^{d-k-1}$, for all $\ell \in \mathcal{L}(t)$.*

Proof. We assume $\varepsilon = +$ (the other case is similar). It suffices to consider the sublattices $L(a, b)$, where u acts as a and ℓ acts as b. Recall that the eigenlattices for ℓ are $ssBW_{2^{d-1}}$ lattices, for which we may use 5.7 to compute the traces for z. Without loss, we may assume that z has nonnegative traces. We get:

sublattice	rank	multiplicity of +1 for z	multiplicity of -1 for z
$L(+1,+1)$	$2^{d-2} + 2^{d-k-2}$	$2^{d-3} + 2^{d-k-2}$	2^{d-3}
$L(+1,-1)$	$2^{d-2} + 2^{d-k-2}$	$2^{d-3} + 2^{d-k-2}$	2^{d-3}
$L(-1,+1)$	$2^{d-2} - 2^{d-k-2}$	$2^{d-3} - 2^{d-k-2}$	2^{d-3}
$L(-1,-1)$	$2^{d-2} - 2^{d-k-2}$	$2^{d-3} - 2^{d-k-2}$	2^{d-3}

$\square$

6 About inherted groups

We continue to use the notations $G := G_{2^d}$, $R := R_{2^d}$. See the ancestor section of [GrPO2^d] for discussion.

Notation 6.1. We use bars for images under restriction $C_G(t) \to O(L^\varepsilon(t))$. As in 5.8, we write $\overline{G_\varepsilon}$ for the image of $C_G(t)$ in $O(L^\varepsilon(t))$ under the restriction homomorphism.

Lemma 6.2. *Suppose that Z is an elementary abelian subgroup of R containing $Z(R)$ and that* $\mathrm{rank}(Z) = s + 1$. *Let L_λ be the eigenlattice for L, defined by the linear character λ of Z, which is assumed to be nontrivial on $Z(R)$. The set $\mathcal{F}$ of such λ has cardinality 2^s.*

There is a finite subgroup of the orthogonal group $O(\mathbb{Q} \otimes L^\varepsilon(t))$ of the form $\prod_\lambda R_\lambda$ with the property that R_λ acts on L_λ as a lower group and acts trivially on L_μ for $\mu \neq \lambda$. We have $|\prod_\lambda R_\lambda| = 2^{s(1+2(d-s))}$.
  (i) *When $s = 1$, $\overline{C_G(Z)} \geq \prod_\lambda R_\lambda$.*
  (ii) *When $s = 2$, $\overline{C_G(Z)} \cap \prod_\lambda R_\lambda$ is an index $2^{2(d-2)}$ subgroup of $\prod_\lambda R_\lambda$ with the property that if $\mathcal{J}$ is any 3-set in $\mathcal{F}$, then the projection of $\overline{C_G(Z)}$ to $\prod_{\lambda \in \mathcal{J}} R_\lambda$ is onto. The kernel of this homomorphism is just $Z(R_\mu)$, where $\mu \in \mathcal{F}$ is the index missing from $\mathcal{J}$.*

Lemma 6.3. *Let $\mathcal{I} \subseteq \mathcal{F}$ be any nonempty collection of characters as in 6.2 and let $J := J(\mathcal{I})$ be the direct summand of L determined by $\mathrm{span}\{J_\nu | \nu \in \mathcal{I}\}$. If $\lambda \in \mathcal{I}$ and $g \in C_G(Z)$ acts trivially on J_λ, then g acts on J as an element of the group $\prod_\lambda R_\lambda$, defined in 6.2.*

Proof. If μ, ν are any two distinct indices so that L_μ and L_ν are stable under $h \in G$, then if h acts trivially on L_μ modulo its first lower twist, then h does the same on L_ν. By considering all distinct pairs of indices $\mu, \nu \in \mathcal{I}$, we deduce that g acts on J as a member of $\prod_{\eta \in \mathcal{I}} R_\lambda$. See [GrPO2^d] □

Corollary 6.4. *Use the notation of 6.3. Assume that $s = 2$, $\mathcal{I}$ has cardinality 3 and $N_{O(J)}(\overline{Z}) = \overline{N_G(Z)} C_{O(J)}(\overline{Z})$. Then $N_{O(J)}(\overline{Z})$ is inherited.*

Proof. 6.3 and 6.2(ii). □

Corollary 6.5. *Let $t \in G$ be an involution and let $Z := Z(C_R(t))$.*
  (i) *Suppose that t is a clean involution of defect 1. Then $N_{O(L^\varepsilon(t))}(\overline{Z})$ is inherited.*
  (ii) *Suppose that t is a split dirty involution of defect 1. Then $N_{O(L^\varepsilon(t))}(\overline{Z})$ is inherited.*

Proof. Note that defect 1 implies that $s = 2$, in the notation of 6.2. (i): This follows from 6.4.

(ii): Let t be such an involution. Let $t = u\ell$ be a UL-factorization 5.11. We define Z_u as $Z(C_R(u))$ and define $Z := Z(C_R(t)) = Z_u \times \langle \ell \rangle$. A character value analysis shows that elements of Z_u have 0 trace on $L^\varepsilon(t)$ and elements of the coset $Z_u \ell$ have nonzero trace 5.12. Therefore, $N_{O(L^\varepsilon(t))}(\overline{Z}_u) \geq N_{O(L^\varepsilon(t))}(\overline{Z})$.

We shall use 6.4 to prove that $N_{O(L^\varepsilon(t))}(\overline{Z}_u)$ is inherited by showing that the latter group induces only $Sp(2,2)$ on Z. Assume that this is false. We have an action of $AGL(2,2) \cong Sym_4$ on Z. Let H be the linear group which $N_{O(L^\varepsilon(t))}(\overline{Z}_u)$ induces on Z_u. The action of $N_{C_G(t)}(Z_u)$ on Z_u preserves the coset $Z_u \setminus Z$ and has orbits modulo $Z(R)$ of lengths 1 and 3. Its orbits on $Z_u \setminus Z$ must have lengths 1,1,3,3 since elements in that coset have nonzero trace on $L^\varepsilon(t)$ so are not conjugate to their negatives. It follows that a Sylow 2-group S of $N_{O(L^\varepsilon(t))}(\overline{Z}_u)$ fixes an element, say ℓ, in this coset. If $x \in Z \setminus Z(R)$, then there exists $g \in S$ so that $x^g = -x$, since we are assuming an action of $AGL(2,2)$ on Z. It follows that $(x\ell)^g = -x\ell$, which is a contradiction since $(x\ell, xu)$ is a UL factorization of t (because $xu \in uZ = u^R$ consists of clean elements). $\square$

7 The split defect 1 cases

7.1 The clean defect 1 case

Definition 7.1. Suppose that M is an integral lattice and X is an SSD lattice. Define $SSD(M, X)$ to be the subgroup of $O(M)$ generated by the SSD involutions associated to sublattices of M which are isometric to X.

This is clearly a normal subgroup of $O(M)$.

We continue to use the notation 3.5. Since the defect is 1, rank$(Z) = 3$. The case $d = 3$ is treated in 4.2, so we assume $d \geq 4$.

Remark 7.2. In the notation of 7.1, if X is SSD and $\det(M) = 1$, then $M \cap X^\perp$ is SSD. This will apply for us when $M \cong BW_{2^d}$ and d is odd.

Lemma 7.3. Suppose that $d > 3$. Let t be a clean involution of defect 1 and positive trace. Then $SSD(L^+(t), ssBW_{2^{d-1}}) = Z$.

Proof. A sublattice X of $L^+(t)$ which is isometric to $ssBW_{2^{d-1}}$ is SSD in the overlattice L. By [GrPO2^d], the associated SSD involution is lower (here, we are using $d > 3$), so lies in $C_R(t)$. Since $Tr_{L^+(t)}(\varepsilon_X) \neq 0$ (see 5.10), $\varepsilon_X \in Z$, the only elements of $C_R(t)$ which have nonzero trace on $L^+(t)$. The action of $C_G(t)$ on Z is that of $O(2k+1,2)$ on its natural module 5.5. Therefore, every element of $Z \setminus Z(R)$ is such an SSD involution. $\square$

Lemma 7.4. Suppose that $d > 3$. Let t be a clean involution of defect 1 and positive trace. Then $O(L^+(t))$ is inherited.

Proof. By 7.3, $\overline{Z}$ is normal in $O(L^+(t))$. Now use 6.4. $\square$

Remark 7.5. If t is a clean involution of defect 1 and positive trace, $L^-(t) \cong ssBW_{2^{d-1}}$, whose automorphism group is known.

7.2 The split dirty defect 1 case

Lemma 7.6. Suppose that $d > 3$ and d is odd. Let t be a split dirty involution of defect 1. Then $SSD(L^\varepsilon(t), ssBW_{2^{d-2}}) = \overline{C_R(t)}$ and $\overline{Z}$ is a subgroup of $Z(SSD(L^\varepsilon(t), ssBW_{2^{d-2}}))$ which is normal in $O(L^\varepsilon(t))$.

Proof. We may suppose that $t = \varepsilon_b$ for a defect 1 midset $b \in RM(2,d)$ and we may assume that $\varepsilon = +$. Since d is odd, an $ssBW_{2^{d-2}}$ sublattice is SSD. Let X be such a sublattice of $L^\varepsilon(t)$. Its associated involution in $O(L)$ is conjugate to an involution of the form $\varepsilon_c \in \mathcal{E}$, where the codeword c is an affine codimension 2 subspace.

Since ε_c acts nontrivially on $L^+(t)$, $c \cap b = \emptyset$ Let b' be the complement of b. We may consider the involution ε_h where h is a hyperplane so that $h \cap b' = c$ (see 4.5). Then ε_c acts on $L^+(t)$ as ε_h, which is a lower involution.

Define K to be the normal subgroup of $O(L^+(t))$ generated by all $\overline{\varepsilon_X}$, where X is an SSD sublattice isometric to $ssBW_{2^{d-2}}$. This is a subgroup of $\overline{C_R(t)}$ which is normal in $\overline{C_G(t)}$ and contains $\overline{\varepsilon_h}$, so is not contained in $\overline{Z(R)}$. The normal subgroups are $\overline{1}, \overline{Z(R)}, \overline{Z}, \langle \overline{Z}, \overline{\ell} \rangle, \overline{C_R(t)}$, where $\ell \in \mathcal{L}(t)$ is any lower part of a UL-factorization. For all such normal subgroups, Y not contained in $Z(R)$, we claim that Z is normal in $N_{O(L^+(t))}(Y)$. This is obvious except when Y is one of the latter two cases. In those cases, $Z(Y) = \langle Z, \ell \rangle$. In the action on $L^+(t)$, the elements of $Z \setminus Z(R)$ have trace 0 and the elements of $Z\ell$ have nonzero trace (see 5.12). The claim follows and so does the lemma since K is normal in $O(L^+(t))$. □

Lemma 7.7. *Suppose that $d > 3$ and d is odd. Let t be a split dirty involution of defect 1. Then $O(L^\varepsilon(t))$ is inherited.*

Proof. Use 7.6 and 6.5(ii). □

This completes the proof of 1.6.

8 The nonsplit defect 1 case

The style of proof here is rather different. The smallest value of d for this case is $d = 2$. Involutions in $BRW^+(2^2) \cong W_{F_4}$ are discussed in A.15. Involutions in $BRW^+(2^3)$ are discussed in 4.2(i), the even rank sublattice cases.

Lemma 8.1. *If t is a nonsplit involution of defect 1, $L/2L$ is a free $\mathbb{F}_2\langle t \rangle$-module, i.e., the Jordan canonical form for t consists of 2^{d-1} blocks of degree 2.*

Proof. The result may be checked directly for $d \leq 2$ since we know $Tel(t)$ and $Tel(t) + 2L/2L$ is the fixed point space for the action of t on $L/2L$ (see [GrPO2^d]). The idea is to use induction on d plus the fact that t leaves invariant the summands of a decomposition $L = L^\pm(u) \oplus L^\pm(v)$, where u,v generate a lower dihedral group which centralizes t. This proves that L is a free $\mathbb{Z}\langle t \rangle$-module, so reduction modulo 2 has the claimed structure. □

Lemma 8.2. *Let t be a nonsplit involution of defect 1. Then $L^\varepsilon(t)$ is doubly even for $d \geq 4$, i.e., $\frac{1}{\sqrt{2}} L^\varepsilon(t)$ is an even integral lattice.*

Proof. When $d = 4$, $L^\pm(u) \cong \sqrt{2} L_{E_8}$, so the property is clearly true. By 8.1, we have $Tel(t) = 2L + [L,t]$. For $x,y \in L$, $(x(t-1), y(t-1)) = (x,y) + (xt, yt) - (x, yt) - (xt, y) = 2(x,y) - 2(x, yt) \in 2\mathbb{Z}$. It follows that $(Tel(t), Tel(t)) \leq 2\mathbb{Z}$. We take $x = y$. We want $(x, xt) \in 2\mathbb{Z}$ to conclude that $x(t-1)$ has norm divisible by 4. This will follow if it is so for a spanning set. Consider the summands of a decomposition $L = L^{\pm}(u) \oplus L^{\pm}(v)$, where u, v generate a lower dihedral group which centralizes t. For $x \in L^{\pm}(u)$, which is a $ssBW_{2^{d-1}}$, $x(t-1)$ has norm divisible by 4 for $d \geq 5$, by induction. □

Lemma 8.3. *Let t be a nonsplit involution of defect 1 in $BRW^+(2^4)$. Then $L^\varepsilon(t) \cong \sqrt{2}BW_{2^2} \perp \sqrt{2}BW_{2^2} \cong \sqrt{2}L_{D_4} \perp \sqrt{2}L_{D_4}$.*

Proof. Let $L = BW_{2^4} \cong L_{E_8}$. We follow the strategy in the proof of 4.2. Then there exists a lower dihedral group $D \leq C_R(t)$. Let u, v be involutions which generate D. Then by 2/4-generation [GrPO2^d], $L = L^{\pm}(u) \oplus L^{\pm}(v)$, all summands are $ssBW_{2^3} \cong \sqrt{2}L_{E_8}$ lattices which are t-invariant and on them t acts like a nonsplit dirty involution. It follows that each $L^{\pm}(w)^\varepsilon(t)$ is isometric to $\sqrt{2}L_{A_1^4}$, for any noncentral involution $w \in D$. Reasoning as in 4.2, we argue that $Tel(t)$ has index 2^8 in L and $\det(Tel(t)) = 2^{16}\det(L) = 2^{24}$. From 8.2, we know that $Tel(t)$ is doubly even. Therefore, each $L^\varepsilon(t)$ is doubly even and has determinant 2^{12}. Therefore, there is an even integral lattice, P, so that $L^\varepsilon(t) \cong \sqrt{2}P$, $\det(P) = 2^4$ and P contains a sublattice Q isometric to $L_{A_1^8}$, of index 4 in P.

Let $r_1, \ldots, r_8$ be an orthogonal basis of roots for Q. Any nontrivial coset of Q in P consists of even norm vectors, so contains an element of shape $\frac{1}{2}\sum_{i \in I} r_i$, where $I \subseteq \{1,2,3,4,5,6,7,8\}$ and $|I| = 4$ or 8 (note that $\exp(P/Q) \neq 4$ since vectors of shape $\sum_{i=1}^{8} \pm\frac{1}{4}r_i$ have norm 1).

Let I, I' be any two 4-sets which arise as above. We claim that they are disjoint. Assume otherwise. Since P is even, $I \cap I'$ is a 2-set. Then P is isometric to $L_{D_6} \perp L_{A_1} \perp L_{A_1}$, whence $C_R(t)$ fixes the unique indecomposable orthogonal summand isometric to L_{D_6}. This is impossible since $C_R(t)$ contains a subgroup of shape 2_+^{1+4}, whose faithful irreducibles have dimension divisible by 4. We conclude that there exists a partition J', J'' of $\{1,2,3,4,5,6,7,8\}$ so that J' and J'' are 4-sets and $P = P' \perp P''$, where $P' := \{x \in P | supp(x) \subseteq J'\}$, $P'' := \{x \in P | supp(x) \subseteq J''\}$ and $P' \cong P'' \cong BW_{2^2} \cong L_{D_4}$. □

Proposition 8.4. *For all $d \geq 2$, if $t \in BRW^+(2^d)$ is a nonsplit dirty involution, then $L^\varepsilon(t) \cong ssBW_{2^{d-2}} \perp ssBW_{2^{d-2}}$.*

Proof. If $d = 2$, this is true by the discussion in A.15. For $d = 3, 4$, we use 4.2, 8.3.

Let $d \geq 5$. Then there exists a lower dihedral group $D \leq C_R(t)$. Let u, v be involutions which generate D. Then by 2/4-generation [GrPO2^d], $L = L^{\pm}(u) \oplus L^{\pm}(v)$, all summands are $ssBW_{2^{d-1}}$ lattices which are t-invariant and on them t acts like a nonsplit dirty involution. By induction, we know the eigenlattices for t on each.

Consider $L^+(u) \perp L^-(u)$. The involution v interchanges the summands and acts trivially on $L/L^+(u) \perp L^-(u)$. The same is therefore true for the actions of v on $L^+(u)^\varepsilon(t) \perp L^-(u)^\varepsilon(t)$ and $L^\varepsilon(t)/L^+(u)^\varepsilon(t) \perp L^-(u)^\varepsilon(t)$.

Since $d - 1 \geq 4$, induction implies that each $L^{\pm}(u)^\varepsilon(t)$ is the orthogonal sum of two orthogonally indecomposable lattices. Furthermore, if S is one of these two indecomposable direct summands of $L^+(u)^\varepsilon(t)$, we deduce that the same is true for the actions of v on $S \perp S^v$ and on $L^\varepsilon(t) \cap (\mathbb{Q} \otimes (S \perp S^v))/S \perp S^v$.

We finish by quoting the uniqueness theorem [GrPO2^d, GrPO2^{d}corr], applied to the containment of $S \perp S^v$ in $L^\varepsilon(t) \cap (\mathbb{Q} \otimes (S \perp S^v))$, for each S. Note that t centralizes a natural $BRW^+(2^{d-2})$-subgroup of $BRW^+(2^d)$ and that it stabilizes S and S^v. □

The main result 1.7 follows.

Appendix About BRW groups

This is an updated and corrected version of Appendix 2 from [GrPO2^d].

Basic theory of extraspecial groups extended upwards by their outer automorphism group has been developed in several places. We shall use [GrEx, GrMont, GrDemp, GrNW, Hup, BRW1, BRW2, B].

Notation A.1. Let $R \cong 2^{1+2d}_\varepsilon$ be an extraspecial group which is a subgroup of $GL(2^d, \mathbb{F})$, for a field $\mathbb{F}$ of characteristic 0. Let $N := N_{GL(2^d, \mathbb{F})}(R) \cong \mathbb{F}^\times . 2^{2d} O^\varepsilon(2d, 2)$. The *Bolt-Room-Wall group* is a subgroup of this of the form $2^{1+2d}_\varepsilon . \Omega^\varepsilon(2d, 2)$. If $d \geq 3$ or $d = 2, \varepsilon = -$, N' has this property. For the excluded parameters, we take a suitable subgroup of such a group for larger d. We denote this group by $BRW^0(2^d, \varepsilon)$ or $\mathcal{D}d$. It is uniquely determined up to conjugacy in $GL(2^d, \mathbb{F})$ by its isomorphism type if $d \geq 3$ or $d = 2, \varepsilon = -$. It is conjugate to a subgroup of $GL(2^d, \mathbb{Q})$ if $\varepsilon = +$. Let $R = R_{2^d}$ denote $O_2(G_{2^d})$. We call R_{2^d} *the lower group* of $BRW^0(2^d, +)$ and call G_{2^d}/R_{2^d} *the upper group* of $BRW^0(2^d, +)$.

For $g \in N$, define $C_{R \bmod R'}(g) := \{x \in R | [x, g] \in R'\}$, $B(g) := Z(C_{R \bmod R'}(g))$ and let $A(g)$ be some subgroup of $C_{R \bmod R'}(g)$ which contains R' and complements $B(g)$ modulo R', i.e., $C_{R \bmod R'}(g) = A(g)B(g)$ and $A(g) \cap B(g) = R'$. Thus, $A(g)$ is extraspecial or cyclic of order 2. Define $c(d) := \dim(C_{R/R'}(g))$, $a(g) := \frac{1}{2}|A(g)/R'|$, $b(g) := \frac{1}{2}|B(g)/R'|$. Then $c(d) = 2a(d) + 2b(d)$.

Corollary A.2. Let L be any $\mathbb{Z}$-lattice invariant under $H := BRW^0(2^d, +)$. Then H contains a subgroup $K \cong AGL(d, 2)$ and L has a linearly independent set of vectors $\{x_i | i \in \Omega\}$ so that there exists an identification of Ω with $\mathbb{F}_2^d$ which makes the $\mathbb{Z}$-span of $\{x_i | i \in \Omega\}$ a permutation module for $AGL(d, 2)$ on Ω.

Proof. In H, let E, F be maximal elementary abelian subgroups and let K be their common normalizer. It satisfies $K/R \cong GL(d, 2)$. Now, let z generate $Z(R)$ and let E_1 complement $\langle z \rangle$ in E and F_1 complement $\langle z \rangle$ in F. The action of K on the hyperplanes of E which complement $Z(R)$ satisfies $N_K(E_1)F = K, N_F(E_1) = Z(R)$. Now consider the action of $N_K(E_1)$ on the hyperplanes of F which complement $Z(R)$. We have that $K_1 := N_K(E_1) \cap N_K(F_1)$ covers $N_K(E_1)/E$. Therefore, $K_1/Z(R) \cong GL(d, 2)$. Let K_0 be the subgroup of index 2 which acts trivially on the fixed points on L of E_1, a rank 1 lattice. So, $K_0 \cong GL(d, 2)$. Let x be a basis element of this fixed point lattice. Then the semidirect product $F_1{:}K_0$ is isomorphic to $AGL(d, 2)$ and $\{x^g | g \in F_1\}$ is a permutation basis of its $\mathbb{Z}$-span. □

Definition A.3. We use the notation of A.1. An element $x \in N$ is *dirty* if there exists g so that $[x, g] = xz$, where z is an element of order 2 in the center. If g can be chosen to be of order 2, call x *really dirty* or *extra dirty*. If x is not dirty, call

x clean.

Lemma A.4. *Let $\mathbb{F}_2^{2d}$ be equipped with a nondegenerate quadratic form with maximal Witt index. The set of maximal totally singular subspaces has two orbits under $\Omega^+(2d, 2)$ and these are interchanged by the elements of $O^+(2d,2)$ outside $\Omega^+(2d,2)$.*

Proof. This is surely well known. For a proof, see [GrElAb]. □

Definition A.5. An involution in $BRW^+(2^d)$ has *defect k* if its commutator space on the Frattini factor of the lower group has dimension $2k$. The defect is an integer in the range $[0, \frac{d}{2}]$. Note that an automorphism of R_{2^d} has even dimensional commutator space on $R_{2^d}/Z(R_{2^d})$ if and only if it is even; see [GrMont], [GrElAb].

Definition A.6. An involution in $BRW^+(2^d)$ is *split* if it centralizes a maximal elementary abelian subgroup of R_{2^d}, and is otherwise *nonsplit*.

Notation A.7 Write $R = D_1 \cdots D_d$ as a central product of dihedral groups, D_i of order 8. The involution $\alpha_{d,r}$ in $Aut(BRW^+(2^d))$, defined up to conjugacy, acts trivially on $d-r$ of the D_i and performs an outer automorphism on the other r of them. When $r = 2k$ is even, $\alpha_{d,2k}$ is represented in $BRW^+(2^d)$ by an involution $\eta_{d,2k,+}$ (see A.13). In case $r = 2k < d$, we define an involution $\eta_{d,2k,-} := \eta_{d,2k,+}z$, where z is a noncentral involution in the above product of the $d-r$ elementwise fixed D_i.

Theorem A.8. *We use the notation of A.1, A.3. Let $g \in N$. Then $\mathrm{Tr}(g) = 0$ if and only if g is dirty. Assume now that g is clean and has finite order. Then $\mathrm{Tr}(g) = \pm 2^{a(g)+b(g)}\eta$, where η is a root of unity. If $g \in BRW(d, +)$, we may take $\eta = 1$. Furthermore, every coset of R in $BRW(d, \varepsilon)$ contains a clean element and if g is clean, the set of clean elements in Rg is just $g^R \cup -g^R$.*

Proof. [GrMont]. □

Lemma A.9. *Suppose that t, u are involutions in $\Omega^+(2d, 2)$, for $d \geq 2$. Suppose that their commutators on the natural module $W := \mathbb{F}_2^{2d}$ are totally singular subspaces of the same dimension, e. Suppose that $e < d$ or that $e = d$ and that $[W, t]$ and $[W, u]$ are in the same orbit under $\Omega^+(2d, 2)$. Then t and u are conjugate.*

Proof. Induction on d. □

Corollary A.10. *Suppose that t, u are clean involutions in H so that $\mathrm{Tr}(t) = \mathrm{Tr}(u) \neq 0$. Then t and u are conjugate in G_{2^d}, if their common defect is less than $\frac{d}{2}$. If the defects are $\frac{d}{2}$, then there are two classes.*

Proof. We may assume that t, u are noncentral. These involutions are not lower and have the same dimension of fixed points on $R/R' \cong \mathbb{F}_2^{2d}$. Let $T, U \leq R$ be their respective centralizers in R. Since both t, u are clean, $[R, t]$ and $[R, u]$ are elementary abelian subgroups of T, U, respectively. From A.9, we deduce that Rt and Ru are conjugate in G_{2^d} if their common defect is less than $\frac{d}{2}$ and there are two possible conjugacy classes in case of common defect $\frac{d}{2}$. We may assume that $Rt = Ru$. Now use A.8 to deduce that t is R-conjugate to u or $-u$. The trace condition implies that t is conjugate to u. □

Remark A.11. The extension $1 \to R_{2^d} \to G_{2^d} \to \Omega^+(2d, 2) \to 1$ is nonsplit for $d \geq 4$. This was proved first in [BRW2], then later in [BE] and in [GrEx] (for both kinds of extraspecial groups, though with an error for $d = 3$; see [GrDemp] for a correction). The article [GrEx] gives a sufficient condition for a subextension $1 \to R_{2^d} \to H \to H/R_{2^d} \to 1$ to be split, and there are interesting applications, e.g. to the centralizer of a 2-central involution in the Monster [Gr72]. A general discussion of exceptional cohomology in simple group theory is in [GrNW].

Lemma A.12. *Let $V = \mathbb{F}_2^{2d}$ have a nonsingular quadratic form, q, of plus type. Let W be an isotropic subspace, $U := W^\perp$. Then every nontrivial coset of U contains singular and nonsingular vectors if $d > 1$.*

Proof. Suppose that $v + U$ is a coset which consists entirely of either singular or nonsingular vectors. Then for all $x, y \in v + U$, $q(x+y) = (x, y) + q(x) + q(y) = (x, y)$. Take $a, b \in U$ so that $a + b = x + y$. Then $(x, y) = q(a+b) = q(a) + q(b) + (a, b)$. Also $(x + a, y + a) = (x, y)$ implies that $0 = (x, a) + (a, y) = (x+y, a) = (a+b, a) = (a, b)$. It follows that for any two elements a, b of U, $(a, b) = 0$. Since U is the annihilator of W, $U = W$. Let $Z := \{x \in W | q(x) = 0\}$, a subspace of W of codimension 0 or 1. Suppose $d > 1$. Let $x \in V \setminus W$. If there is $z \in Z$ so that $(x, z) = 1$, then x and $x + z$ have different values under the quadratic form. If this fails to be so, then $\dim(Z) = 0$, i.e., $d = 2$ and W contains nonsingular vectors. Then x annihilates a nonsingular vector, $w \in W$ and so x and $x + w$ have different values under the quadratic form. □

Lemma A.13. *Let $V = \mathbb{F}_2^{2d}$ and let g be an involution in $\Omega^+(2d, 2)$ so that $[V, g]$ has dimension $r > 1$ and contains nonsingular vectors. There exists a basis of singular vectors $x_1, \ldots, x_d, y_1, \ldots, y_d$ so that $(x_i, y_j) = \delta_{ij}$ and g interchanges x_i and y_i for $i = 1, \ldots, r$ and fixes each x_j, y_j for $j \geq r + 1$.*

Proof. Let W be the codimension 1 subspace of $[V, g]$ which contains all the singular vectors of $[V, g]$. Take a basis u_i, $i = 1, \ldots, 2k$, of $[V, g]$ of nonsingular vectors. For $x \in [V, g]$, let $P(x) := \{v \in V | v(g - 1) = x\}$, a coset of $[V, g]^\perp$. For all x, $P(x)$ contains singular vectors (see A.12). We therefore may take x_1 so that $x_1(g - 1) = u_1$ and we define $y_1 := x_1^g$. We may use induction on $\operatorname{span}\{x_1, y_1\}^\perp$. The only problem might be that we are unable to use A.12 at the last stage in case $r = \frac{d}{2}$. But then we use the fact that V has plus type and the conclusion is forced. □

Lemma A.14. *(i) Suppose that t is a clean upper involution of G_{2^d}. Then the coset tR_{2^d} represents $s + 1$ different conjugacy classes of involutions in G_{2^d}, where s is the number of orbits of $C_{G_{2^d}}(t)$ on the cosets of $[R, t]$ in $C_{R_{2^d}}(t)$ which contain involutions. We have $s = 1$ if $k = \frac{d}{2}$ and $s = 2$ if $k < \frac{d}{2}$. This gives respectively one and two dirty classes of involutions in the coset.*

(ii) If t is $\eta_{d,2k,\pm}$ (so is dirty and nonsplit), the coset tR_{2^d} represents one

class of involutions if $k = \frac{d}{2}$, and two otherwise; all involutions in tR_{2^d} are dirty.

Proof. Exercise. $\square$

Lemma A.15. (i) A defect k involution in $G_{2^d}/R_{2^d} \cong \Omega^+(2d,2)$ is represented in $BRW^+(2^d)$ by an involution, specifically, by either a clean involution of defect k, or the dirty nonsplit involution $\eta_{d,2k,+}$, for a unique integer $k \leq \frac{d}{2}$. Furthermore, for any d and positive $k \leq \frac{d}{2}$, both cases occur and are mutually exclusive.

(ii) An eigenlattice of $\eta_{2,2,+}$ has an orthogonal basis, of norms $2, 4$.

Proof. It is clear from a direct construction (or 3.26) and A.14 that both cases occur and that they are mutually exclusive. Since G_{2^d} contains a natural central product of k natural $BRW^+(2^2) \cong W_{F_4}$ subgroups, it suffices to give a direct construction for the case $k = \frac{d}{2} = 1$, which we now do. Notice that for $d = 2$, $BRW^+(2^2) \cong W_{F_4}$ contains two conjugacy classes of reflection (upper and clean, of defect 1, representing the two classes when $k = \frac{d}{2}$) and a nonsplit involution. Note that the product of two reflections for orthogonal roots has trace 0, so is dirty. There are two orbits of W_{F_4} on orthogonal pairs of roots, distinguished by root lengths, but the resulting products of two reflections represent only two classes: one class (for the pairs of equal length roots) and a second class for the case of unequal root lengths. The latter gives the upper class. For this case, we have an orthogonal set of vectors of norms 2 and 4 in a given eigenlattice, M, corresponding to orthogonal roots of different lengths. $\square$

References

[BW] E. S. Barnes and G. E. Wall, Some extreme forms defined in terms of abelian groups, J. Aust. Math. Soc. 1 (1959), 47-63.

[BRW1] B. Bolt, T. G. Room and G. E. Wall, On the Clifford collineations, transform and similarity groups, I. J. Aust. Math. Soc. 2 (1961/1962), 60-79.

[BRW2] B. Bolt, T. G. Room and G. E. Wall, On the Clifford collineations, transform and similarity groups, II. J. Aust. Math. Soc. 2 (1961/1962), 80-96.

[B] B. Bolt, On the Clifford collineations, transform and similarity groups, III; generators and relations, J. Aust. Math. Soc. 2 (1961/1962), 334-343.

[Bour] N. Bourbaki, Élements de Mathématique, Groupes et algèbres de Lie, Chapitres 2 et 3, Diffusion C.C. L. S., Paris, 1972.

[BE] M. Broué and M. Enguehard, Une famille infinie de formes quadratiques entière; leurs groupes d'automorphismes, Ann. scient. Éc. Norm. Sup., 4eme série, t. 6 (1973), 17-52.

[Car] Roger Carter, Simple Groups of Lie Type, Wiley-Interscience, London, 1972.

[CR] C. Curtis and I. Reiner, Representation Theory of Groups and Associative Algebras, Interscience, 1962.

[Dieud] J. Dieudonne, La Géométrie des Groupes Classiques, Springer, Berlin Heidelberg New York, 1971.

[Gor] D. Gorenstein, Finite Groups, Harper and Row, New York, 1968.

[Gr72] R. L. Griess, Jr., Automorphisms of extra special groups and non-vanishing degree 2 cohomology (research announcement for [5]), in Finite Groups 1972: Proceedings of the Gainesville Conference on Finite Groups, (T. Gagen, M. P. Hale and E. E. Shult, eds.), North Holland Publishing Co., Amsterdam, 1973, 68-73.

[GrEx] R. L. Griess, Jr., Automorphisms of extra special groups and non-vanishing degree 2 cohomology, Pacific J. Math. 48 (1973), 403-422.

[GrDemp] R. L. Griess, Jr., On a subgroup of order $2^{15}|GL(5,2)|$ in $E_8(C)$, the Dempwolff group and $Aut(D_8 \circ D_8 \circ D_8)$, J. Algebra 40 (1976), 271-279.

[GrMont] R. L. Griess, Jr., The monster and its nonassociative algebra, in Proceedings of the Montreal Conference on Finite Groups, Contemp. Math. 45 (1985), 121-157, American Mathematical Society, Providence, RI.

[GrNW] R. L. Griess, Jr., Sporadic groups, code loops and nonvanishing cohomology, J. Pure Appl. Algebra 44 (1987), 191-214.

[GrElAb] R. L. Griess, Jr., Elementary abelian subgroups of algebraic groups, Geometria Dedicata 39 (1991), 253-305.

[Gr12] R. L. Griess, Jr., Twelve Sporadic Groups, Springer, 1998.

[GrPOE] R. L. Griess, Jr, Pieces of Eight, Adv. Math. 148 (1999), 75-104 .

[GrPO2^d] R. L. Griess, Jr., Pieces of 2^d: existence and uniqueness for Barnes-Wall and Ypsilanti lattices, Adv. Math. 196 (2005), 147-192; see also

[GrPO2^{d}corr] R. L. Griess, Jr., Corrections and additions to " Pieces of 2^d: existence and uniqueness for Barnes-Wall and Ypsilanti lattices.", Adv. Math. 211 (2007), 819-824.

[Hup] B. Huppert, Endliche Gruppen I, Springer, Berlin, 1968.

[MS] J. MacWilliams and N. Sloane, The Theory of Error Correcting Codes, North-Holland, 1977.

[McL] J. E. McLaughlin, Some subgroups generated by transvections, Arch. Math. 18 (1969), 108-115.

[MH] J. Milnor and D. Husemoller, Symmetric Bilinear Forms, Ergebnisse der Mathematick und Ihrer Grenzgebiete, Band 73, Springer, New York, 1973.

[Poll] H. Pollatsek, Cohomology groups of some linear groups over fields of characteristic 2, Illinois J. Math. 15 (1971) 393-417.

[Se] Jean-Pierre Serre, A Course in Arithmetic, Springer, Graduate Texts in Mathematics 7, 1973.

DATE DUE

QA 177 .G74 2011

Griess, Robert L., 1945-

An introduction to groups and lattices